TROPICAL NATURE

The Environment in History: International Perspectives

Series Editors: Stefania Barca, *University of Santiago de Compostela*; Kieko Matteson, *University of Hawai'i at Mānoa*; Christof Mauch, *LMU Munich*; Helmuth Trischler, *Deutsches Museum, Munich*

eseh european society for environmental history

Rachel Carson Center
ENVIRONMENT AND SOCIETY

Recent titles:

VOLUME 26
Tropical Nature: Colonial and Post-Colonial Conservation in Africa and Asia
Edited by Guillaume Blanc, Mathieu Guérin and Grégory Quenet

VOLUME 25
Thinking Russia's History Environmentally
Edited by Catherine Evtuhov, Julia Lajus, and David Moon

VOLUME 24
Planting Seeds of Knowledge: Agriculture and Education in Rural Societies in the Twentieth Century
Edited by Heinrich Hartmann and Julia Tischler

VOLUME 23
Environing Empire: Nature, Infrastructure, and the Making of German Southwest Africa
Martin Kalb

VOLUME 22
The Russian Cold: Histories of Ice, Frost, and Snow
Edited by Julia Herzberg, Andreas Renner, and Ingrid Schierle

VOLUME 21
Risk on the Table: Food Production, Health, and the Environment
Edited by Angela N. H. Creager and Jean-Paul Gaudillière

VOLUME 20
The Meanings of a Disaster: Chernobyl and Its Afterlives in Britain and France
Karena Kalmbach

VOLUME 19
Conservation's Roots: Managing for Sustainability in Preindustrial Europe, 1100–1800
Edited by Abigail P. Dowling and Richard Keyser

VOLUME 18
Colonial Seeds in African Soil: A Critical History of Forest Conservation in Sierra Leone
Paul Munro

VOLUME 17
Hazardous Chemicals: Agents of Risk and Change, 1800–2000
Edited by Ernst Homburg and Elisabeth Vaupel

VOLUME 16
Planning for the Planet: Environmental Expertise and the International Union for Conservation of Nature and Natural Resources, 1960–1980
Simone Schleper

VOLUME 15
Changes in the Air: Hurricanes in New Orleans from 1718 to the Present
Eleonora Rohland

VOLUME 14
Ice and Snow in the Cold War: Histories of Extreme Climatic Environments
Edited by Julia Herzberg, Christian Kehrt, and Franziska Torma

VOLUME 13
A Living Past: Environmental Histories of Modern Latin America
Edited by John Soluri, Claudia Leal, and José Augusto Pádua

For a full volume listing, please see the series page on our website: http://berghahnbooks.com/series/environment-in-history.

Tropical Nature

Colonial and Post-Colonial Conservation in Africa and Asia

Edited by
Guillaume Blanc, Mathieu Guérin and Grégory Quenet

Translated by Adrian Morfee

berghahn
NEW YORK • OXFORD
www.berghahnbooks.com

First published in 2025 by
Berghahn Books
www.berghahnbooks.com

Originally published in French as
Protéger et détruire: Gouverner la nature sous les tropiques (XX-XXIe siècle)

Chapters 2, 4, 8, 10 and 12 translated by Adrian Morfee

Library of Congress Cataloging-in-Publication Data

A C.I.P. cataloging record is available from the Library of Congress

Library of Congress Cataloging in Publication Control Number: 2024046832

British Library Cataloguing in Publication Data

A catalogue record for this book is available from the British Library

ISBN 978-1-80539-891-2 hardback
ISBN 978-1-80539-892-9 epub
ISBN 978-1-80539-893-6 web pdf

https://doi.org/10.3167/9781805398912

Contents

Illustrations

Figures

Tables

Maps

Acknowledgments

The three editors of this book conducted their research as part of the ANR (CE 27) PANSER programme 'Patrimoines naturels aux Suds: Une histoire globale à échelle réduite', directed by Guillaume Blanc and hosted by the TEMPORA laboratory (7468) at the University of Rennes 2.

Six chapters have been translated from French to English. The translations were carried out by Adrian Morfee and funded by Guillaume Blanc's Institut universitaire de France (IUF) research programme.

This book is an adjusted version of a volume initially published in France titled *Protéger et détruire: Gouverner la nature sous les Tropiques (xxe–xxie siècle)*, edited by Guillaume Blanc, Mathieu Guérin and Grégory Quenet (Paris: CNRS Éditions, 2022).

Abbreviations

AFW: African Wildlife Foundation

BMEO: British Middle East Office

CGIAR: Consultative Group on International Agricultural Research

CITES: Convention on International Trade in Endangered Species

CTCA: Commission for Technical Cooperation in Africa South of the Sahara

ECOSOC: United Nations Economic and Social Council

FAO: Food and Agriculture Organization

FPS: Fauna Preservation Society (formerly SPFE)

FZS: Frankfurt Zoological Society

IOPN: International Office for the Protection of Nature (renamed IUPN in 1948)

IUCN: International Union for the Conservation of Nature (formerly IUPN)

IUPN: International Union for the Protection of Nature (renamed IUCN in 1956)

MOD: Ministry of Overseas Development

NYZS: New York Zoological Society (renamed WCS in 1991)

ODA: Overseas Development Administration

SPFE: Society for the Preservation of the Fauna of the Empire (renamed FPS in 1950)

UNDP: United Nations Development Programme

UNESCO: United Nations Educational, Scientific and Cultural Organization

WB: World Bank

WCS: Wildlife Conservation Society (formerly NYZS)

WWF: World Wildlife Fund

INTRODUCTION

Protecting Nature in Africa and Asia

Towards a Small-Scale Global History

Guillaume Blanc

This book addresses two histories, both ongoing. The first may be followed over the course of its twelve chapters. These focus on different times and spaces, from the conservation of the imperial forests in Singapore to those in Ali Bongo's Gabon, from Captain Ritchie's colonial career in Africa and Asia to the postcolonial African itinerary of the agronomist Arthur Bunting, without forgetting the mythicised bestiaries of independent India and Tanzania. Each chapter thus acts as a gateway to the global history of tropical nature conservation from the late nineteenth through to the early twenty-first century – in other words, they are thirteen pieces of a far from complete puzzle.

But its pieces have not been positioned by chance. It would have been impossible to retrace these small-scale histories without also apprehending the African and Asian history in which they take place. That is the paradox of the discipline. One has to already know one's field of enquiry to formulate a plausible general hypothesis, but one needs to start from a solid hypothesis in order to correctly investigate one's field. In other words, 'to raise a historical question, one has to already be a historian'.[1] This first chapter therefore goes over a second history, that of our enquiry into global nature conservation in an area covering Asia and Africa. It addresses how it originated, the objectives pursued, the hypotheses envisaged, the resultant reflective framework and our initial findings – which, far from bringing our research project to a crowning close, mark the enthusiastic beginnings of our joint venture.

A Starting Point: From Histories of Encounters to Histories of Circulations

The starting point for this book took place in Ethiopia. It is the history of the 2,500 inhabitants of the village of Gich, who in 2016 were displaced by the wardens of Simien National Park. They settled with their livestock and financial compensation on the outskirts of the little town of Debark, 32 kilometres west of the park. This brought the history of Gich to a close. But where did it start?

First, this displacement occurred during the very contemporary period of global nature governance. Consultants working for the IUCN (International Union for Conservation of Nature) and the WWF (World Wildlife Fund), operating under UNESCO mandate since the 1980s, argued on several occasions that agro-pastoralism posed a threat to the natural values of the Simien mountains, listed as a World Heritage Site. UNESCO thus downgraded the park in 1996, placing it on its List of World Heritage in Danger. Since then many conservation professionals – working for the African Wildlife Foundation based in Tanzania, the German development bank KfW (Kreditanstalt für Wiederaufbau), – or the IUCN – have worked to safeguard the place, providing the technical and financial support needed to displace populations 'voluntarily' and to preserve the *walia ibex*, an endemic species of wild goat. For Ethiopia's leaders, this naturalisation of the places in the park was a way to boost its international profile, which it planned on leveraging to become a leading player in East Africa; it also stoked state capitalism fuelled by the tourist industry, making it easier to control a region of scrubland like Simien Park. Inhabitants' reactions were mixed. On being stripped of the land they had maintained for decades, and deprived of their mutual support networks, most sank silently into poverty, though some sought to cling to the coat-tails of international heritage actors, a potential source of revenue, prestige and power within their community.

The displacement of the Gich villagers also reaches back to the more distant days of national parks. From the mid-1930s through to the late 1970s, the Ethiopian emperor, Haile Selassie, turned to European 'advisers' from neighbouring East African colonies, then to Western 'experts' sent by several international conservation institutions. These biologists, agronomists and ecologists helped create the Ethiopian national parks, including Simien Park. Their guiding principle was to save the Simien's exceptional wildlife, flora and vistas. In an international context conducive to neo-Malthusian theories, this meant evicting populations: this is what East Africa's colonial administrators and then Western experts told Haile Selassie, and it is a policy he sought to implement in the Simien mountains. But the inhabitants resisted. They regularly clashed, rifle in hand, with representatives of the empire; they even tried to kill every single *walia ibex*, whose disappearance would scupper any plans for a 'national park'. The more the state drew on the global heritagisation of nature to build a centralised national administration, the more the inhabitants used nature to resist first the empire, then the nation state.

The displacement of the Gich villagers takes us back, lastly, to the time of game reserves. In early twentieth-century sub-Saharan Africa, the European powers, working in league with colonial foresters and botanists, spread the myth of a *previously* untouched African 'nature' that had since been damaged. This myth legitimised the expropriation of the colonised and the colonial exploitation of resources. It was against this backdrop that Emperor Menelik II of Ethiopia called in German zoologists to create the Simien game reserve, a move serving various purposes. The Ethiopian administration was seeking to leverage international cooperation to obtain a seat at the table of the great powers of the period. Granting a monopoly on game rights exclusively to imperial military officers was also a way of signifying the administration's superiority over the region's local chieftains. To impose their power on Simien communities, the officers granted themselves a monopoly on trophy and game hunting; but it was the emergent nation's representatives in Addis Ababa who henceforth determined land use.

There are two possible interpretations of this Ethiopian history. The first is that of exceptionalism, according to which contemporary Ethiopia, as the only African country not to have been colonised, was following a singular trajectory as the continent's first nation state. This hypothesis is not wholly wrong: each country does indeed follow its own path. However, the North–South circulations marking this Ethiopian history suggest we should consider a larger history: that of the (post)colonial space-times of global nature governance.[2] Rather than focusing on a *before* and *after* of colonisation, Ethiopia suggests we should envisage '*interlocking*' time frames: far from emerging one after the other, the rationales for turning the land into a park accumulated on top of one another.[3] Instead of seeing North–South relations solely through the prism of domination, it is more useful to think of power in terms of circulations, transfers arising from 'perpetually extending the remote links binding ever more people together over an ever vaster space'.[4] And to do that, it is of course necessary to expand the enquiry, that is, to move from individual work on Ethiopia to a group research project.[5] The first stage was to call in three colleagues.

First, research by Mathieu Guérin has shown that throughout South East Asia, heritage invention was marked by worldwide encounters, whose chronology spills across the colonisation–decolonisation watershed.[6] In Malaysia, for instance, nature management changed over the course of negotiations involving different protagonists: colonial administrators then leaders of the Federation of Malaysia, Malay sultans then regional leaders, and colonial conservation experts like Theodore Hubback, some of whom switched to being international experts in the wake of independence, such as James Hislop. From the 1930s to the early 1970s, these conservation practitioners debated, agreed or clashed, especially over the place populations were to have in protected areas.

Above and beyond these relations between (former) imperial metropoles and their colonies, historian Violette Pouillard has examined the circulation of people and ideas within continents.[7] At the turn of the twentieth century, the British hunter and game reserve advocate Edward North Buxton was active in Sudan, Uganda and Somaliland. Then from the 1970s to the 2010s, the British zoologist Kes Hillman Smith, an IUCN consultant employed by the Institut congolais pour la conservation de la nature, worked in parks in Congo, Kenya and Ethiopia. Whether at the beginning or end of the twentieth century, these conservationists seem always to have worked closely with the colonial then national authorities, with imperial then international conservation institutions, as well as with the wardens and inhabitants of protected areas.

Lastly, these nature professionals were at the centre of transcontinental circulations. Archives collected by Grégory Quenet show, for instance, that members of committees regulating hunting in Cochinchina (in southern Vietnam) in the 1920s took direct inspiration from earlier work by some of their members in French West Africa.[8] Other archives indicate that the broad lines of forest conservation policy in Tanzania, the Seychelles and Suriname were all drawn up by John Procter, a British forester and ecologist, over periods stretching from ten years before these countries became independent to several years after.

These hypotheses, stemming from our initial comparison of our fields and archives, led us to devise a research project (PANSER, *PAtrimoines Naturels aux Suds: pour une histoire globale à Échelle Réduite*, or Natural Heritage in the Global South: For a Small-Scale Global History). This project ran from 2018 to 2022, with funding from the Agence française nationale de recherche, and was headed by Mathieu Guérin, Grégory Quenet and me. It sprang from three observations: first, in the twentieth century, nature professionals circulated in and between Africa and Asia; second, the time frames of these circulations followed a chronology other than that of colonisation–decolonisation; and third, these histories of global encounters always took place at the village level. It seemed to us that following the trajectories of these conservation professionals moving from one protected area to another offered a way of exploring the history of global environmental heritage interventions. Studying this history at the heart of the territories where it occurred – from Equatorial Africa to the China Sea – provided a way of envisaging, on the ground, the encounters, negotiations and conflicts that, over the course of the twentieth century, fashioned an Afro-Asian heritage area.

Our Objective: Placing the Global South in Its Specific World History

To produce a history of the Global South from its own perspective, we need to relinquish any opposition between 'colonial' and 'postcolonial' periods, and also move beyond any separation into 'international stakeholders' and 'local populations'. The histories of science, of heritage and of the environment show how hard it is to encompass the various times and protagonists at work in conservation in the Global South in a single account. Concerning colonial times, certain works have revealed the role played by members of a genuine diaspora of scientists working in tropical nature heritagisation,[9] while others have flagged the need to embrace the ways that nature policies are negotiated at a local scale.[10] Conversely, concerning postcolonial times, while some historians have noted how 'local forms of resistance and creativity' may influence the social spaces of natural heritage in the Global South,[11] others have called for greater attention to the role played by the international professionals defining conservationist practices as they moved between North America, Europe, Africa and Asia.[12] The difficulty is thus to reconstitute a genuinely contemporary and truly global history. Doing so entails combining at least two approaches.

A first approach consists in focusing on the 'assemblage vectors' involved in constructing nature policies.[13] Once aspect is to examine how conservation knowledge was built up during missions to explore Africa and Asia, and then standardised in scientific institutions based mainly in the West. Another aspect is to analyse how nature professionals shared this knowledge (at international conferences) and how they disseminated it (through correspondence, scientific journals and official publications). Various works have emphasised the validity of this approach. Several French researchers have thus studied 'imperial trajectories' to throw light on a spatial history of colonial phenomena. This has enabled them to abandon the centre–periphery model to instead envisage colonial territories as functioning in a network, and so apprehend 'the hybridisation of knowledge and of action rationales'.[14] This approach, applied to the history of conservation, seeks to identify nature professionals and to map the networks they built up across Africa and Asia over the course of a long twentieth century.

This idea of hybridisation also encourages us to adopt a second approach. In building up knowledge to define right and wrong usages of nature, foresters, economists and ecologists conducted field studies. Admittedly, their findings were often shackled by their prejudices concerning tropical environments and their inhabitants. Nevertheless, these findings stemmed directly from their interaction with local agents.[15] As several historians of the colonial period have shown, heritage expertise was far from being a closed system of ideas and practices, and we hence need to examine the 'interactions and interpenetration of action and knowledge'.[16] This second line of enquiry leads us to study how, during and after colonisation, heritage knowledge and norms were both imposed and negotiated, constructed and reconstructed on the ground. Conducting situated case studies, from Côte d'Ivoire to Malaysia, throws light on the contacts, negotiations and clashes making heritagised nature a place of conflict and conflicts to build a territory, impose a representation and define rights to access and use resources.

To do this, we need to combine analysis of international archives with in-depth field studies. We hold that such an approach may help show that 'the modern world was not simply born in Europe', while also ensuring that the countries of the South have *their own* contemporary chronology: that of a lengthy twentieth century marked by the history of *their specific* world, driven by both exogenous and endogenous processes.[17]

Our Hypothesis: The (Other) Space-Times of an Afro-Asian Area

In 1950, the Society for the Preservation of the Fauna of the (British) Empire changed its name, becoming the Fauna Preservation Society (FPS). Several international conservation institutions did likewise; new organisations were also founded, such as the World Wildlife Fund. These all emphasised the need to reconcile development and conservation requirements in independent Africa and Asia. Several historians who have studied their statements and programmes argue that the 1960s marked a radical shift in global environmental policies: preservationism – isolating nature under a protective dome and driving out its occupants – is said to have given way to conservationism – protecting resources to use them sustainably.[18] According to this account, the shift really got under way in September 1961, when the IUCN – with the support of UNESCO, the FPS and the FAO (Food and Agriculture Organization) – held a 'Conference for the Conservation of Nature and Natural Resources in Modern African States' in Arusha (Tanzania). This was attended by international experts and leaders of independent African states, and resulted in the founding of the WWF, conceived of by several experts – such as Julian Huxley, the former head of UNESCO, and Max Nicholson, from the (British) Nature Conservancy – as a bank to finance conservation. According to them, and to certain historians, 1961 also marked the end of colonial preservationism and the beginning of developmentalist conservation, backed by international institutions.[19]

But field studies bring a different story to light.[20] International conservation institutions, with the enforced yet opportunistic cooperation of independent leaders, conducted projects in Africa planned by former colonial experts.[21] Research conducted with Mathieu Guérin and Grégory Quenet further suggests that from Africa to South East Asia, similar heritage practices were conducted by the same heritage makers: scientists who had worked in colonies and switched to being international experts, some of whom retained their positions – and even their offices – through to the mid-1970s.

This led us to abandon the chronology that tends to be used to describe societies in Africa and South East Asia: the imperial construction of colonial states through to the 1960s, followed by the formation of nations, then, after the 1990s, the federal recomposition or collapse of neoliberal states. The history of conservation brings a different chronology to light, that of *game reserves* (the 1900s to the 1930s), *national parks* (the 1940s to the 1980s) and *community conservation* (the 1990s to the present day). We shall return to this chronology, but for the moment let us note that it has the advantage of leading us away from a Eurocentric approach. When former colonies acceded to independence, this undeniably marked a historical watershed for imperial metropoles: they lost their empire, and with it the resources and networks bolstering their wealth and power. But any such historical break was less clear-cut for the colonies. Colonised independentists became postcolonial leaders, who both reconfigured and perpetuated policies and power structures, and the 'issue of "colonial legacy" became that of the context of action shaped by the colonial past.'[22]

This intermingling of colonial and postcolonial times is common to all societies in Africa and South East Asia. Interdisciplinary studies have shown how the conservationist model devised at the beginning of the twentieth century was reproduced from country to country,[23] before being generalised by international institutions during the 1950s.[24] However pioneering this approach may be,[25] and though still relevant,[26] it does not explain why policies drawn up in a colonial context continued to be implemented after independence. Yet by following the trail of nature professionals, we may place this Afro-Asian area in the history of the specific world of which it is a part. Whether called 'scientists' (the 1900s to the 1930s), 'experts' (the 1940s to the 1980s) or 'consultants' (the 1990s to the present day), nature professionals were omnipresent; this may well explain why, after decolonisation, conservationist practices continued to be globalised, a phenomenon that in fact only intensified. But because no single scheme can encompass the singularities of the socio-environmental dynamics in, say, Mozambique, the Seychelles or Vietnam, we need to relinquish the dream of a 'totalising' global history,[27] to instead try to conduct a 'small-scale global history'.[28]

A Reflexive Framework for Apprehending Global Encounters

This project would not have been feasible without the extensive body of literature in which to immerse ourselves. Practitioners of the human and social sciences have noted that 'nature' is both an object of study and a way of studying societies: 'Because parks are supposed to be, but aren't, the antithesis of how nature is treated in the rest of society, they end up being very clear expressions of that society.'[29] From this perspective, conservation areas need to be viewed as one of those topics that, though seemingly 'wholly minor, even slightly derisory', in fact reveal dynamics that cannot be otherwise apprehended.[30] By studying how public authorities use the governance of natural things to organise their governance of people, our ultimate purpose is to apprehend how nature is made, and thereby how the colonial then national state was built.

Environmental history has contributed to this history for the United States,[31] Canada[32] and then Britain's former empire in Africa and Asia.[33] We now need to further explore lines of enquiry opened up for Kenya and Tanzania, for example. We also need to expand the field to countries that have been little studied either because they were colonised by a power other than Great Britain (Algeria, Mozambique, Syria and Vietnam) or because their archives are little known (the Seychelles and Malaysia). The purpose of this dual approach is hence also theoretical. As stated earlier, we need simultaneously to examine whether the Afro-Asian area actually forms a single whole, and to look at the continued existence of colonial nature policies in postcolonial times.

A convenient explanation of this would be neo-imperialism conducted by Western institutions. This would however postulate the reproducibility not of a model but of situations, thereby setting to one side the dynamics driving transnational yet always locally grounded circulations.[34] That is why it is important to view conservation agents as mediating between different natures: nature became progressively decontextualised from its inhabitants, and so a single universal and borderless nature emerged, which thereby needed protecting. How could such a definition of nature coexist with solidly rooted social realities? Only because of certain arrangements, and because of a dialogue between European powers, international organisations, (formal and informal) public authorities and local communities making a living from (and living in) the environment. Though asymmetrical, these relations had to benefit several groups in order to last. We thus need to study these connections from the bottom up to detect the transactions and oppositions binding together all those involved in producing nature: agronomists, biologists and ecologists turned administrators, experts and consultants; nature journalists and film-makers; agro-pastoralists, tourist guides and local development brokers; local officials and regional or national leaders. And to do so, we need to go back to the chronology.

The era of game reserves started in the early twentieth century. The Germans were the first to create a reserve in German East Africa, in 1896, followed by the British in the neighbouring colonies, then other European colonial powers. They all convened in London in May 1900 where they drew up the Convention for the Preservation of Wild Animals, Birds and Fish in Africa. The practice of placing wildlife in reserves subsequently spread from colonies in Africa to those in Asia. And in each empire, the process involved hunting elites, botanists or foresters, and colonial officials, as well as colonised subjects. It would be easy to view the former as quite naturally allying against the latter. But quite frequently colonists' interests diverged. Hunters generally sought to appropriate then dominate nature. Many naturalists were inspired by a dual representation, that of a nature that had been lost in Europe and (re)found in the tropics, and that of inhabitants who would destroy this still untouched Eden. As for colonial administrators, they could be convinced both of the benefits of their 'civilising mission' and of the need to govern by force. Shaped by these various contexts and circumstances, alliances were built up around nature. Some viewed it as a tool for controlling unruly spaces,[35] others as a place of knowledge and preservation of the wilderness,[36] and others still as a territory for collecting trophies and becoming a sportsman.[37] And so in Asia and Africa from 1900 to the late 1930s, the rise of game reserves meant that inhabitants were increasingly expropriated and criminalised. This process thus needs to be studied in the light of the institutionalisation of sciences within the state, the invention of an Edenic tropical nature and the role of so-called 'native' protagonists: on what endogenous nomenclature of nature did the authorities draw to learn about and (re)name the landscape and animal species? And did the inhabitants clash with the new wildlife managers, or did they negotiate how to use their territory? These questions provide a way of shedding light on how all these protagonists played a part in the colonial construction of nature, in a context of intensifying contacts between ever closer worlds.

Then the second era started, with reserves being converted into national parks. Once again, the destroyers were also the protectors. The beginnings of colonisation had triggered or at least accelerated deforestation and a drop in wildlife specimens, prompting European and North American hunters to call for game reserves. Faced with the exponential collapse in big game, these hunters-turned-'penitents' urged the colonial states to create wildlife sanctuaries.[38] And once again, the European authorities used nature the better to control inhabitants and exploit the land. Some also truly sought to protect gorillas and large herds of savannah herbivores. The first park was created in the Belgian Congo in 1925 when two game reserves were merged. Then in 1926 the South African Transvaal reserve became Kruger National Park; in 1928 Western conservationists circulating between empires created the International Office for the Protection of Nature (IOPN); and then in 1933, once again in London, the Convention Relative to the Preservation of the Fauna and Flora in their Natural State was drawn up.[39] Since then, in both colonial and postcolonial contexts, heritagisation has led to the introduction of national parks and socio-ecological models to help manage humans and nature, mainly in lands home to semi-nomadic peoples, 'ethnic' minorities or mountain dwellers, all of whom tend to oppose ruling powers.[40] Once again, these spaces were largely outside the states' hold, and nature there was considered less 'spoiled' by mankind. Paradoxically, states therefore set about extending their grasp over these lands so as to protect them. As for the narrative underpinning the process, it was mainly conveyed by the agronomists, ecologists and ethologists who succeeded the botanists and foresters. Dreading erosion in the 1930s, and then alarmed at the prospect of overpopulation in the 1960s, nearly all of them were convinced of the need to expel or at least constrain the occupants of tropical nature, viewed as *previously* untouched but *now* under threat. These men of science thus helped turn these colonies into an experimental laboratory. Then, after independence, they became advisers to national leaders who reappropriated their "Foucauldian" discourse of degradation and turned it to their own ends. Both before and after independence, protecting nature was thus also a way of wielding power.[41]

Lastly, the 1980s saw the beginning of the era of community conservation. In terms of texts, this period is symbolised by great international charters, such as the 1980 World Conservation Strategy drawn up by the IUCN, the WWF, UNESCO and the FAO, or the United Nations Convention on Biological Diversity ratified in Rio in 1992. The spirit behind this new conservation was thus based on sacralising the twin principles of 'parks for people' and 'local community' representation. Local people were cast as saviours who would convert the 'preservationist fortresses' of national parks into zones for protecting biodiversity and boosting economic development.[42] And, alongside ecologists who were still active in protected natural areas, international consultants now played a central role. Trained in 'development', 'international cooperation', 'participatory governance' and 'participatory economics', they asserted their capacity to 'drive the circuits of power where normative frameworks were produced, knowledge conveyed, and resources distributed'.[43] Given the increasingly technical nature of conservation, we need to examine how consultants' actions responded to the societies in which they intervened: were they instrumentalised by the public authorities, with the occupants of protected areas being increasingly coerced by wardens? Or on the contrary, did inhabitants and consultants form mutually beneficial alliances to profit from the heritagisation of nature, perhaps at the expense of national authorities? Answering these questions provides a way of exploring the recent history of 'global green governmentality'.[44] This stems from multiple factors: the recent articulation between the state, society and the market; accelerating contact between ever more varied and numerous actors; and the persistence of colonial representations of an Edenic tropical nature now under threat. In other words, to understand this governmentality, we need to view it through the prism of knowledge agents' increasing involvement in heritage actions.

Findings

This approach guided Mathieu Guérin's, Grégory Quenet's and my investigations between 2018 and 2022, looking at Malaysia, Indonesia, the Seychelles,

Vietnam, Ethiopia and the European headquarters of international conservation institutions. The reason for pooling our research was to identify circulations within and between continents, and then observe how they played out locally. And to do this, it was essential to broaden the enquiry once again. We thus assembled a small team of colleagues to explore the global yet situated histories of the governance of tropical nature.

The sources used vary depending on the periods and the thirteen fields of study: in some cases, the documentation was produced by foresters, veterinary surgeons, biologists or agronomists; in others, the material came from archives left by colonial, national and international administrators; in other cases, the research focused exclusively on correspondence, first-person accounts and photographs. But whatever the nature of our sources, we adopted a common framework of enquiry. First, in order to identify the heritage chronologies that built up over time, we sought to apprehend: for the colonial period, the 'knowledge machinery' fashioned by scientists working in the colonies;[45] for the time of independence, their repositioning as 'foot-soldiers for international conservation institutions';[46] and for the subsequent era of community conservation, the 'atypical relation between the government and the "specialist" – a relationship stemming from the colonial legacy'.[47] Second, to gauge the social depth of the networks and circulations shaping nature governance, we looked for intersections between the situated practices of (colonial or national) state employees and the deterritorialised practices of the representatives of (imperial or international) conservation institutions. Lastly, to assess the role local populations played in building these global bio-policies, we read the archives against the grain to see how the protected areas were, if not contested, then at least negotiated by those living there and in the surrounding areas.

The matter of archives turned out to be crucial for this common framework, and also, ultimately, for each of our individual lines of enquiry. From West Africa to South East Asia, from the headquarters of UNESCO (Paris) to those of the FAO (Rome), the FPS (Kew, in London), the IUCN and the WWF (Gland, in Switzerland), the sheer volume of the archives, and their dispersal, thwarted any attempt to retrace a situated Afro-Asian history of 'the global situation'.[48] Indeed, studying the local situation in depth frequently meant neglecting the international dimension; conversely, extensive analysis of international conservation networks implied neglecting the local level to a certain extent. The only way to make up for this lopsidedness, which is inherent to our historical question, is to conduct collective research over time, punctuated by intermediate conclusive phases. This work is intended as the first such phase.

Our first conclusion relates to the very essence of tropical conservation: this policy is, in itself, contradictory. Richard Grove has already shown how, in the early modern period, ecologism and capitalism emerged concomitantly in tropical islands colonised by Europeans.[49] Yet this link between predation and protection kept on growing stronger. Timothy Barnard points out that in the late nineteenth century, when British colonisers were creating the first forest reserves in Singapore, they sought to combat the deforestation caused by the very farming practices for growing pepper and gambier that they had themselves developed on the islands. The reserves thus served two contradictory objectives: planting trees to protect soil and water resources; and cutting down trees to provide wood for the colony and for export. What mattered was to protect and so exploit, and on other occasions it was a matter of exploiting so as to protect. This was notably the case in East Africa, between 1955 and 1965, during a vast 'wildlife cropping' project devised by biologists and ecologists with the backing of the IUCN and the UK-based Nature Conservancy. This project was implemented in Queen Elizabeth Park (Uganda) and Tsavo Park (Kenya), the former being home to hippopotamuses and the latter to elephants. Among others, Raf de Bont has looked at this 'conservation by culling': raising wild animals, killing some of them, selling their meat to replace cattle in the African diet and thus replacing the sheep- and goat-rearing economy, seen as ecologically destructive, with a supposedly conservationist wildlife economy. This programme failed to produce any tangible results, but it is a particularly striking illustration of the inherent contradiction still affecting conservation policy today.[50] Geographer Johan Oszwald thus notes that in eastern Côte d'Ivoire, during French colonisation and then once Félix Houphouët-Boigny was president, export agriculture led to such over-exploitation that the protected forests of Béki and La Bossémati were the last remaining wooded territories in the region. And so by the early twenty-first century, while no exploitation was authorised within these park spaces, the surrounding land was exhausted in the wake of an influx of farmers who had left the over-exploited countryside looking for any remaining arable land. Protection on the one hand, predation on the other. This paradox is part of the very nature of conservation: it does not exist *alongside* destruction, but *in tandem with* it.

However contradictory it may be, conservation is nevertheless rational. Indeed, tropical nature governance seems constantly driven by the global production and circulation of knowledge about human–nature relations. Raphaël Devred shows this via the history of merino sheep raised at the Rambouillet agricultural station from the late nineteenth century to the 1930s, then sent to Algeria, Madagascar or French West Africa to be crossed with 'native' sheep. This project, backed by the industrial sheep lobby, was devised by livestock experts who theorised the 'degenerative' nature of African sheep and proposed this cross-breeding strategy to return them to 'their former balance'. This operation sought to regenerate African wool-producing breeds; it further sought to perfect the African 'race' by sedentarising nomads and converting them to a value-generating export economy. Managing tropical nature was thus primarily a matter of transforming it, of turning it into what it 'ought' to be, and then conserving it 'as it was'. This absolute confidence in science's capacity to improve nature and tropical societies developed during the colonial period, and lasted after independence, as did the careers of those championing it. Joseph Hodge points this out concerning the career of Arthur Hugh Bunting, a British agronomist and biochemist. From the 1930s to the 1970s, Bunting circulated between Africa, Europe and Asia. As a communist, he was convinced poverty could be solved by societal change, and to his mind science was the best tool for bringing this about. It was with this in mind that he worked for the Colonial Office in Gambia and in Sudan. Then after independence, as a professor in agricultural botany at Reading University, he established a global network of professionals working in rural development and agricultural science. From India to Nigeria, these employees of foreign affairs ministries or United Nations agencies disseminated knowledge and know-how promoting the tropics' agricultural 'development'. Science was thus central to transforming rural societies in general, and conservation areas in particular. In this respect, Simone Schleper emphasises the ever-important role played by technology. In Serengeti Park in Tanganyika, for example, scientists employed various techniques to study the migration of gnus, zebra and gazelles. Over time, this enabled them to assert their authority over wildlife management. But depending on the technologies available, and without necessarily being aware of it, their perceptions of human–nonhuman relations shifted, and they consequently applied different conservation policies. Thus in the 1950s and 1960s, aerial monitoring of ungulates showed seasonal migration, and local hunting that was likewise seasonal. This was deemed reasonable, and authorised. Then in the 1970s, aerial photography captured static landscapes. Where populations used to alternate between hunting and agro-pastoralism following the seasons, conservationists now saw the two activities lasting all year round: this was deemed destructive, and curtailed or sanctioned. Lastly, from the 1980s to the early 2000s, computer modelling pushed park managers to reason in mathematical terms: since the number of ungulates was stable, but human populations were increasing, hunting had to be banned. In the absence of the qualitative studies required to analyse the sociopolitical factors pushing certain populations to practise illegal hunting, conservation policies were now based on computer projections; these led to social exclusion due to the way human–nature relations were perceived and managed.

This link between representations of nature and nature governance leads us, thirdly, to the key influence of discourse and images. Historians of the environment have noted the extent to which imposing 'correct' usage of nature has involved assimilating 'correct' visions of this nature. But because representations engender new ontologies, they are also direct drivers of socio-environmental change. William Beinart demonstrates this in his study of the book and film about Elsa, the Kenyan lion of George and Joy Adamson (the author of *Born Free*). As the colonial era drew to a close in the 1960s, the Adamsons scuttled imperial hunting culture by publicising a human relationship with wild animals based on empathy, affection and non-utilitarian protection. This shift had got under way with the increasing number of African national parks, responding to the need for nature of increasingly urbanised Western societies. But the success of the Adamsons' photographs and tales did not stop there. By fashioning new global attitudes towards the wilderness, they played a part in 'reimagining postcolonial modernity'. The Kenyan government was for that matter involved in the process. Indeed in independent Asia and Africa, after having been placed in the service of colonial power, nature was now put to use for the political ends of new leaders. Grégory Quenet tracks this through the Seychelles archives. From the 1960s to the present day, thus spanning independence in 1976, two parallel processes may be observed. On the one hand, the British deliberately placed the most remarkable parts of the territory under the oversight of autonomous conservationist foundations in order to prevent the Seychelles people from taking control of these conservation sites. But on the other hand, in reaction to this, the authorities of the young country produced and disseminated the idea of a specific form of Seychelles environmentalism, drawing on a French tradition reinterpreted as a Creole figure in which nature was both inhabited and harmoniously exploited. Despite drawing publicly on this representation to combat dominant conservationist ideology, the Seychelles state cooperated with various international conservation institutions to implement it on the ground. Images of nature thus served power, while revealing its mechanisms. Or rather, power struggles, as anthropologist Meera Oommen shows for India, where the postcolonial recycling of British representations of nature has brought two societies into conflict: rural dwellers episodically hunting big game to protect their lives and fields have sparked the ire of town dwellers living far removed from wild animals that they wish to protect at all costs. Thus since the 1980s, tigers, Asian elephants and saltwater crocodiles have played a dual role, acting as 'flagships' for international conservation, yet also as 'battleships' in Indian social conflicts. In concrete terms, their protection now depends on power struggles involving the national administration, nature professionals, animal movements and village groupings. More fundamentally, conserving these species seems to clash with 'correct' usage of nature since, by definition, only one of the two groups has the 'correct' representation.

This global Indian history leads to our fourth conclusion: the need to accept the profound complexity of social relations to nature. Each author contributing to this work would have preferred to present analysis showing conservation policies to be a success. But the archives do not lie: most of the time, conservation fails to attain its objectives. That is why we believe it useful to apprehend the global conservation of tropical nature through the prism not of a model, but rather of situations determined by the eminently complex interlocking of different scales. First, spatial scales. Mathieu Guérin points this out in his study of the career of Captain Archibald Thomas Ayre Ritchie. After fifteen or so years as head of the Wildlife Department in British Kenya, Ritchie was sent by the Colonial Office to Malaya in 1937 to set up its Game Department and to draw up wildlife protection legislation. His intention was to take the conservation policy he had developed in Africa – based on the principles of culling big game to regulate conflicts with reserve inhabitants, and opening reserves to tourists – and transfer it to Asia. But he was hindered by three factors: the political influence of Malay sultans; horizontal power relations between colonial administrators and colonial subjects; and the singularity of European and 'native' hunting practices. So though he did professionalise the new Game Department in his fifteen months in Malaya, he left the colony without having fulfilled his objectives, recognising, despite himself, that above and beyond (or below) an empire and an imperial policy in the singular, there were also colonies and conservation territories in the plural. In this case, the spatial scales overlap; in others, time frames accumulate. The case of Mozambique is a good example. In 2000 a law was passed transforming all the country's protected areas into 'conservation areas for tourist purposes'. Since then, there have been continual clashes between nature managers and local users on the one hand, and national administrators on the other, decrying from the capital that 'use' now wins out over 'conservation'. So while the year 2000 marks a clear break, political scientist Rozenn Nakanabo Diallo shows that to understand it, we need to see how the colonial past weighs heavily on the present. Indeed, the Portuguese left a legacy of many practices and representations. Since independence in 1975, nature has been conserved 'for the benefit of the Mozambican people'; but the narrative of environmental deterioration persists, the state continues to view parks as tools for controlling territories and the private sector remains a major player not in 'enhancing' but in 'developing' them. Thus far from following one after the other, conservation ethics accumulate: exclusion practices specific to the conservation fortress period of the 1930s were overlaid with those from the community era starting in the 1980s, to which have been added, since the 2000s, principles relating to the financialisation of nature. This mixing of conservation times and spaces explains why local dynamics still predominate in global conservation. Human ecology specialist Pamela McElwee and geographer Diana Davis each point this out in their own way in their studies of conservation or, rather, of its partial and biased implementation. In early twentieth-century Vietnam, a series of measures was introduced in response to fears about a collapse in big game numbers in the colony. But these measures seem to have been dictated in Paris, not Hanoi. They were selectively introduced depending on the territory, imposing greater constraints on the Vietnamese population than on European tourists;

more generally, the colonial farming lobby and 'resident' elites put pressure on the colonial administration to prevent regulations drawn up in France being applied in the colony. In this instance, it is the colony–metropole dynamic that explains why conservation was ineffective. The history of Syria and Lebanon enables us to build on this hypothesis. In the interwar period, the new French administration brought in colonial agents from Morocco to implement a forestry policy. As in North Africa, they were convinced by the declinist theory that inhabitants had destroyed previously extensive dense forests. Unlike in Algeria and Morocco where these nature professionals had used reforestation as a tool in their civilising mission, forests in Syria became an instrument for economic development by improving agriculture, while in Lebanon they were used to boost tourism. As in Vietnam, using the land won out over protecting it. Only this time, the lack of conservation stemmed from objectives drawn up in Paris rather than in the colony. For France was not looking to build a colonial society in the Middle East, but above all to increase the profits extracted from the two territories placed under its 'mandate' while awaiting sovereignty. Above and beyond the colony–metropole dynamic, the explanation lies more generally in the tension between the general and the situated.

Global nature conservation in the Global South has changed in tune with the fluctuations in a permanent contradiction between predation and protection. Yet to some extent this contradiction has been toned down by knowledge and images circulating within an Afro-Asian area in which territories have overlapped, time frames accumulated and centre–periphery dynamics been recomposed in various ways throughout the twentieth century.

These are the initial findings of our joint research. They need to be tested in other fields to be confirmed or invalidated, and in any case developed. But there are four firm beliefs underpinning the following chapters that seem, to our minds, to be definitive.

The first is that there is an undeniable gap between standardised imperial and international conservation models and the many ways in which they have been applied in colonial and postcolonial societies. This observation leads us to our second belief: studying the contemporary history of conservation in Africa and Asia always entails revisiting the history of colonisation and decolonisation. This (post)colonial history does not follow a binary logic opposing local societies to empires and then to international institutions: there is a kaleidoscope of situations in which conservation is, to varying degrees, violent and exclusionary, and this violence is associated with the colonial order, its postcolonial legacy and the context in which these take shape. Our third belief, then, is that to understand conservation policies we need to go beyond an approach focusing solely on protected areas, which are only one way of governing people and things. It is thus important to consider the entire set of laws, practices and representations seeking to order the natural world by regulating hunting, establishing and turning tracts of land into parks and managing access to so-called natural spaces, together with agronomy, the domestication of animals and the process by which culture fashions the categories of town and nature, of 'local' and 'global'. Lastly, however indispensable it may be, a historical approach cannot suffice on its own. This book, devised as the first step in an ongoing enquiry, hopes to demonstrate this over the course of its thirteen chapters, combining history, anthropology, human ecology, political science and geography. It is only by being open to other social sciences that we may truly decipher and understand the relations binding human societies to the nonhuman world on which they depend.

Guillaume Blanc is Associate Professor of Contemporary History at the University of Rennes 2 and member of the Institut universitaire de France (2021–26), and works on the global governance of nature and people in Ethiopia and contemporary East Africa. He directs the 'Environmental History' series for Éditions de la Sorbonne and has recently published *La nature des hommes* (La Découverte, 2024) and *The Invention of Green Colonialism* (Polity, 2022).

Notes

1. Antoine Prost, *Douze leçons sur l'histoire* (Paris: Seuil, 1996), 80.
2. In French historical scholarship, the idea of 'space-times' (*espaces-temps*) has been used since the 1970s to emphasise the historically and geographically situated nature of global phenomena.
3. Achille Mbembe, *De la postcolonie: Essai sur l'imagination politique dans l'Afrique contemporaine* (Paris: Karthala, 2000), 36.
4. Gérard Noiriel, *Introduction à la socio-histoire* (Paris: La Découverte, 2006), 31.
5. See in particular: Guillaume Blanc, 'Governing Nature and Ethiopia: Struggles around World Heritage, Nation-Building and Ecologies (1963–2012)', *Northeast African Studies* 18(1–2) (2018), 137–64.
6. See in particular: Mathieu Guérin, 'Conserver la faune sauvage de la péninsule malaise: De la Malaya britannique à la Malaisie indépendante', *VertigO* 17(1) (2017), http://journals.openedition.org/vertigo/18503.
7. See in particular: Violette Pouillard, *Histoire des zoos par les animaux: Impérialisme, contrôle, conservation* (Seyssel: Champ Vallon, 2019), 250.
8. See in particular: Grégory Quenet and Jan Synowiecki, 'Ce que conserver veut dire: Praxis et historicité de la nature (1770–1810)', *Annales historiques de la Révolution française* 399(1) (2020), 97–121.
9. Paul Basu and Vinita Damodaran, 'Colonial Histories of Heritage: Legislative Migrations and the Politics of Preservation', *Past and Present* 226(10) (2015), 240–71.
10. James McCann, *Green Land, Brown Land, Black Land: An Environmental History of Africa, 1800–1990* (Portsmouth/Oxford: Heinemann/James Currey, 1999), 47–48.

11. William Beinart, 'African History and Environmental History', *African Affairs* 99(395) (2000), 269–302.
12. Dominique Guillaud, Chayan Vaddhanaphuti and Olivier Evrard (eds), *Mobility and Heritage in Northern Thailand and Laos: Past and Present* (Chiang Mai: Good Print/ Institut de recherche pour le développement, 2013).
13. Bruno Latour, *La science en action: Introduction à la sociologie des sciences* (Paris: Gallimard, 1995), 515.
14. Hélène Blais, Florence Deprest and Pierre Singaravélou (eds), *Territoires impériaux: Une histoire spatiale du fait colonial* (Paris: Publications de la Sorbonne, 2011), 8.
15. Isabelle Surun, 'L'exploration de l'Afrique au XIXe siècle: Une histoire pré-coloniale au regard des *postcolonial studies*', *Revue d'histoire du XIXe siècle* 32 (2006), 21–39.
16. William Beinart, Karen Brown and Daniel Gilfoyle, 'Experts and Expertise in Colonial Africa Reconsidered: Science and the Interpenetration of Knowledge', *African Affairs* 108(432) (2000), 413–33, here 432.
17. Edmund Burke and Kenneth Pomeranz (eds), *The Environment and World History* (Oakland: University of California Press, 2009), 9.
18. Yannick Mahrane et al., 'De la nature à la biosphere: L'invention politique de l'environnement global, 1945–1972', *Vingtième siècle: Revue d'histoire* 113(1) (2012), 127–41.
19. Anna-Katharina Wöbse, '"The World After All Was One": The International Environmental Network of UNESCO and IUPN, 1945–1959', *Contemporary European History* 20(3) (2011), 331–48.
20. Thomas Lekan, *Our Gigantic Zoo: A German Quest to Save the Serengeti* (Oxford: Oxford University Press, 2020).
21. William Adams and David Hulme, 'Conservation and Communities: Changing Narratives, Policies and Practice in African Conservation', in Institute for Development Policy and Management (ed.), *Community Conservation Research in Africa* (Manchester: University of Manchester, 1998), 1–31.
22. Jean-François Bayart and Romain Bertrand, 'De quel "legs colonial" parle-t-on?', *Esprit* 12 (2006), 134–60, here 145.
23. Corey Ross, *Ecology and Power in the Age of Empire: Europe and the Transformation of the Tropical World* (Oxford: Oxford University Press, 2017), 395.
24. William Adams, 'Nature and the Colonial Mind', in William Adams and Martin Mulligan (eds), *Decolonizing Nature: Strategies for Conservation in a Post-Colonial Era* (London: Earthscan, 2002), 16–50.
25. Marie-Christine Cormier-Salem et al. (eds), *Patrimonialiser la nature tropicale: Dynamiques locales, enjeux internationaux* (Paris: IRD Éditions, 2002).
26. Marie-Christine Cormier-Salem et al. (eds), *Ambivalences patrimoniales au Sud: Mises en scène et jeux d'acteurs* (Paris: IRD Éditions/Karthala, 2016).
27. Christophe Charle (ed.), *Histoire sociale, histoire globale?* (Paris: MSH, 1993).
28. Francesca Trivellato, *Corail contre diamants: De la Méditerranée à l'océan indien au XVIIIe siècle* (Paris: Seuil, 2016), 36.
29. Alan MacEachern, 'Writing the History of Canadian Parks: Past, Present and Future', *History Publications* 1 (2008): 1–9, here 7.
30. Pierre Bourdieu, *Réponses: Pour une anthropologie réflexive* (Paris: Seuil, 1992), 191.
31. Roderick Nash, *Wilderness and the American Mind* (New Haven, CT: Yale University Press, 1967).

32. Janet Foster, *Working for Wildlife: The Beginning of Preservation in Canada* (Toronto: University of Toronto Press, 1978).
33. John MacKenzie, *The Empire of Nature: Hunting, Conservation and British Imperialism* (Manchester: Manchester University Press, 1988); David Arnold and Ramachandra Guha (eds), *Essays on the Environmental History of South Asia* (New Delhi: Oxford University Press, 1995).
34. Roy MacLeod, 'Nature and Empire: Science and the Colonial Enterprise. Introduction', *Osiris* 15 (2000), 1–13.
35. See in particular Bernhard Gissibl, *The Nature of German Imperialism: Conservation and the Politics of Wildlife in Colonial East Africa* (New York: Berghahn, 2016).
36. Peder Anker, *Imperial Ecology: Environmental Order in the British Empire, 1895–1945* (Cambridge, MA: Harvard University Press, 2001).
37. David Anderson and Richard Grove, 'The Scramble for Eden: Past, Present and Future in African Conservation', in David Anderson and Richard Grove (eds), *Conservation in Africa: People, Policies and Practice* (Cambridge: Cambridge University Press, 1987), 1–12.
38. Richard Fitter and Peter Scott, *The Penitent Butchers: The Fauna Preservation Society 1903–1978* (London: Collins, 1978).
39. Guillaume Blanc, *The Invention of Green Colonialism* (Cambridge: Polity, 2022), 35–37.
40. Roderick Neumann, 'The Postwar Conservation Boom in British Colonial Africa', *Environmental History* 7(1) (2002), 22–47.
41. Melissa Leach and Robin Mearns, 'Challenging Received Wisdom in Africa', in Melissa Leach and Robin Mearns (eds), *The Lie of the Land: Challenging Received Wisdom on the African Environment* (Oxford/Portsmouth: James Currey/Heinemann, 1996), 1–33, here 8.
42. David Hulme and Murphree Marshall, 'Communities, Wildlife and the "New Conservation" in Africa', *Journal of International Development* 11(2) (1999), 277–85.
43. Birgit Müller, 'Comment rendre le monde gouvernable sans le gouverner: Les organisations internationales analysées par les anthropologues', *Critique internationale* 54(1) (2012), 9–18, here 10.
44. Timothy Luke, 'Environmentality as Green Governmentality', in Éric Darier (ed.), *Discourses of the Environment* (Oxford: Blackwell, 1998), 121–51.
45. Helen Tilley, 'African Environments & Environmental Sciences', in William Beinart and Joan McGregor (eds), *Social History & African Environments* (Oxford/Portsmouth: James Currey/Heinemann, 2003), 109–30, here 113.
46. Jonathan Adams and Thomas McShane, *The Myth of Wild Africa: Conservation without Illusions* (Oakland: University of California Press, 1996), 91.
47. Anderson and Grove, 'The Scramble for Eden', 3.
48. Anna Tsing, 'The Global Situation', *Cultural Anthropology* 15(3) (2000), 327–60.
49. Richard Grove, *Green Imperialism: Colonial Expansion, Tropical Island Edens and the Origins of Environmentalism, 1600–1860* (Cambridge: Cambridge University Press, 1995).
50. Raf de Bont, 'Abattre pour conserver: Protéines, organisations internationales et faune sauvage africaine (1955–1965)', in Guillaume Blanc, Mathieu Guérin, and Grégory Quenet (eds), *Protéger et détruire: Gouverner la nature sous les Tropiques (20e–21e siècle)* (Paris: CNRS Éditions, 2022), 189–211.

Bibliography

Adams, Jonathan, and Thomas McShane. *The Myth of Wild Africa: Conservation without Illusions*. Oakland: University of California Press, 1996.

Adams, William. 'Nature and the Colonial Mind', in William Adams and Martin Mulligan (eds), *Decolonizing Nature: Strategies for Conservation in a Post-Colonial Era* (London: Earthscan, 2002), 16–50.

Adams, William, and David Hulme. 'Conservation and Communities: Changing Narratives, Policies and Practice in African Conservation', in Institute for Development Policy and Management (ed.), *Community Conservation Research in Africa* (Manchester: University of Manchester, 1998), 1–31.

Anderson, David, and Richard Grove. 'The Scramble for Eden: Past, Present and Future in African Conservation', in David Anderson and Richard Grove (eds), *Conservation in Africa: People, Policies and Practice* (Cambridge: Cambridge University Press, 1987), 1–12.

Anker, Peder. *Imperial Ecology: Environmental Order in the British Empire, 1895–1945*. Cambridge, MA: Harvard University Press, 2001.

Arnold, David, and Ramachandra Guha (eds). *Essays on the Environmental History of South Asia*. New Delhi: Oxford University Press, 1995.

Basu, Paul, and Vinita Damodaran. 'Colonial Histories of Heritage: Legislative Migrations and the Politics of Preservation'. *Past and Present* 226(10) (2015), 240–71.

Bayart, Jean-François, and Romain Bertrand. 'De quel "legs colonial" parle-t-on?' *Esprit* 12 (2006), 134–60.

Beinart, William. 'African History and Environmental History'. *African Affairs* 99(395) (2000), 269–302.

Beinart, William, Karen Brown and Daniel Gilfoyle. 'Experts and Expertise in Colonial Africa Reconsidered: Science and the Interpenetration of Knowledge'. *African Affairs* 108(432) (2000), 413–33.

Blais, Hélène, Florence Deprest and Pierre Singaravélou (eds). *Territoires impériaux: Une histoire spatiale du fait colonial*. Paris: Publications de la Sorbonne, 2011.

Blanc, Guillaume. 'Governing Nature and Ethiopia: Struggles around World Heritage, Nation-Building and Ecologies (1963–2012)'. *Northeast African Studies* 18(1–2) (2018), 137–64.

———. *The Invention of Green Colonialism*. Cambridge: Polity, 2022.

Bourdieu, Pierre. *Réponses: Pour une anthropologie réflexive*. Paris: Seuil, 1992.

Burke, Edmund, and Kenneth Pomeranz (eds). *The Environment and World History*. Oakland: University of California Press, 2009.

Charle, Christophe (ed.). *Histoire sociale, histoire globale?* Paris: MSH, 1993.

Cormier-Salem, Marie-Christine, Dominique Juhé-Beaulaton, Jean Boutrais and Bernard Roussel (eds). *Patrimonialiser la nature tropicale: Dynamiques locales, enjeux internationaux*. Paris: IRD Éditions, 2002.

Cormier-Salem, Marie-Christine, Dominique Juhé-Beaulaton, Yves Girault and Dominique Guillaud (eds). *Ambivalences patrimoniales au Sud: Mises en scène et jeux d'acteurs*. Paris: IRD Éditions/Karthala, 2016.

De Bont, Raf. 'Abattre pour conserver: Protéines, organisations internationales et faune sauvage africaine (1955–1965)', in Guillaume Blanc, Mathieu Guérin and Grégory Quenet (eds), *Protéger et détruire: Gouverner la nature sous les Tropiques (20ᵉ–21ᵉ siècle)* (Paris: CNRS Éditions, 2022), 189–211.

Foster, Janet. *Working for Wildlife: The Beginning of Preservation in Canada*. Toronto: University of Toronto Press, 1978.

Gissibl, Bernhard. *The Nature of German Imperialism: Conservation and the Politics of Wildlife in Colonial East Africa*. New York: Berghahn, 2016.

Grove, Richard. *Green Imperialism: Colonial Expansion, Tropical Island Edens and the Origins of Environmentalism, 1600–1860*. Cambridge: Cambridge University Press, 1995.

Guérin, Mathieu. 'Conserver la faune sauvage de la péninsule malaise: De la Malaya britannique à la Malaisie indépendante'. *VertigO* 17(1) (2017), https://doi.org/10.4000/vertigo.18503.

Guillaud, Dominique, Chayan Vaddhanaphuti and Olivier Evrard (eds). *Mobility and Heritage in Northern Thailand and Laos: Past and Present*. Chiang Mai: Good Print/Institut de recherche pour le développement, 2013.

Hulme, David, and Murphree Marshall. 'Communities, Wildlife and the "New Conservation" in Africa'. *Journal of International Development* 11(2) (1999), 277–85.

Latour, Bruno. *La science en action: Introduction à la sociologie des sciences*. Paris: Gallimard, 1995.

Leach, Melissa, and Robin Mearns. 'Challenging Received Wisdom in Africa', in Melissa Leach and Robin Mearns (eds), *The Lie of the Land: Challenging Received Wisdom on the African Environment* (Oxford/Portsmouth: James Currey/Heinemann, 1996), 1–33.

Lekan, Thomas. *Our Gigantic Zoo: A German Quest to Save the Serengeti*. Oxford: Oxford University Press, 2020.

Luke, Timothy. 'Environmentality as Green Governmentality', in Éric Darier (ed.), *Discourses of the Environment* (Oxford: Blackwell, 1998), 121–51.

MacEachern, Alan. 'Writing the History of Canadian Parks: Past, Present and Future'. *History Publications* 1 (2008), 1–9.

MacKenzie, John. *The Empire of Nature: Hunting, Conservation and British Imperialism*. Manchester: Manchester University Press, 1988.

MacLeod, Roy. 'Nature and Empire: Science and the Colonial Enterprise. Introduction'. *Osiris* 15 (2000), 1–13.

Mahrane, Yannick, Marianna Fenzi, Céline Pessis and Christophe Bonneuil. 'De la nature à la biosphere: L'invention politique de l'environnement global, 1945–1972'. *Vingtième siècle: Revue d'histoire* 113(1) (2012), 127–41.

McCann, James. *Green Land, Brown Land, Black Land: An Environmental History of Africa, 1800–1990*. Portsmouth/Oxford: Heinemann/James Currey, 1999.

Mbembe, Achille. *De la postcolonie: Essai sur l'imagination politique dans l'Afrique contemporaine*. Paris: Karthala, 2000.

Müller, Birgit. 'Comment rendre le monde gouvernable sans le gouverner: Les organisations internationales analysées par les anthropologues'. *Critique internationale* 54(1) (2012), 9–18.

Nash, Roderick. *Wilderness and the American Mind*. New Haven, CT: Yale University Press, 1967.

Neumann, Roderick. 'The Postwar Conservation Boom in British Colonial Africa'. *Environmental History* 7(1) (2002), 22–47.

Noiriel, Gérard. *Introduction à la socio-histoire*. Paris: La Découverte, 2006.
Pouillard, Violette. *Histoire des zoos par les animaux: Impérialisme, contrôle, conservation*. Seyssel: Champ Vallon, 2019.
Prost, Antoine. *Douze leçons sur l'histoire*. Paris: Seuil, 1996.
Quenet, Grégory, and Jan Synowiecki. 'Ce que conserver veut dire: Praxis et historicité de la nature (1770–1810)'. *Annales historiques de la Révolution française* 399(1) (2020): 97–121.
Ross, Corey. *Ecology and Power in the Age of Empire: Europe and the Transformation of the Tropical World*. Oxford: Oxford University Press, 2017.
Surun, Isabelle. 'L'exploration de l'Afrique au XIXe siècle: Une histoire pré-coloniale au regard des *postcolonial studies*'. *Revue d'histoire du XIXe siècle* 32 (2006), 21–39.
Tilley, Helen. 'African Environments & Environmental Sciences: The African Research Survey, Ecological Paradigms & British Colonial Development 1920–1940', in William Beinart and Joan McGregor (eds), *Social History & African Environments* (Oxford/Portsmouth: James Currey/Heinemann, 2003), 109–30.
Trivellato, Francesca. *Corail contre diamants: De la Méditerranée à l'océan indien au XVIIIe siècle*. Paris: Seuil, 2016.
Tsing, Anna. 'The Global Situation'. *Cultural Anthropology* 15(3) (2000), 327–60.
Wöbse, Anna-Katharina. '"The World After All Was One": The International Environmental Network of UNESCO and IUPN, 1945–1959'. *Contemporary European History* 20(3) (2011): 331–48.

CHAPTER 1

Laissez-Faire Conservation

Nature Protection in Colonial Vietnam

Pamela McElwee

This chapter begins with a puzzle, namely the limited scope of nature conservation efforts undertaken in French colonial Vietnam (the colonies then known as Tonkin, Annam and Cochinchina). No significant protected areas were created during the decades when it was under French control (1860s–1954), in contrast to British and Dutch measures in other parts of Asia, and even compared to France's overseas colonies elsewhere. From 1900 onward, colonial empires and countries around the world attended early global environmental events to draw up conservation treaties; but though some of these gatherings were hosted by France, it in fact implemented no comprehensive conservation plans for its own Asian empire. This was despite Vietnam's wealth of exotic fauna and bounty of economically valuable flora, all of which were under pressure from hunting, agricultural expansion and economic exploitation. Yet when French colonial rule collapsed for good in 1954, Tonkin, Annam and Cochinchina did not have a single national park designated for conservation.

Given the rise of environmental activities elsewhere – in France, its other colonies and worldwide – together with the clear need for action as species declines were documented, how are we to explain this neglect of nature conservation in French colonial Vietnam? This chapter addresses this question by examining how global discussions about species and nature shaped the decisions taken for French Indochina only to a limited extent.

In some respects, Vietnam was typical of colonial attempts to manage species: just as the first conservation treaties focused on protecting game rather than on conservation or ecosystems per se, so too did the first laws in Cochinchina. Measures to regulate hunting focused on important game species, not those that were most endangered; they differentiated between European hunters and locals; and their main goal was to extract greater profit from hunting (through permits and taxes) rather than stop it altogether. Species were not treated as having any intrinsic value or ecosystem function, but viewed as providing economic benefits and sport. Such measures reflected elements of the 1900 London Convention for the Preservation of Wild Animals, Birds and Fish in Africa, which France had helped draft.

From 1900 to the 1920s hunting reserves were set up in many other colonies, with more protection-focused parks eventually being created in the 1920s and 1930s in South Africa, Malaya, India and elsewhere. But Vietnam remained an outlier. It was only in the late 1920s that a patchwork of flora and fauna regulations and restrictions focusing more closely on conservation started to emerge. This largely resulted from pressure by scientists and organisations in the metropole, who urged local authorities in Vietnam to consider the merits of a more organised system of protected areas and stricter measures to conserve species. Yet though studies were done, papers written and maps drawn up, there was never enough follow-through to create the types of national parks or reserves that were rapidly expanding elsewhere.

Overall, the lack of any real conservation effort under French colonial rule reflected the metropole's general attitude towards the colony, viewing it as a source of revenue and resources to be unevenly shared with local residents, and only rarely as warranting any sustained interest and investment.[1] This created a situation where the economic benefits of exploitation often took precedence over conservation.

There was also little sustained conservation campaigning back in France, unlike in other metropoles, such as lobbying by the Society for the Preservation of the Fauna of the Empire in Britain.[2] Only a few individuals called for better conservation for nature's sake, and they were often at odds with other lobbying groups, such as hunting societies and agricultural concessionaries whose words carried greater weight with local officials. The paucity of knowledge about flora and fauna also acted as a barrier: state officials often had no idea of trends in species numbers, nor did they have the expertise to manage ecosystems for conservation rather than exploitation.

This chapter recounts the history of nature conservation efforts in Indochina, using a variety of sources generated by colonial administrators and global gatherings, as well as by individual scientists, hunters and local people. These sources provide contrasting viewpoints on the role the colonial state ought to play in managing natural resources, conveying different values as to which natural elements were important. Colonial officials' attitudes often differed depending on how close they were to Paris, with the bottom rungs often resisting mandates and laws emanating from the Ministry of Colonies or from Hanoi. Thus the sources used here should not be taken as a definitive record, but as a partial sample of the different opinions and approaches held by a range of protagonists.

Colonial Conservation Trajectories

Historians of empire have produced a rich body of scholarship on the links between colonialism and nature. Colonial powers implemented regulations on the use of natural resources, from game to timber. They did so for many reasons, including fear of declining resource availability, the desire to dedicate reservations for the sole use by colonists and the wish to control newly colonised peoples.[3] Richard Grove has argued that the roots of modern environmentalism reach back to this period, with its focus on scientific knowledge and management, driven by fears of degradation.[4]

New scientific societies were founded in metropoles like London and Paris, and colonial powers were closely involved in the emergence of new science facilities such as botanical gardens and experimental stations.[5] Formal 'science' was often invoked to justify the regulation of local practices deemed deleterious.[6] In the colonies, new provisions were introduced for forestry, game and water use. The land was mapped, reservations were created and harvesting or trading in natural produce was regulated, with local resource use being criminalised.[7]

The perceived depletion in game in African colonies was one of the root reasons for the first international treaties on wildlife protection in the early years of the twentieth century.[8] Excessive hunting for sport, food and trade lent urgency to a meeting in London in May 1900 that resulted in the Convention for the Preservation of Wild Animals, Birds and Fish in Africa – which though signed by most colonial powers, including France, did not come into effect. Nevertheless, professional scientific and campaigning organisations kept these initiatives alive thanks to nascent conservation-minded networks.[9] The 1900 London meeting was followed by a Convention for the Protection of Birds Related to Agriculture, signed in 1902, an International Congress for the Protection of Landscapes in 1909, the first and second International Congresses for the Protection of Nature in 1923 and 1931, and the 1933 International Conference for the Protection of the Fauna and Flora of Africa.[10]

Even though France played a major role in these early nature conservation events, even hosting several in Paris, the country was ambivalent over how best to appreciate and manage nature. For example, French officials watered down the 1900 London Convention by rejecting restrictions on trade in feathers and higher tariffs on animal parts due to the demand for wild bird plumes from milliners in Paris.[11]

France also had only a small number of conservation champions, mostly associated with the Muséum national d'histoire naturelle in Paris and the Société zoologique d'acclimatation (SZA), which had been founded to encourage the exchange of plants and animals with the colonies so as to 'enrich' natural landscapes, as well as to exploit new economic possibilities.[12] Other like-minded organisations included the Ligue française pour la protection des oiseaux and the Société pour la protection des paysages de France. But these groups did not agree on what to protect, nor was their influence comparable to that of the Society for the Protection of the Flora and Fauna of the Empire and the Royal Society for the Protection of Birds in Britain.[13] Indeed, Anglo-American ideas of nature dominated international conservation conferences throughout the early to mid-twentieth century, with France playing a less prominent role.[14]

As others have noted, this reduced influence was partly because interest in conservation developed more slowly in France than elsewhere in Europe or America.[15] Different groups remained divided as to the importance of protecting nature and how best to achieve it, leading to successive and at times competing impulses. Debates took place between those who supported preservation – especially of spectacular landscapes and cultural heritage, rather than species and nature per se – and those moved by utilitarian sentiments, for whom nature existed to provide humans with benefits, such as timber, water or game.[16] Despite a 1906 law allowing for the protection of 'sites and natural monuments of an artistic character', there were few protected areas in France itself until a large reserve was set up in the Camargue in 1927, championed by the SZA.[17] The lack of experience with protected areas in France itself, combined with the absence of any strong interest in colonial nature conservation in colonies in South East Asia, resulted in less pressure for protected areas there.[18]

Hunting as a Driver of Conservation in Indochina

Wild animals in Indochina were seen as a resource to be managed: for economic profit from tourism in big game hunting, and to control depredation of agricultural fields by birds and small mammals. By the 1920s Indochina had become a popular destination for tourists seeking exotic game hunts, with most foreign visitors touring the southern part of Annam and Cochinchina. Elephants served both as quarry and as transportation to find other targets such as tigers and panthers. Other frequently hunted species included wild boar, deer, serow and wild cattle such as gaur and banteng, along with large birds such as marabou storks and pelicans (see Map 1.1).

The Didier and Defosse Safari Service and many related businesses (taxidermists, rifle rentals, refrigeration specialists) sprang up in Saigon to help promote hunting as a reason to visit. Tourists were authorised to bring shotguns and sporting rifles with two hundred cartridges per gun, while photographic safaris were organised for the more squeamish. These adventures were not the sole preserve of the rich. While it cost around USD 1,500 per person for a month-long first-class safari in the 1920s, including porters and elephant

Map 1.1. Game found in Indochina in the 1920s. From J. Bordeneuve, *Les grandes chasses en Indochine* (Saigon: Albert Portail, 1925). Archive.org, public domain.

Figure 1.1. Advertisement for Didier and Defosse, 1937. From Bureau officiel du tourisme indochinois, *Les grandes chasses en Indochine* (Saigon, 1937). Photo © BnF/Gallica, reproduced with permission.

transport, an eight-day tent safari with the promise of a tiger cost only 330 USD. Adjusted for inflation, such a trip would be around 5,000 USD today, an affordable luxury for the upper middle classes.[19]

Affordable tourism in the Far East meant that even inexperienced hunters could join; for many Americans it was cheaper and easier to get to than Africa, and, unusually, women were allowed to hunt in Indochina. The Tourism Office published lists of famous people – Hollywood stars, political scions, minor European royalty – who went hunting in the colonies.[20] But many professional hunters of the day disdained these amateurs and tourists, preferring to see their own activities as more noble and ethical, and requiring far more skill.[21]

The king of the safari was undoubtedly the tiger, and tour operators promoted their availability as a major draw. There were estimated to be up to eighty thousand tigers throughout Indochina at the turn of the century; Khánh Hòa province was even said to be 'peopled with tigers'.[22] These widespread magnificent beasts were considered a 'pest', so killing them was regarded as a favour to local residents (Hunting tigers was never restricted as they were perceived as dangerous to humans). François Josef Defosse, the founder of the Didier and Defosse Safari service, was unofficially known as 'the General Exterminator of Indochina's tigers' due to his aggressive efforts to hunt them (see Figure 1.1). Bagging a tiger to take home was advertised as practically certain if one were to visit Indochina: 'If you have piasters and a gun, you will

Figure 1.2. Tiger attacking the Marquis de Barthélemy, *Le petit journal* (17 February 1907). Photo © BnF/Gallica, reproduced with permission.

get a tiger'.[23] Mathieu Guérin notes that practically every hunting story from Indochina has a 'tiger episode'.[24]

Safari companies took elaborate steps to ensure a successful hunt, ranging from building permanent hides from which tigers could be viewed and shot, to burning open glades to create tiger habitat, and putting out freshly killed buffalo and then lying in wait for a tiger to arrive.[25] Despite these efforts to make it easier to find a tiger, hunts could still be dangerous, and there are numerous tales of animals attacking hunting parties. Some foreigners were even killed on hunts, while others, like the Marquis de Barthélemy, survived attacks and went on to pen best-selling books about their experiences (see Figure 1.2).

Even as tourism authorities were promoting safari travel to the region, officials back in the metropole, at the Ministry of Colonies, began to grow concerned about the need to protect the empire's flora and fauna from over-exploitation. This stemmed less from an interest in nature conservation per se than from a wish that countries continue to attract overseas visitors: 'We have been aware of the immense advantage likely to be drawn from the hunting resources contained in our overseas territories: in Indochina, and especially in Africa' began a 1922 report from the Commission permanente de la chasse coloniale,[26] which had been set up in 1911 under the auspices of the Ministry to ensure that hunting remained viable for sportsmen throughout the empire.

However, by the early 1900s it was already clear that certain birds such as egrets and ostriches were being over-exploited due to demand for their feathers driven by global fashion trends: import records indicate that in 1905 alone British consumers used the feathers of thirty million birds on their hats, with global demand running as high as three hundred million birds a year. Marabou storks, great egrets, little egrets and grey egrets all suffered, with Rach Gia in Kien-Giang province in Indochina '[being] known for its fan feathers'.[27]

The Lacey Act had been passed in the US in 1900 in response to the fast disappearance of certain bird species. It prohibited the sale and import of illegally hunted species, and other countries were under pressure to enact similar restrictions. The 1900 London Convention introduced further measures intended to halt the destruction of economically valuable animals, including: lists of species classified by their level of endangerment, with various restrictions on harvesting; hunting rules pertaining to types of equipment and seasons, and the issuing of permits; spatially limited game reserves; and regulations on trade and sale.

This utilitarian approach focused on the protection of game animals and waterfowl over non-game species; and on the conservation of prized prey rather than on the preservation of habitat.[28] Significantly, the Convention noted that reserves were needed to build up game populations, and a number were created across Africa in its wake. Yet such plans were largely neglected in Asia, leading the Marquis de Barthélemy, a former enthusiast of tiger hunts in Indochina, to assert: 'It is necessary to extend the effects of the [London] Convention, and after study, probably to make it applicable in all our Colonies. Big game is of tremendous value to our overseas possessions. It must therefore be protected, not in some but in all.'[29]

The first hunting laws in Indochina had focused on equipment, such as restricting the use of gunpowder to French colonists, though this probably reflected fears of rebellion more than any concern for animals. The first regulation specifically targeting particular species was introduced in 1889 in response to the decimation of egrets for the millinery trade; there were also strongly utilitarian motives, as egrets were viewed as useful for eliminating crop pests.

By the 1910s it was clear that more extensive approaches were needed. In 1909 revised laws specified where hunting could take place, with what equipment and at what times and seasons, with the sale of live game being prohibited out of season. In 1914, responding to pressure from Paris to tighten these rules, officials at the Service des eaux, forêts, et chasses proposed creating 'hunting reserves' in Cochinchina, where hunting pressure had been greatest and declines in game the most obvious.[30] The trade in egrets and marabou stork feathers from Cochinchina, once a major supplier, was completely prohibited by 1915, as was the indiscriminate hunting of other birds beneficial to agriculture due to their being 'great destroyers of rats and snakes'.[31]

In 1917, the local Forest Service office in Langbian (around present-day Dalat) classified the entire territory into zones designating where hunting could take place and how: in the reserved zone, only permit holders were allowed to hunt specific species, while in other areas restrictions were introduced on total amounts of game that could be bagged.[32] In the north, Tonkin banned elephant hunting as of 1917, for population numbers were significantly lower there.[33] Yearly laws were also passed in Tonkin on when the hunting season began and how much could be bagged, although the scientific justifications for these were often estimates rather than being based on formal studies.[34]

Further pressure to improve environmental management came from the Ministry of Colonies in response to rising global interest, particularly since France had hosted the 'Premier congrès international pour la protection de la nature, faune et flore, sites et monuments naturels' in Paris in June 1923. French scientific societies co-sponsored the meeting, including the SZA and the Museum of Natural History in Paris, and after the congress the Minister of Colonies asked the Museum to set up a new Comité national permanent pour la protection de la faune coloniale in 1925. This panel presented numerous suggestions for new restrictions to be applied throughout the empire; for example, Madagascar's first protected areas (nature reserves) were proposed and implemented at this time.[35] Yet while a list of possible reserves was proposed for Indochina as well (Table 1.1), none of them actually came into force, as discussed below.

Table 1.1. National Parks proposed for Vietnam by the Comité national permanent pour la protection de la faune coloniale, 1925. From Abel Gruvel, 'Projets soumis au Ministre des Colonies le 8 Avril 1925 pour la protection de la faune colonial', in *Rapports, voeux, réalisations, revus et annotés: Congrès international pour la protection de la nature* (Paris: Guillemot et de Lamothe, 1926), 355. Cornell University, public domain.

Location	Reason	Proposed size
Tonkin		
Lao-Kay-Chapa	Collarless pheasants, deer	Small
Halong Bay	Monkeys, goats	Medium
Babe and Langson lakes region	Ring-necked pheasants, spotted deer	Large
Annam		
Dalat	Eld deer, elephant	Medium
Darlac	Elephants, deer	Large
Vinh-Nape	Elephants, deer	Medium
Col des Nuages	Edwards' pheasants	Small
Quang-Tri-Donghoi	Sambar and Eld deer, oxen, elephants, birds	Medium
Cochinchina		
Trang-Bom Forest	Not specified	Medium
Camao peninsula	Waders and waterfowl	Large
Cap Saint-Jacques	Small birds	Small
Poulo Condore island	Not specified	Whole island

Small: 50,000 hectares and above; medium: 500,000 hectares and above; large: 2,000,000 hectares and above.

Conservation or 'Civilising' the Natives?

The Commission permanente de la chasse coloniale also considered the benefits of advocating more aggressive conservation measures. In a meeting in 1922, the Vicomte de Poncins noted:

> the protection of widely distributed species and increasing [species population] numbers will be viewed favourably by amateur expeditions to our colonies, who have currently become attached to foreign East Africa. With them, these expeditions will bring not only money, but, better, the complete exploration of the distant parts of our possessions, their scientific study, and

> friendly and frequent contact with the natives who are essential for hunting expeditions, as well as for facilitating the work of civilisation.[36]

Coupling species conservation with a civilising mission sought to reduce the impact of 'native' hunting in the colonies. Despite the clear consequences both of colonial hunters themselves and of demand in London and Paris for animal products such as feathers and ivory, it was very common for officials to blame destruction on locals. One report from a delegate to the Colonial Hunting Commission estimated that:

> 997 out of 1,000 elephant tusks sold in the London market are of native hunting origin – only 3 of this number can be attributed to the rifle of a European! . . . But civilisation has exerted a constantly growing demand even in the remotest parts of the dark continent. The taste for profit has developed proportionally in the native, resulting in excessive destruction . . . destruction against which our country alone has not yet taken comprehensive and truly effective measures.[37]

Blaming locals was common in Indochina. Fernand Millet, a forest inspector and professional hunter in Annam – who bragged of having personally killed more than a hundred tigers – expressed concern that game numbers were declining as a result of native hunting. Millet was particularly incensed by the native practice of 'fire beats [*battues au feu*]' in which '100 to 150 deer, of which many were does and fawns' could perish; he likewise denounced the use of pits, which were indiscriminate in catching 'a brood of partridges or quail, the barely born youngling, the fawn which is still suckling its mother, the pregnant doe, large male or female cattle, and even the imprudent elephant'. He argued that 'the number of big game killed or crippled by traps exceeds in a single year the total kills that could be achieved in ten years by all the European hunters in the country'.[38] Another report noted that local practices were leading to a decline in tiger populations, which had started 'to become scarce in Cochinchina, as its main food, deer, are being slaughtered indiscriminately by the disastrous practice of lantern hunting'.[39]

The preferred solution was to restrict local use while allowing for tourism together with revenue from hunting taxes and permits. Many of the laws reflected outright discrimination against native hunting, stating that only French citizens could perform certain deeds. Other laws favoured European methods of hunting, namely in large parties with guns and dogs, as against native methods, such as hunting with lanterns at night, with snares or using fire.[40] Any justification for such distinctions between the impact of different hunting methods was dubious, for they were never clearly measured through scientific study. Instead, regulations often stemmed from conceptualisations of cultural superiority and a vision of a virtuous French sportsman hunter.[41]

When a new requirement to pay for hunting permits came into effect in Cochinchina in 1914, French colonists and tourists were able to afford such fees, but locals often could not. Foreign tourists were largely unaffected by the hunting restrictions, as they could hunt any time of year provided they paid for a permit.[42] A *Permis A* cost 80 piasters and allowed hunters to bag one elephant in Cochinchina and two in Annam, three and five gaurs respectively, three and four wild buffaloes and three and six wild bulls; though it was forbidden to shoot any rhinos in Cochinchina, one rhinoceros could be killed in Annam per permit. Once dead, a head tax was assessed: 60 piasters per male elephant and rhinoceros, and 40 piasters per gaur, wild buffalo or bull (a *Permis B* allowed hunting of any other kind of unprotected animal). Tourists were allowed and even encouraged to take trophies, while at the same time laws often restricted the sale of live game – to the detriment of locals.

French gentlemanly hunting societies, such as the Saint Hubert Club Haiphonnais, complained to state administrators that locals largely ignored hunting laws, continuing to hunt out of season and at night, and to sell game unimpeded.[43] A 1932 letter from the *Résident Supérieur* in Tonkin to the police department complained that locals were flouting edicts of 1909 and 1914 by selling game in all seasons 'in very large quantities' in front of the Hotel Metropole in downtown Hanoi. The matter had been brought to the *Résident*'s attention by a representative of the Indochinese travel and tourism agency who was concerned that the large quantity of game on sale meant fewer hunting opportunities for others.[44]

Provincial chiefs were urged to crack down on hunting in their territories so as to prevent game from reaching these urban markets. A letter from the *Résident* in Thái Nguyên Province to district officials complained that 'surveillance of poaching is non-existent in the province despite the numerous instructions I have given on this matter. . . . I ask the Canton Chiefs to step up their activity to prevent the complete destruction of game, some of which are useful for agriculture.'[45] Other local *Résidents* were not so sure that restrictions on game were appropriate, or that bans were effective. Some *Résidents* stated that they were unable to keep tabs on the remote villages under their administration, and that villages could not be asked to monitor their own game levels.[46] There was also concern that too strict an enforcement of hunting laws might lead to local unrest and discontent. In a 1931 report the *Résident Supérieur* of Tonkin noted that 'villages draw appreciable resources from capturing passing birds and waterfowl, wood pigeons, turtledoves, snipe, wild geese, ducks, and teals. I would be prepared to tolerate the trapping and the sale of these species only for the villages which are in these conditions, and therefore, if necessary, you will send me the list. This tolerance will allow part of the population to be able to obtain game and will probably not cause any discontent among hunters'.[47]

Hunting Reserves and Conflicts over Management

After recommendations from the Commission for the Protection of Colonial Fauna in 1925, the Ministry of Colonies proposed tightening hunting laws and setting up reserves. A new decree on the Regulation of Hunting in Indochina signed by the French president in 1927 allowed for stricter hunting measures and the establishment of national parks (*parcs nationaux de refuge*) where 'hunting and disturbance will be prohibited to allow wildlife to continue to live in peace, sheltered from any development, tourism, or hunting'.[48]

Despite this 1927 decree there were no moves by local administrators in Indochina to delimit and designate national parks. Indeed, in one of the few areas to have adopted zoned hunting regulations (the Langbian plateau, where to all intents and purposes a 58,000 ha game reserve had been established in 1917), the system had been dismantled by 1927 due to the expansion of agricultural concessions in the province.[49]

Game reserves had been set up in colonies in South Africa, Sudan and German East Africa even before 1900, and other reserves focusing on conservation and tourism eventually emerged.[50] For example, a 1926 National Parks Act in South Africa introduced a colony-wide system of protected areas.[51] In other French colonies reserves were also created or expanded in the 1920s, including seventeen in French West Africa, thirteen in Algeria and twelve in Madagascar, along with other natural areas in Tunisia and Equatorial Africa.[52] These reserves served varying functions: some were for tourism, some for restoration and reforestation and some for preservation and scientific research.[53] In Madagascar in particular, reserves were placed under the scientific oversight of the Muséum national d'histoire naturelle, which introduced a ban on killing many types of bird and sea turtle, while several reserves focused on lemurs.[54]

Elsewhere in Asia, the Dutch East Indies had brought in protection for some forests by 1909 and introduced species protection laws.[55] In British Malaya, a number of reserves were set up as early as 1903, when the Chior Game Reserve was established to protect gaur populations, with ten more reserves being added by the 1920s.[56] Yet Indochina remained stubbornly resistant to more game reserves and to national parks. A 1931 review of conservation laws around the world included detailed sections on Malaya, Japan and the Dutch Indies, but just one sentence on French Indochina, noting that there were 'plans' for protected areas and some restrictions on hunting, but nothing more.[57]

The 1927 hunting decree referred both to national parks and to temporary hunting reserves, and this latter, less strict option was eventually taken up. In 1932, the *Résident Supérieur* of Tonkin (RST) decided to encourage local implementation: in each province, at least 5 per cent of the land was to be enclosed in temporary hunting reserves for three years. The areas would rotate, with the first reserved area completely closed for three years, then reserve number 2 closed for three years, and so on.[58] The RST noted that the reserves should be based on boundaries marked by roads, canals or rivers, so that hunters would know where each zone began and ended (see Map 1.2).[59]

Because the reserves were based on the successful repopulation of game animals, the schedule could be changed, and if game levels did not recover, the reserves could be closed indefinitely.[60] While this was a step towards the type of game reserves set up in Africa and other parts of Asia in previous de-

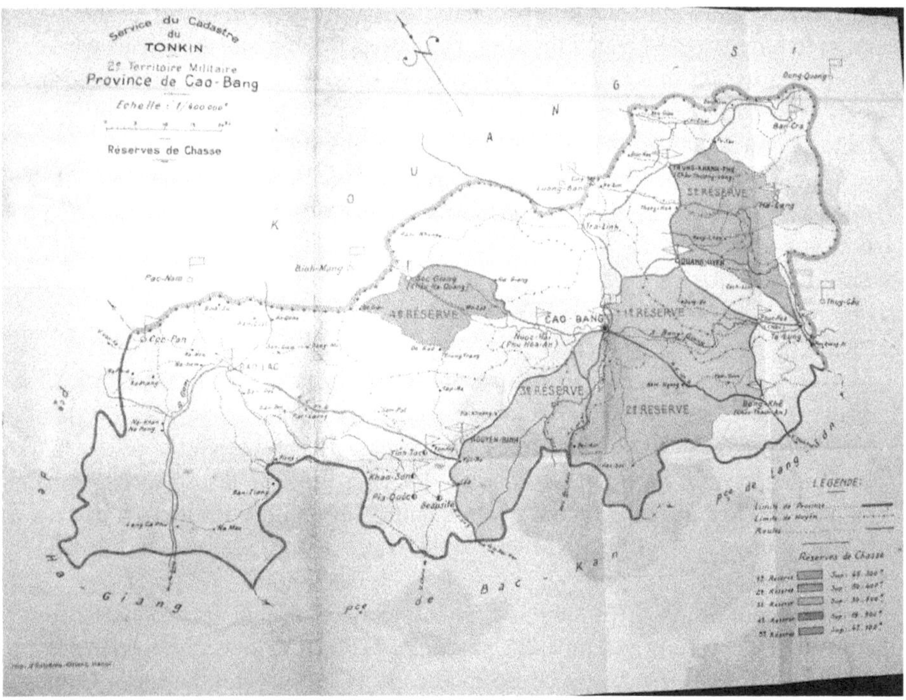

Map 1.2. Example of proposed hunting reserve schedule in Cao Bang province. From Résidence supérieure du Tonkin, *Recueil des textes relatifs à la réglementation de la chasse* (Hanoi: Imprimerie d'extrême-orient, 1936). © BnF/Gallica, reproduced with permission.

cades, their temporary nature compared unfavourably to permanent reserves elsewhere.[61]

In addition to these new temporary reserves, increased attention was paid to the particular needs of certain species. The 1927 decree had established the categories of *reserved game*, *protected game* and *free game*, to be defined by the Governor-General, while enforcement was entrusted to the local forest services, which were already under-staffed.[62] A 1935 regulation from the Ministry of Colonies attempted to address this problem by allowing for 'hunting lieutenants' (who had to be French citizens) to act as technical advisers to local authorities. The advisors were to be consulted on all questions relating to game tourism, the protection of fauna, the organisation of hunting and the monitoring of reserves.[63]

An additional measure by the Ministry of Colonies in 1936 required more stringent lists of species for the highest level of protection; this was the last major hunting regulation in Indochina (Table 1.2). The law not only applied across all five divisions of Indochina but also expanded the number of reserved species that could not legally be hunted, which now included rhinoceroses, bears, tapirs, sloths, pangolins, antelopes, gibbons, langurs, monkeys, lorises, flying squirrels and many birds. Yet the shortage of local rangers to police such measures meant these restrictions were rarely enforced.

Flora and Landscapes in an Expanding Conservation System

While hunting pressure was the clear impetus for the evolution of game reserves, over-exploitation of forests spurred interest in botanical reserves. A Forest Service (*Service forestier*) had been established for each division of the Indochinese Union at the turn of the century, along with local forest laws. The Forest Service created different types of 'reserved' and 'protected' forests, which though often justified on environmental grounds were not in fact conservation areas. Rather, they were subject to different levels of methodical treatment through rotational logging plans, with rights to timber being sold at public auction – as for state forests in France – and the tax revenues going to colonial authorities. Forest regulations also sought to restrict local forest use, to prevent the use of fire and swidden agriculture and to keep populations out of the reserves for which the French Forest Service had exclusive authority.[64] In some areas, such as the Mekong Delta, swamp forests dominated by cajeput (*Meleleuca* spp.) trees were treated essentially as 'wastelands' and managed by the Department of Public Works, not the Forest Service. In such cases locals were often encouraged to convert the trees to more economically valuable charcoal, while dredging operations were carried out to clear canals through the peatlands.[65]

Table 1.2. Major laws and restrictions on hunting in Indochina, 1887–1936. Table data from: J. Bordeneuve, *Les grandes chasses en Indochine* (Saigon: Albert Portail, 1925); Résident Supérieur of Tonkin, *Recueil des textes relatifs à la réglementation de la chasse* (Hanoi: Imprimerie d'extrême-orient, 1936). © Pamela McElwee.

Year	Key regulations	Where	To whom did the law apply	To what species did the law apply
1887	*Restrictions on equipment*: – gunpowder for hunting restricted to five grams	All colonies	All 'natives' or 'Asian foreigners'	
1889	*Restrictions on who can hunt*: – Hunting of egrets is limited	Cochinchina	Hunting is forbidden to foreign Europeans (non-French) and to Asians. It can only be practised by French citizens who can hunt with the assistance of Annamese	Egrets
1909	*Restrictions on timing of hunts*: – 1 October to 15 March only *Restrictions on sale of game*: – can only be sold during open season	Tonkin	Applies to all	Hares and sedentary game birds (partridges, wild hens, pheasants, peacocks, woodcock, snipe)
1914	*Restrictions on type of hunting*: – Beating with nets, calls, decoys and traps prohibited from 1 February to 1 June – Hunting with horns and cries, shooting and hounds authorised throughout the year *Restrictions on sale of game*: – live game sales forbidden year-round	Cochinchina	Applies to all	Deer, wild boars, hares, peacocks, wild hens, pheasants, partridges, quail
1914	*Restrictions on where hunts can take place*: – Creation of hunting reserves for certain species *Restrictions on who can hunt*: – Hunter must have hunting license and pay a fee *Restrictions on numbers of animals:* – limits to number that can be hunted	Cochinchina	Applies to all	Pachyderms and ruminants, including elephants, rhinos, buffaloes, deer of all kinds

Year	Key regulations	Where	To whom did the law apply	To what species did the law apply
1915	*Restrictions on species*: – Ban on the hunting and capture of egrets and marabou stork and on the sale of their remains anywhere	Cochinchina	Applies to all	Egrets and marabou storks
1917	*Restrictions on where hunts can take place*: Province of Lang-Bian is divided into three different zones: 1st Reserved area (880 km^2): requires permit for hunting. Fee for permit: 200 piasters 2nd Protected area of Dalat: No hunting of gaur, female elephants, Eld deer or female deer (*con-nai*) Hunting, trapping and sale of partridges, quail, hares prohibited from 15 March to 15 September 3rd Free zone: Hunting allowed at all times except for females of Eld deer, elephant and wild cattle. Permit (30 piasters) for hunting 1 male gaur.	Langbian province (Annam)	Customary rights of natives to hunt with native weapons applies, with the exclusion of trapping tools. Game that would cause crop damage may be hunted.	Permit allows hunting of: 1 gaur, 1 elephant and 3 wild oxen (*Bo-rung*). All other game except tigers/panthers not allowed.
1917	*Restrictions on species*: – No elephant hunting except in case of authorisation by the Resident Superior	Tonkin	Applies to all	Elephants
1925	*Restrictions on timing of hunts*: – All hunting in the colony to be restricted to seasons *Restrictions on who can hunt*: – All hunters need a permit for guns *Restrictions on where hunts can take place:* – No one to have the right to hunt on the property of others without owner's consent *Restrictions on type of hunting*: Lantern hunting prohibited Punishments: fine of 150 to 250 Fr and/or hunting ban for a period of one year	Cochinchina	To get a hunting licence: must be listed on a village roll, cannot have criminal offenses including rebellion, must be older than 16 Tourists and naval officers may purchase permits A permit is not required for natives who hunt by day with spears, crossbows or bows	Destructive animals (to humans or crops) allowed without a permit

(*continued*)

Table 1.2. *Continued*

Year	Key regulations	Where	To whom did the law apply	To what species did the law apply
1927 General law on hunting in Indochina	*Restrictions on who can hunt*: – Permits required for firearms; hunting permits of different kinds to be set up and re gularised across Indochina: *grand permis de chasse*, *permis ordinaire*, and *licence de destruction (permis de capture scientifique* also allowed) *Restrictions on where hunts can take place:* Limits to amounts bagged to depend on local decrees. No hunting allowed in *réserves de chasse* or *parcs de refuge* *Restrictions on types of hunting*: No lantern hunting at night (because of accidents)	All Indochina (from Minister of Colonies)	Free personal permits may be issued by Local Administration heads to officials and European agents No permit required for hunting with bows, arrows or spears	Local authorities asked to set up lists of *gibier réservé*, *gibier protégé* and *gibier libre*. Animals harmful to livestock and agriculture may be hunted at all times (e.g. tigers, panthers, leopards)
1931	*Restrictions on where hunts can take place:* Establishing restrictions on hunting across colonies *Restrictions on species*: No hunting of protected species during mating season in Cochinchina	Cambodia and Laos: free zones for hunting Tonkin, Annam and Cochinchina: hunting territories that require permits and have restrictions		Different restrictions across colonies for *gibier protégé* Cochinchina: waterfowl, wild chick-ens, partridges, peacocks, cranes, quail, hornbills, storks, deer, pelicans, ibis, herons, antelopes Annam: elephants, gaur, wild cattle, muntjacs, deer, banteng, gaur Tonkin: pheasants and partridges
1935	*Restrictions on timing of hunts*: – Hunting season from 1 October to 1 February for protected species; free species may be hunted 1 August to 1 May. No hunting of *gibier réservé*			Removal of wild boars from list of protected species due to crop damage *Gibier réservé*: Peacock, egrets and elephants

Year	Key regulations	Where	To whom did the law apply	To what species did the law apply
1935 (*continued*)				Harmful animals that may be hunted at any time: tigers, panthers, leopards, bears, civets, wild cats, foxes, weasels, polecats, birds of prey (eagles, vultures, kites, etc)
1935	*Restrictions on types of hunting:* – Aerial hunting is prohibited	All Indochina		
1935	*Restrictions on where hunts can take place:* – Selected temporary hunting reserves (*réserves temporaires de chasse*) to be established to repopulate species (5 per province, rotating every 3 years) – Owners of agricultural concessions have the right to make their own hunting restrictions	All provinces in Tonkin		
1936	*Restrictions on species:* – New lists of species that are reserved and protected for all colonies	All Indochina		*Gibier réservé* in all hunting areas: rhinoceroses, bears, tapirs, sloths, pangolins, antelopes, gibbons, langurs, monkeys, lorises, flying squirrels, storks, egrets, cranes, ibis, herons, pheasants, peafowl, cuckoos, drongos, weavers, pythons and other non-venomous snakes in Cochinchina Restriction on number to be hunted: elephants, gaur, wild buffalo and cattle Restrictions on males only: list above plus deer, muntjacs Minimum size requirements: elephants and gaur

Many scientists and observers pointed out that greater protection for plants was needed, as existing forest reserves were solely about economics, not 'the biological necessities of protecting fauna and flora'.[66] Botanists Henri Lecomte and Paul Alfred Pételot argued for setting up botanical reserves to save 'virgin nature', though they grounded this in utilitarian purposes, such as the future discovery of economically valuable species.[67] Other prominent scientists such as Auguste Chevalier from the Muséum national d'histoire naturelle in Paris, who had previously worked as a taxonomist in Indochina, told the 1923 Congrès international pour la protection de la nature that French colonies needed better nature protection measures. He urged officials to inventory as soon as possible natural sites 'where the vegetation and the fauna – with the exception of venomous snakes and possibly certain large predators – will be preserved intact in their natural state'.[68]

In 1932 the Governor-General of Indochina eventually proposed that the Forest Service conduct a study to determine the measures to be taken to protect flora. One park for the protection of both flora and fauna was proposed for Chapa (current day Sapa), while Cochinchina proposed a small number of plant reserves, including a '*réserve de Palétuvier*' for mangroves in Baria province.[69]

In addition to botanical reserves, there was also interest in protecting picturesque locations and landscapes, similar to parks in France that had been founded for reasons of tourism and aesthetics. Where these sites were threatened with development, creating a park or reserve could forestall such dangers. One example played out in 1927, when the *Résident* in Đồng Hới province wrote to the *Résident Supérieur* of Annam (RSA) with a dilemma: he was happy to report that mining by the Phosphates Company of Tonkin was rapidly expanding in the area, but the company had made an application to exploit an area called '*phong nha*' (wind and teeth), which included the entrance to the Cu Lake cave, 'one of the most remarkable of Indochina'. The *Résident* asked, 'what are the texts on which I can rely for the possible protection of this site and what restrictions are needed relating to mineral exploration in such cases', particularly as the site was already recognised as a picturesque location in local guidebooks to Annam.[70]

The *Résident* had come up with an ingenious though temporary solution to halt the mining: having the caves listed as 'historic monuments', because they contained Cham artifacts as well as spectacular nature. He noted however that the 1925 historical preservation law he was relying on did not clearly allow for the classification of caves as protected historical monuments, and that the law was inadequate because it would need to protect not just one 'building' (the cave) but also to prohibit any drilling nearby that might imperil the geological formations. The RSA forwarded the request to the Governor-General, who stated that the local areas could decide on any measures deemed appropriate, allowing the RSA to approve Phong Nha as a protected landscape. It is now recognised as Phong Nha-Ke Bang National Park, the world's widest limestone cave and a UNESCO World Heritage Site.

These ad hoc protections for landscapes could not address the growing pressure on resources in Vietnam caused by development, whether for mining, agriculture or other products. Just as groups like the SZA and hunting clubs had called for more species protections, new groups sprang up to campaign against the loss of flora and forests, with the Groupe d'action contre le déboisement de l'Indochine being founded in 1936.[71]

The Ministry of Colonies again asked the Governor-General in Hanoi to draw up expanded lists of endangered flora and fauna in Indochina, not just for game species, and to consult expert botanists if necessary to detail the loss of flora. The Governor in turn wrote to the *Résidents* of Tonkin, Annam and Cochinchina in 1936, asking them to compile such lists, adding that one impetus for such information was a forthcoming international conference in London for the protection of Asian fauna and flora (which did not in fact take place).[72] The lists were to contain any animal and plant species threatened with extinction that could benefit from special protection measures.

In response, the RST noted that this kind of in-depth scientific assessment was difficult if not impossible, as the Chef des services agricoles in Tonkin did not have sufficient expertise, and suggested that the education department should look into the possibility of calling in outside scientists.[73] The RST also noted that reserved game had been legally protected from hunting since 1931. However, officials also noted that regulations alone had proved ineffective, and that 'it would be difficult, without the creation of nature reserves [*réserves naturelles*] or parks [*parcs de refuges*] to increase the protection these animals actually receive'.[74] Accordingly, the RST proposed (again) creating a park for the protection of flora and fauna in Chapa, and parks for the protection of fauna in the provinces of Yen Bay, Phu Tho, Tuyen Quang, Bac Kan and Thai Nguyen Quang, though apparently nothing was ever finalised.

In 1939 there was another flurry of communication between the RST and the Governor-General's office regarding these unfinished plans for 'national parks', but by then it was too late – the Japanese occupation of Indochina in 1940 ended all hopes of expanding protected areas. It was not until 1959 that South Vietnam declared the first national park at the former French arboretum at Trang Bom, while North Vietnam established Cuc Phuong National Park in 1962.[75]

What can we make of the story of French colonial conservation efforts in Vietnam? One of the justifications for French colonial rule was the need for scientific management of resources and the prevention of 'native destruction':

as one observer noted, 'in Cochinchina and in Cambodia, all that is wood has long been devastated in all accessible areas, either for the needs of the nation, or by the destructive habits of the natives'. He went on to ask: 'Have we improved this state of affairs? Have we made it worse?'[76] The answer is probably the latter, as the so-called scientific management of game and timber was almost never based on empirical studies, but rather on concerns about degradation raised far away in the metropole.

Early game regulations around the world eventually gave way to more conservation-oriented international treaties, and in other countries to protected territorial areas. But in Indochina change was begrudging and slow, often in response to what Paris wanted without much enthusiasm in local provinces. In fact, global treaties for stricter nature conservation like the London Convention were watered down by France, then these watered-down approaches were transferred outward from the Ministry of Colonies in Paris, before being even further watered down in the colonies. This process can be seen in the calls for national conservation parks for Indochina that never came to fruition, with only 'temporary' and weak game reserves being put in place that were rarely seriously enforced.

The primary regulations in place for most of the early twentieth century were minimal hunting laws, aimed not at protecting nature but rather at maintaining economic and social interests via the continued presence of game for tourists and sportsmen. Further, many of these regulations reflected colonial notions of race and culture and applied unevenly to natives and colonists. Pressure from advocacy groups and scientists eventually forced some hesitant steps towards conservation, such as the numerous calls for better-protected reserves, but these never got off the ground. The continual 'studying' of conservation reserves that were never actually implemented indicates the lack of enthusiasm on the part of many local administrators.

In the end, France failed to leave a legacy of protected areas in Vietnam, and only a small number of currently important parks (especially Phong Nha) can be attributed to conservation from this period. In this, Vietnam stands alone among former colonies in Asia, giving a sense of France's failure to provide clear priorities in its colony regarding nature's value to humans.

Pamela McElwee is Professor of Human Ecology in the School of Environmental and Biological Sciences at Rutgers University (New Brunswick, NJ). She earned a joint PhD in anthropology and forestry at Yale and since 1996 has conducted research in Vietnam. In 2017, the European Association for Southeast Asian Studies (EUROSEAS) awarded the prize for best social science book on South East Asia to her book *Forests are Gold: Trees, People and Environmental Rule in Vietnam*.

Notes

1. Pierre Brocheux and Daniel Hémery, *Indochina: An Ambiguous Colonization, 1858–1954* (Berkeley: University of California Press, 2011).
2. Richard Fitter and Peter Scott, *The Penitent Butchers: The Fauna Preservation Society 1903–1978* (London: Fauna Preservation Society, 1978).
3. William Beinart and Lotte Hughes (eds), *Environment and Empire* (Oxford: Oxford University Press, 2007); Kalyanakrishnan Sivaramakrishnan, *Modern Forests: Statemaking and Environmental Change in Colonial Eastern India* (Stanford, CA: Stanford University Press, 1999); John MacKenzie (ed.), *Imperialism and the Natural World* (Manchester: Manchester University Press, 1990).
4. Richard Grove, *Green Imperialism: Colonial Expansion, Tropical Island Edens, and the Origins of Environmentalism 1600–1800* (Cambridge: Cambridge University Press, 1995).
5. Marie-Noëlle Bourguet and Christophe Bonneuil, 'Présentation: De l'inventaire du monde à la mise en valeur du globe. Botanique et colonisation (fin 17e siècle–début 20e siècle)', *Revue française d'histoire d'outre-mer* 322–23 (1999), 7–38.
6. Kalyanakrishnan Sivaramakrishnan, 'State Sciences and Development Histories: Encoding Local Forestry Knowledge in Bengal', *Development and Change* 31(1) (2000), 61–89; Richard Grove, 'Scottish Missionaries, Evangelical Discourses and the Origins of Conservation Thinking in Southern Africa 1820–1900', *Journal of Southern African Studies* 15(2) (1989), 163–87; William Beinart, 'Soil Erosion, Conservationism and Ideas about Development: A Southern African Exploration, 1900–1960', *Journal of Southern African Studies* 11(1) (1984), 52–83.
7. Greg Barton, *Empire Forestry and the Origins of Environmentalism* (New York: Cambridge University Press, 2002); Nancy Peluso and Peter Vandergeest, 'Genealogies of the Political Forest and Customary Rights in Indonesia, Malaysia, and Thailand', *Journal of Asian Studies* 60(3) (2001), 761–812.
8. John MacKenzie, *The Empire of Nature: Hunting, Conservation, and British Imperialism* (Manchester: Manchester University Press, 1997).
9. Raf de Bont, Simone Schleper and Hans Schouwenburg, 'Conservation Conferences and Expert Networks in the Short Twentieth Century', *Environment and History* 23(4) (2017), 569–99.
10. Mark Cioc, *The Game of Conservation: International Treaties to Protect the World's Migratory Animals* (Athens: Ohio University Press, 2008), 40–42; Raf de Bont, *Nature's Diplomats: Science, Internationalism, and Preservation, 1920–1960* (Pittsburgh: University of Pittsburgh Press, 2021).
11. Cioc, *The Game of Conservation*, 36–38.
12. Michael Osborne, 'The Société Zoologique d'Acclimatation and the New French Empire: Science and Political Economy', in Patrick Petitjean, Catherine Jami and Anne-Marie Moulin (eds), *Science and Empires: Historical Studies about Scientific Development and European Expansion* (Boston: Springer, 1992), 299–306.
13. David Prendergast and William Adams, 'Colonial Wildlife Conservation and the Origins of the Society for the Preservation of the Wild Fauna of the Empire (1903–1914)', *Oryx* 37(2) (2003), 251–60.
14. De Bont et al., 'Conservation Conferences', 569–99.

15. Carolyn Ford, *Natural Interests: The Contest over Environment in Modern France* (Cambridge, MA: Harvard University Press, 2016), 114–15; Yannick Mahrane, Frédéric Thomas and Christophe Bonneuil, 'Mettre en valeur, préserver ou conserver? Genèse et déclin du préservationnisme dans l'empire colonial français (1870–1960)', in Charles-François Mathis and Jean-François Mouhot (eds), *Une protection de l'environnement à la française (XIXe–XXe siècles)* (Seyssel: Champ Vallon, 2013), 62–80.
16. Ford, *Natural Interests*, 127.
17. Henri Jaffeux, 'La longue et passionnante histoire des parcs nationaux français', *Revue du comité d'histoire du ministère de l'écologie* 9 (2010), 138–62.
18. Mathieu Guérin, 'France, the IUCN and Wildlife Conservation in Cambodia: From Colonial to Global Conservationism', *Journal of Southeast Asian Studies* 53(1–2) (2022), 183–208.
19. Office central du tourisme indochinois, *Les grandes chasses en Indochine* (Saigon, 1937).
20. Office central du tourisme indochinois, *Glimpses of a Great Tourist Country: French Indochina at the Beginning of 1938* (Saigon, 1938).
21. Shaun K. Malarney, 'Defining the True Hunter: Big Game Hunting, Moral Distinction, and Virtuosity in French Colonial Indochina', *Comparative Studies in Society and History* 62(3) (2020), 651–80.
22. Albert Sallet, 'L'Annam: Richesses touristiques', *Bullletin des amis du Vieux Hué* 18 (1931), 65.
23. Roland Dorgelès, *On the Mandarin Road* (New York: Century, 1926).
24. Mathieu Guérin, 'Européens et prédateurs exotiques en Indochine, le cas du tigre', in Jean-Marc Moriceau and Philippe Madeline (eds,) *Repenser le sauvage grâce au retour du loup: Les sciences humaines interpellées* (Caen: Presses Universitaires de Caen, 2010), 211–24.
25. Harry A. Franck, *East of Siam: Ramblings in the Five Divisions of French Indo-China* (New York: Century, 1926).
26. Maurice Rondet-Saint (ed.), *L'Organisation des colonies françaises au point de vue cynégétique* (Paris: Agence Générale des Colonies, 1922).
27. Joanny Guillard, *Au service des forêts tropicales: Histoire des Services forestiers français outre-mer* (Nancy: École nationale du génie rural, des eaux et des forêts, 2010), 12.
28. Mark Cioc-Ortega, 'Hunting, Agriculture, and the Quest for International Wildlife Conservation during the Early Twentieth Century', paper presented at the Yale 'Agrarian Studies' seminar, 20 September 2008.
29. Rondet-Saint, *L'Organisation des colonies françaises*, 37.
30. Reserves were proposed for the provinces of Tây Ninh, Biên Hòa and Thủ Dầu Một: Jean Bordeneuve, *Les Grandes Chasses en Indochine* (Saigon: Albert Portail, 1925), 130.
31. Circulaire du Gouverneur de la Cochinchine du 16 juin 1923 (quoted by Bordeneuve, *Les grandes chasses en Indochine*, 130).
32. Fernand Millet, 'La chasse sur les hauts plateaux moïs du Langbian', *La Revue Indochinoise* 26 (1916), 11–12 and 417–22.
33. Bordeneuve, *Les grandes chasses en Indochine*, 128–29.
34. 'Arrêté fixant les périodes d'ouverture de la chasse', Archives nationales d'outre-mer (ANOM), résidence supérieure du Tonkin (RST), nouveau fonds (NF) 2905.

35. Ford, *Natural Interests*, 121; Caroline Ford, 'Nature, Culture and Conservation in France and Her Colonies 1840–1940', *Past and Present* 183 (2004), 173–98.
36. Rondet-Saint, *L'Organisation des colonies françaises*, 40.
37. Report by M. René Puaux, 'La réglementation de la chasse pour les indigènes', in ibid., 20.
38. Fernan Millet, *Les Grands Animaux sauvages de l'Annam, leurs moeurs, leur chasse et leur tir* (Paris: Plon, 1929), 366–67.
39. Georges Maspero, *Un empire colonial français, l'Indochine*, vol. 1 (Paris: G. Van Oest, 1929), 267.
40. 'Réglementation de la chasse en Indochine', ANOM, RST, NF 2907.
41. Malarney, 'Defining the True Hunter'.
42. ANOM, RST, NF 2910.
43. ANOM, RST, NF 02911.
44. ANOM, RST, NF 02912.
45. ANOM, RST, NF 02911.
46. Letter from the Chief of the *Division Forestière* of Cao Bang to the Commander of the 2nd Military Territory, ANOM, RST, NF 02912.
47. Résidence supérieure du Tonkin, *Recueil des textes relatifs à la réglementation de la chasse* (Hanoi: Imprimerie d'extrême-orient, 1936), 7.
48. Francis Evrard, 'Biogéographie indochinoise et réserves naturelles', in André d'Aubréville (ed.), *Contribution à l'étude des réserves naturelles et des parcs nationaux* (Paris: Paul Lechevalier, 1937), 257–62.
49. Guillard, *Au service des forêts tropicales*, 303.
50. Cioc, *The Game of Conservation.*
51. G. A. Brouwer, *The Organization of Nature Protection in the Various Countries* (New York: Special Publication of the American Committee for International Wildlife Protection, 1938 [1931]), 87.
52. Elizabeth Hone, *African Game Protection* (New York: Special Publication of the American Committee for International Wildlife Protection, 1933); Stephanie Pincetl, 'Some Origins of French Environmentalism: An Exploration', *Forest & Conservation History* 37(2) (1993), 80–89.
53. Ford, *Natural Interests*, 114–37.
54. Brouwer, *The Organization of Nature Protection*, 79.
55. Robert Cribb, 'Birds of Paradise and Environmental Politics in Colonial Indonesia, 1890–1931', in Peter Boomgaard, Freek Colombijn and Davis Henley (eds), *Paper Landscapes: Explorations in the Environmental History of Indonesia* (Leiden: KITLV Press, 1997), 379–407; Paul Jepson and Robert Whittaker, 'Histories of Protected Areas: Internationalisation of Conservationist Values and their Adoption in the Netherlands Indies (Indonesia)', *Environment & History* 8(2) (2002), 129–72.
56. Robert Aiken, 'Peninsular Malaysia's Protected Areas' Coverage, 1903–92: Creation, Rescission, Excision, and Intrusion', *Environmental Conservation* 21(1) (1999), 49–56.
57. Brouwer, *The Organization of Nature Protection.*
58. CAOM RST NF 02925, *Plans des réserves temporaires de chasse des provinces de Bac Giang, Ha Dong, Hai Duong, Lao Cai, Ninh Binh, Phu Yen, Phu tho, Son Tay, Thai Nguyen, Vinh Yen, Cao Bang*, 1935.

59. Résidence Supérieure du Tonkin, *Recueil des textes relatifs à la réglementation de la chasse.*
60. Letter from the *Résident Supérieur* of Tonkin to all provincial chiefs, 17 September 1935, ANOM, RST, NF 2926.
61. Brouwer, *The Organization of Nature Protection*, 109.
62. 'Reserved game' could not be captured or killed at any time. 'Protected game' had restrictions on when it could be hunted, the sex that could be taken and the number allowed. 'Free game' could be hunted without restrictions.
63. The only *Inspecteur des chasses* ever designated in Annam was Henri de Monestrol, who later went to work for Emperor Bao Dai, for whom he managed hunting in the Pays Montagnard du Sud, a so-called autonomous zone set up in 1946: Henri de Monestrol, *Chasses et faune d'Indochine* (Saigon: Édition A. Portail, 1952).
64. Pamela McElwee, *Forests are Gold: Trees, People and Environmental Rule in Vietnam* (Seattle: University of Washington Press, 2016); Frédéric Thomas, 'Protection des forêts et environnementalisme colonial: Indochine 1860–1945', *Revue d'histoire moderne & contemporaine* 56(4) (2009), 104–36.
65. David Biggs, 'Managing a Rebel Landscape: Conservation, Pioneers and the Revolutionary Past in the U Minh Forest, Vietnam', *Environmental History* 10(3) (2005), 448–76.
66. Evrard, 'Biogéographie indochinoise', 258.
67. Thomas, 'Protection des forêts'.
68. Auguste Chevalier, 'Propose diverses mesures générales pour la protection des sites et de la nature dans les Colonies Françaises', in *Rapports, voeux, réalisations, revus et annotés: Congrès international pour la protection de la nature* (Paris: Guillemot et de Lamothe, 1926), 303–4.
69. Province de Baria, service Forestier, National Archives of Vietnam (NAV), 2 IIB 55/256.
70. Protection du site de la grotte de Cu lac, Phong Nha, ANV, 4 RSA/HC 1892.
71. 21 February 1936, ANV, 1 RST 75.420.
72. 'Au sujet de la protection de la faune et de la flore du Tonkin', 1936, ANV, 1 RST N90 75.437.
73. Frédéric Thomas has noted a similar lack of scientific research and expertise within the Forest Service: Frédéric Thomas, *Histoire du régime et des services forestiers française en Indochine de 1862 à 1945* (Hanoi: Gioi, 1999).
74. 'Au sujet de la protection de la faune et de la flore du Tonkin', 1936, ANV, 1 RST N90 75.437.
75. Võ Quý, 'Rare Species and Protection Measures Proposed for Vietnam', in J. W. Thorsell (ed.), *Conserving Asia's Natural Heritage: The Planning and Management of Protected Areas in the Indomalayan Realm* (Geneva: International Union for the Conservation of Nature, 1985), 98–104.
76. Evrard, 'Biogéographie indochinoise', 257.

Bibliography

Aiken, Robert. 'Peninsular Malaysia's Protected Areas' Coverage, 1903–92: Creation, Rescission, Excision, and Intrusion'. *Environmental Conservation* 21(1) (1999), 49–56.

Barton, Greg. *Empire Forestry and the Origins of Environmentalism*. New York: Cambridge University Press, 2002.

Beinart, William. 'Soil Erosion, Conservationism and Ideas about Development: A Southern African Exploration, 1900–1960'. *Journal of Southern African Studies* 11(1) (1984), 52–83.

Beinart, William, and Lotte Hughes (eds). *Environment and Empire*. Oxford: Oxford University Press, 2007.

Biggs, David. 'Managing a Rebel Landscape: Conservation, Pioneers and the Revolutionary Past in the U Minh Forest, Vietnam'. *Environmental History* 10(3) (2005), 448–76.

Bourguet, Marie-Noëlle, and Christophe Bonneuil. 'Présentation. De l'inventaire du monde à la mise en valeur du globe. Botanique et colonisation (fin 17e siècle–début 20e siècle)'. *Revue française d'histoire d'outre-mer* 322–23 (1999), 7–38.

Brocheux, Pierre, and Daniel Hémery. *Indochina: An Ambiguous Colonization, 1858–1954*. Berkeley: University of California Press, 2011.

Cioc, Mark. *The Game of Conservation: International Treaties to Protect the World's Migratory Animals*. Athens: Ohio University Press, 2008.

Cioc-Ortega, Mark. 'Hunting, Agriculture, and the Quest for International Wildlife Conservation during the Early Twentieth Century'. Paper presented at the Yale 'Agrarian Studies' seminar, 20 September 2008, 1–31.

Cribb, Robert. 'Birds of Paradise and Environmental Politics in Colonial Indonesia, 1890–1931', in Peter Boomgaard, Freek Colombijn and Davis Henley (eds), *Paper Landscapes: Explorations in the Environmental History of Indonesia* (Leiden: KITLV Press, 1997), 379–407.

De Bont, Raf. *Nature's Diplomats: Science, Internationalism, and Preservation, 1920–1960*. Pittsburgh, PA: University of Pittsburgh Press, 2021.

De Bont, Raf, Simone Schleper and Hans Schouwenburg. 'Conservation Conferences and Expert Networks in the Short Twentieth Century', *Environment and History* 23(4) (2017), 569–99.

Ford, Caroline. 'Nature, Culture and Conservation in France and her Colonies 1840–1940'. *Past and Present* 183 (2004), 173–98.

———. *Natural Interests: The Contest over Environment in Modern France*. Cambridge, MA: Harvard University Press, 2016.

Grove, Richard. 'Scottish Missionaries, Evangelical Discourses and the Origins of Conservation Thinking in Southern Africa 1820–1900'. *Journal of Southern African Studies* 15(2) (1989), 163–87.

———. *Green Imperialism: Colonial Expansion, Tropical Island Edens, and the Origins of Environmentalism 1600–1800*. Cambridge: Cambridge University Press, 1995.

Guérin, Mathieu. 'Européens et prédateurs exotiques en Indochine, le cas du tigre', in Jean-Marc Moriceau and Philippe Madeline (eds), *Repenser le sauvage grâce au retour du loup: Les sciences humaines interpellées* (Caen: Presses universitaires de Caen, 2010), 211–24.

———. 'France, the IUCN and Wildlife Conservation in Cambodia: From Colonial to Global Conservationism', *Journal of Southeast Asian Studies* 53(1–2) (2022), 183–208.

Guillard, Joanny. *Au service des forêts tropicales: Histoire des Services forestiers français outre-mer*. Nancy: École nationale du génie rural, des eaux et des forêts, 2010.

Jaffeux, Henri. 'La longue et passionnante histoire des parcs nationaux français', *Revue du comité d'histoire du ministère de l'écologie* 9 (2010), 138–62.

Jepson, Paul, and Robert Whittaker. 'Histories of Protected Areas: Internationalisation of Conservationist Values and Their Adoption in the Netherlands Indies (Indonesia)', *Environment & History* 8(2) (2002), 129–72.
MacKenzie, John (ed.). *Imperialism and the Natural World*. Manchester: Manchester University Press, 1990.
———. *The Empire of Nature: Hunting, Conservation, and British Imperialism*. Manchester: Manchester University Press, 1997.
Mahrane, Yannick, Frédéric Thomas and Christophe Bonneuil. 'Mettre en valeur, préserver ou conserver? Genèse et déclin du préservationnisme dans l'empire colonial français (1870–1960)', in Charles-François Mathis and Jean-François Mouhot (eds), *Une protection de l'environnement à la française (XIX e–XXe siècles)* (Seyssel: Champ Vallon, 2013), 62–80.
Malarney, Shaun K. 'Defining the True Hunter: Big Game Hunting, Moral Distinction, and Virtuosity in French Colonial Indochina'. *Comparative Studies in Society and History* 62(3) (2020), 651–80.
McElwee, Pamela. *Forests are Gold: Trees, People and Environmental Rule in Vietnam*. Seattle: University of Washington Press, 2016.
Osborne, Michael. 'The Société Zoologique d'Acclimatation and the New French Empire: Science and Political Economy', in Patrick Petitjean, Catherine Jami and Anne-Marie Moulin (eds), *Science and Empires: Historical Studies about Scientific Development and European Expansion* (Boston: Springer, 1992), 299–306.
Peluso, Nancy, and Peter Vandergeest. 'Genealogies of the Political Forest and Customary Rights in Indonesia, Malaysia, and Thailand'. *Journal of Asian Studies* 60(3) (2001), 761–812.
Pincetl, Stephanie. 'Some Origins of French Environmentalism: An Exploration'. *Forest & Conservation History* 37(2) (1993), 80–89.
Prendergast, David, and William Adams. 'Colonial Wildlife Conservation and the Origins of the Society for the Preservation of the Wild Fauna of the Empire (1903–1914)', *Oryx* 37(2) (2003), 251–60.
Quý, Võ. 'Rare Species and Protection Measures Proposed for Vietnam', in J. W. Thorsell (ed.), *Conserving Asia's Natural Heritage: The Planning and Management of Protected Areas in the Indomalayan Realm* (Geneva: International Union for the Conservation of Nature, 1985), 98–104.
Sivaramakrishnan, Kalyanakrishnan. *Modern Forests: Statemaking and Environmental Change in Colonial Eastern India*. Stanford, CA: Stanford University Press, 1999.
———. 'State Sciences and Development Histories: Encoding Local Forestry Knowledge in Bengal'. *Development and Change* 31(1) (2000), 61–89.
Thomas, Frédéric. *Histoire du régime et des services forestiers française en Indochine de 1862 à 1945*. Hanoi: Gioi, 1999.
———. 'Protection des forêts et environnementalisme colonial: Indochine 1860–1945', *Revue d'histoire moderne & contemporaine* 56(4) (2009), 104–36.

CHAPTER 2

Setting up a Wildlife Department

Kenyan Expertise in Malaya

Mathieu Guérin

As is well known, Britain's colonies in Africa played a central role in the emergence of the contemporary wildlife conservation movement between the early nineteenth century and the Second World War.[1] As Britain expanded its colonies across the continent, especially in Southern and East Africa, it established protected areas and set up conservation agencies. The Convention for the Preservation of Wild Animals, Birds and Fish in Africa, signed in London in 1900, laid the groundwork for the conservation movement in the tropics. Three years later the Society for the Preservation of the Wild Fauna of the (British) Empire (SPFE) was founded, focusing particularly on preserving African big game. The SPFE – nowadays Fauna & Flora International – became one of the twentieth century's major protagonists in wildlife protection.[2] The 1933 London conference resulted in an inter-imperial agreement to protect African flora and wildlife. The impact of these major conferences and the SPFE's action was soon felt outside the African continent.[3]

The circulation of experts within the Empire was meant to establish common practices for wildlife conservation based on experiences in Africa. In 1937, the Colonial Office sent the head of Kenya's Game Department, Archibald Ritchie, to Malaya to establish a similar department there. Study of the British colonial archives in Kew and of correspondence and reports by Kenya's and Malaya's Game Departments, held at their national archives in Nairobi and Kuala Lumpur, enable us to understand: first, why Kenya seemed a model for conservation; second, why the Colonial Office relied on Ritchie to export this model to South East Asia; and, third, how and to what extent he was able to influence Malayan and then Malaysian wildlife conservation policies. Ritchie's mission raises questions about what defined a professional wildlife protector in the days when game departments were being set up in the colonies. These went on to become wildlife departments with staff in charge of implementing the conservation agenda of the newly independent states a few decades later.

The Specificity of Conservation in 1930s Kenya

Although South Africa had a wildlife protection policy well before Kenya, and though the Kruger National Park, created in 1926, was the first the British founded in Africa, Kenya showed a remarkable capacity to preserve large wildlife populations, particularly large mammals. Since the late nineteenth century, Westerners had come to systematically associate representations of Kenya with wildlife, a phenomenon only heightened with the publications of such mainstream works as Karen Blixen's *Out of Africa* and Ernest Hemingway's *Green Hills of Africa*.[4]

Nevertheless, throughout the first decades of the twentieth century, Kenya's Game Department remained incapable of properly carrying out its principal missions. After the appointment of a first ranger, Arthur Percival, in 1901, Kenya's Game Department was officially established in 1907. Four European agents had to enforce hunting laws across the entire colony, to protect game reserves and to combat the illegal trade in ivory, primarily with neighbouring Somalia under Italian rule. The department ceased operating during the First World War. Then in 1923 Archibald Ritchie was named as its head.[5]

Archibald Thomas Ayres Ritchie came from a wealthy Irish family. After studying zoology at Oxford, he joined the French Foreign Legion at the beginning of the war. He afterwards joined the Grenadier Guards with the rank of lieutenant in 1915, and was wounded in combat on several occasions, finishing his military career with the rank of captain. According to his friend and colleague Peter Jenkins, Ritchie arrived in Kenya in 1920, where he worked for the East African Land Company before joining the Game Department.[6]

With Ritchie at its head, the department expanded slightly from three European agents in 1923 to five in 1936, namely the game warden, three conservation assistants and one 'control' assistant, that is, a person tasked with shooting animals posing a threat to humans, cattle or crops. They had sixty-four African scouts under their orders. The African scouts were chosen for their capacities to track down animals, and often came from hunting communities, such as the Kamba. As Ritchie himself admitted, 'it is I must admit no exaggeration to say that the staff of the Game Department is a joke and a poor joke at that'.[7] For historian Nora Kelly, 'the game department was the Protectorate and Colony's Cinderella, hardworking and impoverished'.[8] The colonial government recognised the need to protect wildlife but refused to provide the necessary means. Ritchie was tasked with 'professionalising' the department, which he sought to do by introducing military procedures and discipline.[9]

He introduced several significant changes in the way the Kenyan Game Department operated. To make up for the shortage of staff, the department started using honorary wardens as of 1921 – unpaid volunteers selected from among the European colonists to help enforce hunting laws with the same powers as Game Department agents. By the mid-1930s there were eighty such wardens.

Furthermore, Ritchie expanded game 'control' in response to complaints from white settlers – and to a lesser extent African farmers – about crop destruction and attacks on people. He thus sought to win over colonists and the administration's support in a period when agricultural activities were being developed. The department had specialised agents – though it also called on (at times inexperienced) European hunters – to destroy animals deemed to be 'pests', such as wild pigs, baboons, hyenas, porcupines and so on. Each year they were tasked with slaughtering thousands of animals classed as game, such as zebras, gazelles, elephants and rhinoceroses, when these were accused of destroying fields or of competing with domestic herds for access to water or pasture.[10]

After the rapid success of Kruger National Park,[11] the SPFE campaigned to open national parks in the rest of the Empire, starting with Africa. These were intended to protect vast spaces and the animals living there 'in perpetuity', and were the preferred option among the conservation circles to which Ritchie belonged. Since 1900, Kenya had been home to two large game reserves outside the zones settled by European colonists. The level of protection was limited, and the existence even of these reserves was threatened by the expansion of farmland. In 1930, the governor of Kenya organised a conference on wildlife protection, in which the SPFE took part. A subcommittee chaired by Ritchie was tasked with studying the desirability of establishing a national park in Kenya.[12] Ritchie contributed to and closely monitored the 1933 London conference, one of whose emblematic proposals was the creation of national parks.[13] By the time Ritchie was seconded to Malaya in 1937, the process had been launched, though no national park was as yet established in Kenya.

What set Kenya apart from the rest of the Empire was the importance the colony had acquired for big game hunters, dating from the 1909 expedition by former US president Theodore Roosevelt, which was extensively covered in the press. Kenya became a fashionable destination for rich American and European tourists, especially British aristocrats. The Game Department was thus regularly called on to extend the best possible welcome to prestigious guests who were looked after by professional hunters, operating under the auspices of the East African Professional Hunter's Association, founded in 1934 by Ritchie and Charles Pitman, a game warden from neighbouring Uganda, to monitor tourists' hunting activities more closely.[14] Hunting safaris, increasingly followed as of the 1930s by photographic safaris, became a major source of revenue for the colony. It is reckoned that in the early 1930s, game brought Kenya £150,000 per year.[15]

Conservation Policy in Malaya

British Malaya in the 1930s was divided into crown colonies, known as the Straits Settlements, and protectorates, including the sultanates of Perak, Selangor, Negri Sembilan and Pahang that constituted the Federated Malay States (FMS), plus unfederated sultanates. The wildlife conservation policy was developed mainly where British influence was strongest, namely the Straits Settlements and the FMS. In 1921, the Wild Animals and Birds Protection Enactment introduced Game Departments in the FMS. Each Malay state was to have a European game warden, assisted by Malay rangers, and a few honorary rangers and wardens. In 1936 the FMS had three salaried European wardens, six honorary wardens and twenty or so Malay rangers. European wardens and Malay rangers were selected from among hunters who knew how to use a shotgun or a rifle.

The hunting regulations and creation of a wildlife management corps owed much to the planter and former big game hunter Theodore Rathbone Hubback, who was adviser to the colonial government. Hubback, an honorary game warden in Pahang, became an indefatigable champion of the wildlife preservation cause. In 1930, with SPFE backing, he was appointed chair of a commission of enquiry into wildlife protection regulations in Malaya, looking at damage to crops and how to remedy it after a number of elephants and deer were slaughtered in the peninsula in 1929. In the 1920s, large areas of forest were cut down, especially to make room for rubber plantations and the first palm oil plantations. Elephants could destroy the former by rubbing their skin on the bark, and were drawn to the latter since the fruit is a delicacy for them. The colossal enquiry he conducted was published in 1932, and sent by the SPFE to the Secretary of State for the Colonies.[16] Hubback became the main authority on the issue of wildlife conservation in Malaya, recognised as such both in situ and in London.[17]

Drawing on the results of his commission of enquiry, Hubback wished to unite Malaya's game wardens into a single corps with authority over the FMS, and even the whole of British Malaya. The sultans and colonial administration were opposed, but a compromise solution was reached in 1933. Hubback was named game warden in each of the Federated Malay States, and received the title of Chief Game Warden for the entire FMS, meaning he could coordinate the various Game Departments.[18] Hubback additionally maintained personal ties with the unfederated Malay states. The economic crisis of the 1930s led to severe cuts to the administration's budget, and Hubback – to his wrath – had to run his various departments with very insufficient funding, even though he received no pay for his work.

In 1936, Malaya was visited by Thomas Comyn-Platt, sent by the SPFE with Colonial Office support to assess actions to protect wildlife. Comyn-Platt reported that 'poaching, trapping, unlicensed shooting, and the indiscriminate killing of game is becoming less frequent as the law is becoming more effectively administered'. He found that hunting by colonised populations was more closely monitored. The priority, to his mind, was not to expand the Game Department, for the protected spaces were far too vast in any case, covering nearly 8,000 km^2, but rather to step up education and propaganda. He reckoned that reserves had insufficient legal protection: 'these areas – to put it crudely – are allotted to wild life just so long as they are not required for anything else!'[19]

Hubback's response to the weak status of reserves was a grand plan to set up a national park in Malaya devoted to protecting wild species in perpetuity, based around the Gunung Tahan reserve at the heart of the peninsula. In 1934, newly appointed High Commissioner Shenton Thomas, who had spent most of his career in Africa, lent his backing to the scheme, as did the SPFE.[20] At the proposal of the SPFE and its patron the Prince of Wales, the sultans of Terengganu, Kelantan and Pahang agreed to the creation of a park, to be named after King George V to mark the Silver Jubilee in 1935.

From this point on, the national park became a major political issue in both Malaya and the United Kingdom. Comyn-Platt visited the park and appreciated its splendid landscapes and abundant wildlife: 'I was much impressed by all I saw and formed the opinion, as to which there is no doubt, that as a game sanctuary it will prove invaluable'.[21] Nevertheless, for the moment the park only existed by proclamation of the sultans, and had no reality on the ground or in local legislation. Hubback's vision of what a national park should be was inspired by the model of Yellowstone in the United States: a space totally devoid of people. He tried to constrain the local populations who used its resources. The inhabitants then appealed to the sultans and won their backing – together with that of some of the colonial administration – to safeguard some of their traditional user rights.[22]

Hubback's short temper, aggressive letters and intransigence ended up alienating his supporters, including those at the SPFE, and in 1936 he was pushed into retirement. Two events brought about his downfall. In December 1935 he went to London and attended an SPFE meeting where he met officials from the Colonial Office, including Edward Gent[23] and the Undersecretary of State Lord Plymouth, hoping to convince them that he had been allocated insufficient means.[24] These manoeuvres were perceived by Shenton Thomas as a betrayal, especially as Hubback gave an interview with the *Straits Times* in which he questioned the Malayan administration's conservation policy.[25]

A few months later, Comyn-Platt returned from his mission in Malaya alarmed by Hubback's attitude. He appended a confidential note to his report, which was sent to the Colonial Office, in which he observed:

> During many conversations he told me, to begin with, that as regards Fauna Preservation he was continually at loggerheads with the governor, and had

> on more than one occasion given him a 'piece of his mind'! That the residents knew nothing about game and cared less – a fact that he had more than once reminded them! That the Malayans were little less than savages! – and finally that the whole policy of the Government, as regards game, was futile and those responsible for carrying it out inept! As to the Planter, if his Estate suffered from the depredations of wild animals, he had only himself to blame for settling in a game area![26]

Hubback's fate was sealed once he lost the support of the SPFE. In summer 1936, it was decided to ask Hubback, aged 64 at the time, to leave the Game Department. Fearing he might kick up a scandal, Gent suggested that the Secretary of State for the Colonies, William Ormsby Gore, a major SPFE supporter, write a letter praising Hubback's work, emphasising that it was time for him to retire so as to allow an experienced younger game warden to pursue his work.[27] The Colonial Office started casting around for an official whose legitimacy Hubback could not dispute, and alighted on Captain Ritchie.

Thus what was presented as a wish to professionalise the Malayan Game Department and enhance its expertise in fact resulted from a settling of accounts between the High Commissioner and his subordinate, and a rejection of the intransigent conservation policy advocated by Hubback, which had triggered tensions with the planters, sultans and local populations.[28]

Captain Ritchie's Secondment to Malaya

Ritchie was chosen to replace Hubback during a meeting at the Colonial Office attended by Shenton Thomas, Edward Gent and George Tomlinson. The latter, who headed the Colonial Office's personnel department, put forward two names, Ritchie and Pitman, for their expertise in running a Game Department. Ritchie's name was retained.[29] He was a respected authority on wildlife conservation matters. In his report on the 1933 London conference, he could thus write:

> We may claim without immodesty to have taken no small part in the preparation of the Convention which was placed before the Conference and formed there the basis of discussion and agreement. The Convention was indeed modelled in its exposition of the theory and practice of Game Preservation, very largely on the existent Kenya conception.[30]

All accounts portray him as a warm and affable person, and, unlike Hubback, measured in word and manner. Merwyn Cowie, the first head of Kenya's national parks, and far younger than Ritchie, described him as follows:

> My friend and hero, Captain Archie Ritchie, Game Warden of Kenya – a man with immeasurable knowledge and ability, with a head of white hair like the mane of a lion, and with a tremendous sense of humour; a man to whom you could run for advice on any subject; a man who would always befriend you in any trouble; the custodian of our wild animals and the very embodiment of everything that was fine and splendid.[31]

Ritchie had been awarded the Most Excellent Order of the British Empire for his action in Kenya, and the Military Cross and the *Légion d'honneur* for his gallantry in action during the First World War. Even several decades after having left the army he was systematically called Captain. His aura stretched far beyond Kenya's borders. Excerpts from his annual reports were regularly reprinted in the SPFE journal, and he was sometimes mentioned in the press. A children's magazine presented him as 'a sort of Mowgli grown up and translated to Africa', and accounts of his hunting exploits even figured in a minor Alsatian newspaper.[32]

His education at Magdalen College, Oxford, time in the army, remarkable service, refined manners and capacity for living outdoors meant he resembled a stereotype of the colonial gentlemen, equal parts adventurer, man of culture and *bon viveur*.

Furthermore, Ritchie could draw on exceptional contacts, thanks to his time at university and in the army, and built up further when head of the Game Department in Kenya. He knew King Edward VIII, having been at Oxford at the same time as the then Prince of Wales. They went on to serve together as young officers in the Grenadier Guards during the war. From 1928 through to his abdication, Edward had been the royal patron of the SPFE. He had taken part in two safaris to East Africa, in 1928 and 1930 – accompanied by the famous hunters Denys Finch-Hatton and Bror von Blixen – when he had seen Ritchie again.[33] Ritchie's position as head of the Game Department had led him to receive celebrities, monarchs, businessmen and senior government officials passing through Kenya on safari. He seemed heaven-sent to counter Hubback's influence, especially as Kenyan wildlife management had led to the development of safaris, a profitable activity that also enhanced the social capital of colonists and the colonial authorities.

Ormsby Gore wrote to the Governor of Kenya to ask his permission to offer Ritchie the position of Chief Game Warden of Malaya, with an attractive salary. He presented Ritchie's post as 'to a certain extent a new one' and insisted on the attention he would need to pay to the King George V National Park.[34]

After receiving the proposal from the Colonial Office, Ritchie contacted Hubback.[35] He was reluctant to leave Kenya for good, and negotiated a one-year secondment to get a feel for the position in Malaya. In a letter to his friend

Charles Hobley, who had just retired as secretary to the SPFE, he explained that he had decided to accept the post in Malaya:

> I should not have considered it – I am far too fond of Kenya – had it not been for health reasons. Since getting back here I have had a bad time; in July and August I was for some time on the verge of a complete mental breakdown, in addition to being seriously ill physically. I am far from fit now, and the only chance of getting right again is, I am certain, a complete change to a lower altitude. Whether I shall stay in Malaya or return here I don't know.[36]

Beyond his image as an adventurer, Ritchie was a former combatant who was psychologically and physically weakened, while Hubback, who had been forced into retirement because of his age, still spent long days in the forest. Ritchie arrived in Malaya to take up his post on 10 January 1937. His arrival was prepared by the administration, with Shenton Thomas having announced Ritchie's appointment in November 1936, describing him as 'one of the most expert authorities on game preservation methods in the Empire'.[37]

On his arrival, another glowing press release detailed his biography. Ritchie was presented as Malaya's 'first stipendiary [hence professional] game warden', implicitly diminishing Hubback's importance, who, having carried out his duty without pay for over fifteen years, was now downgraded to the status of a dilettante.[38]

Ritchie's Action in Malaya

It took Ritchie two months to settle in. In early March 1937, he was appointed warden in the various states in his jurisdiction. In April, he wrote a memorandum for Shenton Thomas setting out his plan of action, which involved expanding the Game Department and significantly increasing its budget.[39] Ritchie presented the work of the Game Department as falling into three main categories: 'routine', the national park and 'the prevention of depredation by game animals'.

Given the economic recovery, the high commissioner was broadly favourable to providing what he requested: 'we must do all we can to meet his views. Only thereby shall we be able effectively to meet any criticism that may be levied against us in respect of game preservation.'[40] Indeed, both Shenton Thomas and Ritchie were under pressure. As soon as Ritchie's appointment was announced, the *Straits Times* published an anonymous article denouncing the ousting of Hubback and the 'very ungenerous treatment that has been meted to Mr. Hubback', despite his being active in his work and highly knowledgeable about protecting Malaya's wildlife. Ritchie, for his part, was presented as 'someone who is entirely inexperienced so far as the purely Malayan aspects of wild life conservation are concerned'.[41] The attacks continued throughout his time in Malaya. Hubback held him to know nothing about the country, its inhabitants, language or wildlife. 'He is an expert only by virtue of his office.'[42] These were the same arguments as those found in the press.[43] Ritchie nevertheless set to work. One year after having presented his recommendations to Shenton Thomas, he drew up his 1937 annual report, allowing us to see what he had managed to put in place.[44]

By 'routine', Ritchie meant issuing hunting permits, combating poaching, supervising rangers, gathering information, and actions to raise awareness. He could draw on the support of the two game wardens of Perak and Pahang. He requested a third post for Negri Sembilan and Selangor, and that these three agents be provided with an office, telephone and administrative assistant to receive the public and issue permits. For Hubback, the bulk of a warden's job was to be out in the field, and after their offices were shut due to the 1934 budget cuts wardens had to use a room at home.[45] Ritchie suggested that the agents should move regularly, which posed practical problems given that the game wardens were employed and paid by each of the states. He emphasised that, unlike in Kenya, he could not draw on a reserve of European hunters in case of need. He additionally requested a budget to pay informants about poaching.

His request for additional staff was granted, with two posts being created for European wardens, one for Selangor and Negri Sembilan, to which a former honorary game warden was appointed in 1937, and a national park superintendent.[46] In parallel, the number of honorary game wardens, nearly all of whom were European, was increased from six to fourteen, while that of rangers from local populations rose from about twenty or so to thirty-seven. However, he was not allocated the offices, administrative assistants or culling agents that he had requested. The only real office was that of the Chief Game Warden, which was transferred to Kuala Lumpur, in the museum building.

Any unregulated hunting was designated poaching, yet it was difficult to put anti-poaching measures in place. In arguing for new positions to be created in Selangor, Ritchie explained that 'to deal with this vast army of licensed and unlicensed shooters and trappers we have two game rangers, [which] is token, but little more'.[47] The major mismatch between the number of permits, the number of kills declared to the authorities and the volume of munitions sold in the country gives an idea of the scale of poaching. In 1937, 555 deer and big game licences were issued, and 1,474 bird game licences. While it is not known how many birds were killed, 238 deer, four gaurs and one elephant were declared. In parallel, nearly 1.4 million shotgun cartridges were sold that year, together with 330,000 for .22 long rifles (mainly used for birds and small cervids) and 20,800 cartridges for large-bored rifles. Many Europeans, Chinese and Malays hunted without a permit and did not declare the game they killed.[48] Nevertheless, only forty infringements were recorded for the year, of which eighteen cases went to trial and twenty-two were settled out of court. Enforcing hunting regulations was impossible despite the increase in staff numbers, which remained insignificant given the size of the country.

The key concern for the Colonial Office and the SPFE was setting up the national park. Unlike Hubback, who had been out exploring the forest for two decades, Ritchie recognised that his 'knowledge of the fauna content is very slight'. He felt the priority was to inventory the park wildlife and to cut paths to open up parts of it. He felt that a European agent should be specifically assigned to the park. This was also Hubback's opinion, who volunteered for the position after being removed as head of the Game Department, but he was turned down. The first year was thus spent opening paths, looking for rhinoceroses and monitoring a herd of gaurs living near the park entrance.[49]

Ritchie launched works to open the park to visitors. To his mind, taking Kruger as his reference, the main point of the park was to be open to the public. In Nairobi he had thus initiated discussions to create a national park just outside the city, so that anyone could go and see the wildlife and especially the lions.[50] In Malaya, he endeavoured to facilitate access to the peak of Gunung Tahan to allow people to admire the landscape. Some of the colonial press published enthusiastic articles about a 'holiday resort at Gunung Tahan',[51] which was in fact never built.

Hubback totally disagreed with his successor on this matter. For him, the prime function of the park was to preserve wildlife and forest ecosystems: 'The National Park will just be a National Farce unless it is handled with care and firmness and the idea of making it into a picnic place before it has been made into a real sanctuary for wild life should be strongly resisted.'[52] Differences in landscape explain why tourism failed to develop in Malaya. Whereas Kenya's savannah meant anyone could easily observe herds of grazing animals and even watch carnivores hunting, the dense forest around Gunung Tahan could hide an elephant at fewer than twenty paces, as a British administrator remarked. To spot animals one needed to be used to stalking or to lying in wait.[53]

The Colonial Office and Shenton Thomas thus asked Ritchie to introduce legislation via which the sultans would make their land available, and which would determine how the park was to operate. The three sovereigns, especially Sultan Abu Bakar of Pahang, refused to sign the ordinances unless inhabitants' rights were guaranteed.[54] The park boundary was not even demarcated. Ritchie's main achievement for the national park was to have switched tack away from intransigence, enabling discussions with the sultans and local communities.

In his yearly report he listed the damage to crops and dwellings caused by wild animals, on occasions resulting in deaths. Drawing on his experience in Kenya, Ritchie felt that the priority for the Game Department was to 'control' animals, particularly elephants, even though they were not the same species as in Africa. Elephants had been protected in the FMS since 1911. This was briefly lifted in 1929 due to the many complaints from European planters and local farmers, before being re-established under pressure from the SPFE and lobbying by Hubback. Elephants could only be shot in 'self-defence'.

Ritchie wished to ground his action in the principle of separating those spaces where elephants were to be authorised from those where they were to be eliminated, drawing on policies implemented in East Africa.[55] He thus envisaged exterminating elephants to the west of the mountain range dividing the Malay Peninsula, that is, to the west of the states of Perak, Selangor and Negri Sembilan. To this end he intended to implement a battle plan based on a military approach:

> Just as in war all schemes of defence must be based on such knowledge as can be gathered of the offensive intentions of the enemy, so in Elephant control no plan can anywhere reasonably be evolved without detailed information of their numbers, habitats and movements.[56]

He suggested employing two European hunters specially tasked with killing 'unjustifiable animals' and with training local rangers to kill elephants. When this was turned down, he had to make do with calling on hunters as and when needed.

For the other elephants, including those living in reserves and the national park, he proposed adopting the African policy of educating them.[57] In August 1937, he wrote a memorandum to the British Adviser in Kelantan setting out his theory. Ritchie deemed elephants to be nomadic and highly conservative. Conditioned by what he termed their 'inherited racial memory', they follow the same routes for centuries, heedless of any changes brought about by men. They thus needed to be taught where not to go, by associating certain places with disagreeable experiences, such as being shot at or group members being killed.[58] This theory of using rifles to educate elephants was not specific to Ritchie, being supported by a Pahang Resident and by Charles Hobley, a former secretary to the SPFE.[59]

Ritchie's experience in Malaya came to an end with his secondment. He had encountered systematic hostility from Hubback and his supporters. In his final report he noted: 'Game preservation in Malaya has been the subject of much controversy in the past. Indeed statements, arguments, and contentions, have at times reached a degree of acrimony which is as unfortunate as it should be foreign to so just a cause.'[60] Personally, he never ceased recognising the major contribution made by Hubback, 'one of the greatest authorities on game preservation of all time'.[61] In December 1937, Ritchie's brother-in-law was killed and his wife injured in a plane crash when travelling from Singapore to England.[62] Ritchie then requested to return to Kenya. He left on 15 April 1938, having spent a little over fifteen months in Malaya, and was replaced in October by an employee with the Indian Forest Service, Edward Oswald Shebbeare. After the failure of Ritchie's secondment, India seemed another promising British colonial model for export to Malaya, even though Shebbeare had no experience in wildlife management.

The Impact of Kenyan Expertise on the Malay Game Department

For the Colonial Office, 'Capt Ritchie was appointed to Malaya to organise the Game Department and to prepare legislation for game preservation'.[63] In fact, everything had been put in place by Hubback. Ritchie just strengthened the department without substantially altering its structure. It was undersized given its missions. Paradoxically, while 'professionalisation' involved recruiting a paid Chief Game Warden, it also entailed using a larger number of honorary wardens and rangers. Even though Ritchie was very reasonable in what he asked for, he received only some of the means requested. No marked improvement may be observed in the Game Department's action to regulate hunting or combat poaching, despite this being its core purpose. Even though Ritchie had originally studied zoology, no systematic study of wildlife was undertaken, and agents were not provided with any training. The mission to raise public awareness never got off the ground.

Despite having a more conciliatory character than his predecessor, Ritchie did not manage to finalise the creation of the national park, though he did support the work to negotiate with the various communities, sultans and British authorities. The ordinances creating the park were issued in July 1938 for Kelantan and in 1939 for Pahang and Terengganu. In accepting that local populations not be totally excluded from the park, as Hubback had wished, he helped give it a more effective role in protecting wildlife in the long run, and contributed to its continued existence after the colonial period.[64] Contrary to his wish that the park be an attraction open to the public, it remained primarily a conservation area for species and ecosystems. It was decades before it opened to tourists. In 1948, it was visited by a grand total of four groups. It was not until 1965 that developing park tourism was embraced as part of the country's overall development policy.[65]

Ritchie's influence is discernible in the management of human–animal conflict. While Hubback had been reluctant to allow animals to be shot to protect fields, Ritchie, drawing on his experience in Africa, felt that the security of inhabitants and their crops prevailed. Controlled shooting of elephants became systematic throughout the colonial period and beyond, so that by 1964 there were only six elephants identified in Selangor, down from 166 in 1921.[66]

Ritchie's experience in Malaya raises questions about the professionalism of the Game Department and the effects of imperial circulations. Ritchie was sent to Malaya 'to create an up and running wildlife conservation service there. Among other things, it underscored the influence that conservation in East Africa was having elsewhere',[67] but his impact was in fact limited. Local dynamics continued to prevail. His main contribution was to legitimise the discourse and action of those in favour of shooting animals – especially elephants – deemed dangerous to people. Ritchie is still perceived as the person who initiated the Kenya Wildlife Service,[68] but in Malaysia it is Hubback who is presented as 'the "father" of the present Department of Wildlife and National Parks'.[69]

Studying British game wardens' life courses in Malaya – and then Malaysia – shows that they had a similar profile: big game hunters chosen for their hunting skill and with no aptitude in managing wild species populations. Malaya was no exception.[70] On being asked in 1941 about the qualities of a good Game Department head, Charles Hobley, former secretary to the SPFE, responded:

> Character, as in most posts, is the first essential; in addition, administrative ability is most necessary, for he must have the faculty of carrying people, both black and white, with him in his aims, and to do this successfully, he must possess a judicial mind, and not ignore the grievances of farmers against the depredations of wild life on crops. He must have organising gifts, and be able to state his case fearlessly and with some eloquence. He should by taste and upbringing be a field naturalist or he will probably lack devotion to his task. It is, perhaps, better that he should not be a biological specialist, for if he is, his passion for his pet branch will divert him from the broad outlook which is so necessary. Finally, he must be physically tough, and have an optimistic and resourceful type of mind.[71]

The only technical skills listed by Hobley are administrative. His definition of a 'good' game warden is that of a colonial gentleman hunter. In Malaysia, it was not until 1971 that the Wildlife Department was placed in the hands of someone who had trained for the task, when Mohamed Khan bin Momin Khan – who had studied at the US Department of Fish & Game and the African College of Wildlife Management, founded in Tanzania in 1963 – was appointed as its head. It was he who put an end to culling elephants and tigers. He established solid links with international conservation networks, especially the IUCN, worked with scientists and incorporated research on animal ethology into conservation practices. He played a central role in protecting ecosystems, maintaining protected zones and opening the national park –now renamed Taman Negara – to tourism, and was involved in raising public awareness.[72]

Ritchie's secondment to Malaya shows how the SPFE and Colonial Office sought to propel an imperial model of wildlife conservation based on that developed in Africa. This model shows the importance accorded to hunters turned 'penitent butchers',[73] and the appeal of utilitarian conservation based on developing tourism while protecting settlers and – whenever possible given game departments' limited resources – local populations from animal attacks on their lives and property. Ultimately, it was the local colonial authorities who had the last word, for it was they who negotiated with the authorities in the country and listened to the complaints of colonists and local populations. Thus the policies, and especially the ways they were implemented, differed markedly from one British possession to another.

Colonial protagonists such as Hubback and Ritchie no doubt laid the groundwork and managed to preserve vast areas, but it was not until significantly later that wildlife departments capable of managing this heritage were actually up and running. Although men, including many locals, devoted their careers to wildlife in the first decades of the twentieth century, the professionalisation of these departments still lay some way in the future when the colonial era came to an end. For the colonial administration, professionalism was less a matter of skills than of discipline and status, with paid civil servants doing the job. The use of honorary game wardens and rangers remained widespread, and the level of technical expertise low, unlike in forestry services.[74] Professional conservation incorporating scientific recommendations and conducted by the IUCN, the FAO, UNESCO and major conservation NGOs only arose well after the Second World War, even though the colonial legacy has continued to mark the practices of the wildlife departments in Kenya and Malaysia. The latter is thus far less militarised than the former.

Mathieu Guérin is Professor of History of South East Asia at the Institut national des langues et civilisations orientales (INALCO). His research focuses on the social and economic history of Cambodia in the nineteenth and twentieth centuries, with particular attention to rural societies and the margins. He also examines the history of wildlife conservation in South East Asia since the colonial era, particularly in Cambodia and Malaysia.

Notes

1. John MacKenzie, *The Empire of Nature: Hunting, Conservation and British Imperialism* (Manchester: Manchester University Press, 1988); William Beinart and Lotte Hughes, *Environment and Empire* (Oxford: Oxford University Press, 2007).
2. Tim Knight and Mark Rose, *With Honourable Intent: A Natural History of Fauna & Flora International* (London: William Collins, 2017).
3. Raf de Bont, Simone Schleper and Hans Shouwenburg, 'Conservation Conferences and Expert Networks in the Short Twentieth Century', *Environment and History* 23(4) (2017), 569–99.

4. Karen Blixen, *Out of Africa* (London: Putnam, 1937); Ernest Hemingway, *Green Hills of Africa* (New York: Scribner, 1935).
5. Nora Kelly, 'In Wildest Africa: The Preservation of Game in Kenya 1895–1933' (PhD diss., Simon Fraser University, 1978); Thomas Ofcansky, *Paradise Lost: A History of Game Preservation in East Africa* (Morgantown: West Virginia University Press, 2002); Edward Steinhart, *Black Poachers, White Hunters: A Social History of Hunting in Colonial Kenya* (Nairobi: EAEP, 2006).
6. Lieutenant Archibald Thomas Ayres Ritchie, Grenadier Guards, 1915–1921, UK National archives (UKNA), WO 339/38758; Peter Jenkins, 'Three Stalwarts', in Ian Parker and Stan Bleazard (eds), *An Impossible Dream: Some of Kenya's Last Colonial Wardens Recall the Game Department in the British Empire's Closing Years* (Kinloss: Librario, 2001), 32–38; Kelly, 'In Wildest Africa', 152; Steinhart, *Black Poachers*, 158–63.
7. Letter from the game warden to the Colonial Secretary, 16 November 1936, Kenya National Archives (KNA), KW/27/2.
8. Kelly, 'In Wildest Africa', 154.
9. Steinhart, *Black Poachers*, 158–59.
10. 'Game and vermin control 1935–1950', KNA, KW/5/4; Kelly, 'In Wildest Africa', 158–61, 208–37 and 262–90; Steinhart, *Black Poachers*, 159–73.
11. Jane Carruthers, 'Creating a National Park, 1910 to 1926', *Journal of Southern African Studies* 15(2) (1989), 188–216.
12. Richard Hingston, *Mission to Africa on Behalf of Society for the Preservation of the Empire Fauna*, 1930, IE TCD MS 10476, Trinity College Dublin, Digital Collections, https://digitalcollections.tcd.ie/concern/works/k3569616t?locale=en, retrieved 23 March 2022; KNA, VF/2/2; Kelly, 'In Wildest Africa', 297–98.
13. KNA, KW/8/29; Kenya Game Department report, 1932–1934, KNA, KW/23/160.
14. Robert M. Maxon and Thomas P. Ofcansky, 'East African Professional Hunter's Association', in *Historical Dictionary of Kenya* (Lanham, MD: Scarecrow Press, 2000).
15. Kelly, 'In Wildest Africa', 281; Steinhart, *Black Poachers*, 113–37.
16. Malaysia National Archives (MNA), 1957/0265735.
17. Jeyamalar Kathirithamby-Wells, *Nature and Nation: Forests and Development in Peninsular Malaysia* (Honolulu: University of Hawai'i Press, 2005), 165–226; Mathieu Guérin, 'Conserver la faune sauvage de la péninsule malaise: De la Malaya britannique à la Malaisie indépendante', *VertigO* 17(1) (2017).
18. MNA, 1957/0280032, 0534170.
19. Thomas Comyn-Platt, 'A Report on Fauna Preservation in Malaya', *Journal of the Society for the Preservation of the Fauna of the Empire* 30 (1937), 45–52.
20. At the end of his career Shenton Thomas was vice-president of the SPFE.
21. Comyn-Platt, 'A Report', 50.
22. Mathieu Guérin, 'Protecting the Forest and Its Fauna Against Local Residents in British Malaya', *Journal of the Malaysian Branch of the Royal Asiatic Society* 95(323) (2022), 57–82.
23. During the 1930s and 1940s, Gent closely monitored the affairs of the Malayan Game Department. On being appointed High Commissioner after the Second World War, he played a major role in reorganising how wildlife conservation was administered. Guérin, 'Conserver la faune'.

24. Minutes, letters and meetings about the King George V National Park at the Colonial Office, 1935–1936, UKNA, CO 717/113/2.
25. Memorandum from Gent to Lord Plymouth, 12 March 1936, UKNA, CO 717/116/6.
26. 'Mr. Hubback and Fauna preservation', confidential appendix to the Comyn-Platt report, 1936, UKNA, CO 717/116/6.
27. Memorandum from Gent, 25 July 1936, UKNA, CO 717/116/6.
28. For analysis of local populations' agency vis-à-vis the British conservation agenda, see Guérin, 'Protecting the Forest'.
29. UKNA, CO 717/116/6.
30. Kenya Game Department report, 1932–1934, KNA, KW/23/160.
31. Merwyn Cowie, *I Walk with Lions: The Story of Africa's Great Animal Preserves, the Royal National Parks of Kenya, as Told by Their First Director* (New York: Macmillan, 1961), 56.
32. *The Children's Newspaper*, 517, 16 February 1929; 'Ein schönes Jagdabenteuer [A fine hunting adventure]', *Gebweiler neueste Nachrichten*, 8 September 1933.
33. Patrick R. Chalmers, *Sport and Travel in East Africa: An Account of Two Visits 1928 and 1930 Compiled from the Private Diaries of H.R.H. the Prince of Wales* (London: P. Allan, 1934); Steinhart, *Black Poachers*, 125–27; Trevor Mark Simmons, 'Selling the African Wilds: A History of the Safari Tourism Industry in East Africa, 1900–1939' (PhD diss., University of Texas at Austin, 2015), 401.
34. Draft letter from the Secretary of State for the Colonies to the Governor of Kenya, 17 August 1936, UKNA, CO 717/116/6.
35. Letter from Ritchie to Eric Percy-Smith, 23 September 1936, KNA, KW/27/2.
36. Letter from Ritchie to Charles Hobley, 18 November 1936, KNA, KW/27/2.
37. 'New Chief Game Warden, One of the Empire's Leading Experts', *Straits Times*, 8 November 1936.
38. *Straits Times*, 31 December 1936; *Singapore Free Press and Mercantile Advertiser*, 11 January 1937.
39. Ritchie, memorandum, 22 April 1937, MNA, 1957/0534420.
40. High Commissioner's records, 23 April 1937, MNA, 1957/0534420.
41. 'Dropping the Pilot', *Straits Times*, 9 November 1936.
42. Letter from Hubback to the Pahang Resident, 25 March 1938, MNA, 1957/0538315.
43. See for example: 'The Life That Makes Us Wild', *Straits Times*, 25 February 1937; 'Wild Life vs Farmers in Kenya: Problems Left Behind by Malaya's Chief Game Warden', *Straits Times*, 16 March 1937; 'Pity the Wild Life', *Straits Times*, 7 April 1937.
44. 'Annual Report on the Game Department, Federated Malay States', 1937, MNA, 1957/0289414.
45. Letters from Hubback, MNA, 1957/0280032.
46. Game Department correspondence, MNA, 1957/0289057; 1937 annual report, MNA, 1957/0289414.
47. Chief Game Warden to the Secretary of the Federated Malay States, 6 June 1937, MNA, 1957/0289057.
48. Hubback had already pointed out this anomaly in 1934; MNA 1957/0280032.
49. 1937 yearly report, MNA 1957/0289414; 'Preservation of Wild Life', 1937, UKNA, CO 717/123/4.
50. He carried out this project on returning to Kenya. Nairobi National Park was the first

to be created in the country in 1946; KNA, KW 23/160, VF/2/2; Kelly, 'In Wildest Africa', 293–98.

51. *Malaya Tribune*, 15 July 1938; *Morning Tribune*, 15 July 1938.
52. Letter from Hubback to the Pahang Resident, 11 March 1938, MNA, 1957/0538315.
53. British Adviser in Terengganu, 'Note on the present position of the King George V National Park', 21 August 1938, MNA, 1957/0520096. As experienced by the author in Kenya and Malaysia.
54. Guérin, 'Protecting the Forest'.
55. Kenya Game Department report 1932–1934, KNA, KW 23/160; Richard Hingston, 'Memorandum on the Elephant', in *Mission to Africa*; 1937 annual report, MNA, 1957/0289414; Mathieu Guérin, 'L'homme contre l'éléphant, une guerre de cent ans: Conservation et abattage de l'Afrique de l'Est à la péninsule malaise', in Philippe Madeline, Élisabeth Ridel-Granger and Sylvain Skora (eds), *Les campagnes dans l'histoire: Mélanges offerts à Jean-Marc Moriceau* (Caen: Maison de la Recherche en Sciences Humaines, 2023), 579–89.
56. Report by Ritchie, 12 April 1938, MNA, 1957/0538315.
57. Steinhart, *Black Poachers*, 163.
58. Ritchie, 'Elephant control', 24 August 1937, MNA, 1957/0538315.
59. Memorandum from Pahang Resident Cowgill, 19 May 1937, MNA, 1957/534420; Charles W. Hobley, 'Wild Life Conservation in Its Wider Aspects', *Journal of the Royal Society of Arts* 89(4589) (1941), 468–69.
60. Report by Ritchie, 12 April 1938, MNA, 1957/0538315.
61. 1937 annual report, MNA, 1957/0289414.
62. *Malaya Tribune*, 6 December 1937; *Straits Times*, 7 December 1937.
63. Note from the Undersecretary of State for the Colonies for the conference of Malaya Residents about the chief game warden, 23 August 1938, MNA, 1957/0538315.
64. Guérin, 'Protecting the Forest'.
65. 'First Malaysia Plan, 1966–1970', 1965, 189, MNA, 1995/0003057.
66. MNA, 1957/0215521; Lord Medway, 'Game Statistics for Malaya: 1960–1963', *Malayan Nature Journal* 19 (1965), 223–34.
67. Jenkins, 'Three Stalwarts', 33.
68. Interview with a Nairobi National Park Ranger, 26 October 2020; Parker and Bleazard, *An Impossible Dream*; Stephen Spawls and Glenn Mathews, *Kenya, a Natural History* (London: T & AD Poyser, 2012), 404–6.
69. Mohamed Khan bin Momin Khan, 'Animal Conservation Strategies', in Earl of Cranbrook (ed.), *Key Environments, Malaysia* (Oxford: Pergamon Press, 1988), 252. The expression is in fact taken from James Hislop, the last British chief game warden in Malaysia: 'Protection of Wild Life in the Federation of Malaya', in J. Wyatt-Smith and P. R. Wycherley (eds), *Nature Conservation in Western Malaysia, 1961* (Kuala Lumpur: MNS, 1961), 140.
70. MacKenzie, *The Empire*.
71. Hobley, 'Wild Life Conservation', 471.
72. Guérin, 'Conserver la faune'.
73. The expression 'penitent butchers' was popularised by Fitter and Scott and refers to hunters from the colonial period who went on to become supporters of species preservation and played an increasingly significant role in imperial conservation: Richard

Fitter and Peter Scott, *The Penitent Butchers: The Fauna Preservation Society* (London: HarperCollins, 1978).

74. Peter Vandergeest and Nancy Lee Peluso, 'Empires of Forestry: Professional Forestry and State Power in Southeast Asia', *Environment and History* 12(1) (2006), 31–64.

Bibliography

Beinart, William, and Lotte Hughes. *Environment and Empire*. Oxford: Oxford University Press, 2007.

Carruthers, Jane. 'Creating a National Park, 1910 to 1926'. *Journal of Southern African Studies* 15(2) (1989), 188–216.

De Bont, Raf, Simone Schleper and Hans Shouwenburg. 'Conservation Conferences and Expert Networks in the Short Twentieth Century'. *Environment and History* 23(4) (2017), 569–99.

Guérin, Mathieu. 'Conserver la faune sauvage de la péninsule malaise: De la Malaya britannique à la Malaisie indépendante'. *VertigO* 17(1) (2017), https://doi.org/10.4000/vertigo.18503.

———. 'L'homme contre l'éléphant, une guerre de cent ans: Conservation et abattage de l'Afrique de l'Est à la péninsule malaise', in Philippe Madeline, Élisabeth Ridel-Granger and Sylvain Skora (eds), *Les campagnes dans l'histoire: Mélanges offerts à Jean-Marc Moriceau* (Caen: Maison de la Recherche en Sciences Humaines, 2023), 579–89.

———. 'Protecting the Forest and Its Fauna Against Local Residents in British Malaya'. *Journal of the Malaysian Branch of the Royal Asiatic Society* 95(323) (2022), 57–82.

Kathirithamby-Wells, Jeyamalar. *Nature and Nation: Forests and Development in Peninsular Malaysia*. Honolulu: University of Hawai'i Press, 2005.

Kelly, Nora. 'In Wildest Africa: The Preservation of Game in Kenya 1895–1933'. PhD diss., Simon Fraser University, 1978.

Khan bin Momin Khan, Mohamed. 'Animal Conservation Strategies', in Earl of Cranbrook (ed.), *Key Environments, Malaysia* (Oxford: Pergamon Press, 1988), 251–72.

Knight, Tim, and Mark Rose. *With Honourable Intent: A Natural History of Fauna & Flora International*. London: William Collins, 2017.

MacKenzie, John. *The Empire of Nature: Hunting, Conservation and British Imperialism*. Manchester: Manchester University Press, 1988.

Maxon, Robert M., and Thomas P. Ofcansky. 'East African Professional Hunter's Association', in *Historical Dictionary of Kenya* (Lanham, MD: Scarecrow Press, 2000), 15.

Ofcansky, Thomas. *Paradise Lost: A History of Game Preservation in East Africa*. Morgantown: West Virginia University Press, 2002.

Simmons, Trevor Mark. 'Selling the African Wilds: A History of the Safari Tourism Industry in East Africa, 1900–1939'. PhD diss., University of Texas at Austin, 2015.

Spawls, Stephen, and Glenn Mathews. *Kenya, a Natural History*. London: T & AD Poyser, 2012.

Steinhart, Edward. *Black Poachers, White Hunters: A Social History of Hunting in Colonial Kenya*. Nairobi: EAEP, 2006.

Vandergeest, Peter, and Nancy Lee Peluso. 'Empires of Forestry: Professional Forestry and State Power in Southeast Asia'. *Environment and History* 12(1) (2006), 31–64.

CHAPTER 3

Imperial Forests and Nature Reserves in Singapore, 1883–1959

Timothy P. Barnard

Nathaniel Cantley, the Superintendent of the Singapore Botanic Gardens, submitted a report to the Legislative Council of the Straits Settlements in 1883. The document outlined a serious environmental problem that existed in many British territories and had become particularly evident in Singapore, at the centre of their imperial presence in South East Asia. The island was massively deforested. This process had begun six decades earlier, following the establishment of East India Company authority over an island with fewer than a thousand inhabitants in 1819. In just a few decades, as imperial power secured sovereignty and developed the harbour into an important node of global trade, the cultivation of cash crops transformed the landscape beyond the port.[1]

This began when migrants arrived to work the forest. They planted a variety of species, many of which were unable to flourish in the island's acidic soils and tropical climate. Nutmeg was an early success, but disease soon ravaged the trees. Planters quickly switched to black pepper and gambier. Grown together – with the vines symbiotically supporting each other – this spice and herb were tremendously fruitful. By the 1840s, gambier and pepper became the dominant export crops from Singapore, which continued to function mainly as an entrepôt. The authorities encouraged further cash-crop cultivation and plantations spread throughout the island. Agriculture flourished, forests were felled and Singapore became a modern export centre. The success was fleeting, however, as plantations aged and declined in productivity, leaving behind an island suffering from 94 per cent total deforestation by the 1880s.[2]

In his report Cantley admonished the colonial administration for actively supporting the devastation of the tropical landscape. These decisions, made over decades and reflecting larger imperial policy, had pushed Singapore to the edge of an environmental catastrophe. Temperatures were rising, both anecdotally and as recorded by newly installed instruments, while timber resources for cooking and construction had become scarce.[3] To further exacerbate matters, water resources were also drying up. The lack of water was of particular importance, as the port not only supported one hundred thousand residents by the 1880s but also functioned as a supplier of goods and services for ships plying the China to India trade route that was vital to British imperial economic policy. The environment had been the foundation for the development of the colony and a devastated landscape transformed for agricultural export now threatened its existence. As Cantley emphasised, basic environmental management in Singapore had pushed the island and colony to the edge of catastrophe. It was, as he expressed in his laconic British manner, contrary 'to the rules of good forest conservancy'.[4]

Cantley, however, had a solution for these problems. He proposed the development of a massive tree planting programme in Singapore to re-establish a vibrant biosphere, which would protect the water supply and even provide some relief from the tropical heat. The main proposal was to create a massive forest reserve in the centre of the island, along with a few scattered coastal zones. Singapore may have only had 6 per cent forest cover, but this plan would address the matter and avoid an environmental collapse. Singapore would have managed forests through an application of imperial science and administration, which would herald the benefits of British rule in South East Asia.[5]

The imperial forests of Singapore, and how they grew out of a devastated landscape, is a tale that is not unique. Deforestation and reforestation took place in many territories throughout the British Empire during the nineteenth century. Colonial officials initially viewed indigenous forests throughout the world as timber mines, useful for exploitation and revenue as the land was converted into agricultural zones that supported profitable cash-crop plantations. Systematic forest management only arose when the massive exploitation of timber resources had created a series of observable crises, ranging from soil erosion to threats to water supplies.[6]

By the beginning of the twentieth century, officials within the British Empire managed imperial forests following precepts developed earlier in Germany and France and actively applied by the Indian forest department. The replanting of the forests may have alleviated much of the environmental devastation that imperialism had initially wrought, but it also reinforced the idea that these regions were sites of remunerative productivity as they were overseen and cultivated on the principle of continuing to contribute to the colonies' and metropoles' material prosperity. After the First World War, realising that the unique ecosystems under imperial authority varied in their ability to fulfil the principles of conservation, foresters eventually began to advocate a more ecological agenda in which preservation of biodiversity soon took precedence over utilitarian concerns.[7]

The management of imperial forests and nature reserves globally is clearly reflected in Singapore. As Richard Grove has highlighted, islands are important for understanding environmental change because of their limited size and isolation (although Singapore is not quite isolated), which helps focus our attention on the impacts and mechanisms of environmental policies and their histories.[8] The deforestation of Singapore, as well as its subsequent reforestation and management, all took place within a relatively small, self-contained space of 500 square kilometres; it helps clarify many of these developments, not because of any innovative programme officials enacted to protect its landscape, but because it clearly reflects how policies often originating in Europe, and in this case practised in India, were subsequently applied throughout the world.[9] It is a case study for governing nature in the tropics.

Developing Forest Reserves

Following the submission of the *Report on the Forests of the Straits Settlements*, the colonial government approved the creation of a 'small Forest Department' under Cantley's supervision in late 1883. The remit of this new institution was to ensure 'proper protection', as well as 'working and general management of the forests' throughout the colony. The passage of this legislation was a mere formality as a year earlier Cantley had already identified potential forest reserves on the island – there were initially eight, consisting of more than 10 per cent of the entire island – and began supervising workers on the planting of tree boundaries. As part of these efforts, he also developed nurseries in the Singapore Botanic Gardens, and later the Bukit Timah Reserve, 'for the propagation of trees and shrubs', from which over a hundred thousand saplings would be transplanted yearly into these reserves. This occurred quickly, as Cantley proudly proclaimed that the establishment of forest reserves 'had been carried out to a greater or lesser extent' by 1886.[10]

The forest reserves of Singapore contained a mixture of species, chosen for two distinct purposes, which reflected the reforestation programme's ecological realities as well as its conservation goals. The first type of desired trees was those that grew quickly. Their purpose was to create shade that would choke off native grasses, known locally as *lalang* (*Imperata cylindrica*), which covered most of the abandoned plantations and inhibited the spread of desirable forest flora. Among the species used in these efforts was the American Rain Tree (*Samanea saman*), which was appealing due to its extensive canopy and ability to grow quickly. It also reflected the networks of forestry and botany that stretched from South America to Europe, and then out to the wider world.[11] Although the American Rain Tree and a few other non-native species became established in the Singaporean landscape, officials soon became dependent on native species such as Tembusu (*Fagraea fagrans*), Meranti (g. *Shorea*) and those of the *Eugenia* genus to create forest cover to counter the extensive fields of *lalang*. By the mid-1880s nurseries under the Forest Department were propagating 150,000 plants a year, of which over 110,000 were sent for planting in the reserves, with the primary purpose of re-establishing forest cover in Singapore.[12] While the ability of these trees to displace weedy grasses was an important factor in the development of the Singapore forest reserves in the late nineteenth century, the primary objective of officials was to cultivate utilitarian species, which could provide timber for export as well as high-quality materials for construction needs in the town. To fulfil these needs, the nursery at the Singapore Botanic Gardens also nurtured and then transplanted into the deforested interior twenty thousand tropical hardwoods – primarily teak (*Tectona grandis*) – in the first year of operation and well into the 1890s.[13]

These initial attempts to reforest the island with economically valuable timber reflected many of the early efforts of forest reserves throughout the empire. On the one hand these sought to promote conservation, grounded in a belief that scientific principles should be applied to address environmental damage; on the other, they sought to ensure that the forests would continue to provide economic value to society. This tension between preserving (or re-establishing) the forest and utilising it for public profit was a dichotomy running through the rhetoric of forest management for much of the late nineteenth century.[14] While the contrasting nature of these approaches arose out of the writings and observations of Americans and continental Europeans, forestry officials in Britain quickly adopted and enacted them throughout their empire, and with particular rigour in their Indian colony.[15]

Following these principles of conservation, the forest reserves in Singapore developed and addressed two separate issues. The first was the environmental crisis of a transforming climate and reduced water resources. One reserve accomplished this task by shielding the water supply contained in artificial reserves in the centre of the island. The remaining forest was to provide utilitarian worth to the colony through the provision of limited amounts of timber for general use.[16] Cantley was simply appealing to British administrators when he enacted a similar solution to a local environmental crisis in the Straits of Malacca.

The influence of these mid-nineteenth-century imperial networks and experts, and the bifurcated policies of imperial forest management, can be seen in Cantley's background, and that of his successor, Henry Nicholas Ridley, who arrived to oversee the forests following Cantley's death from ill health in 1887. Both Ridley and Cantley were products of the Royal Botanic Gardens, Kew, reflecting networks of imperial botany that encircled the British empire.[17] Cantley had previous experience working in the herbarium at Kew and had been the Assistant Director of Pamplemousses Botanic Garden in Mauritius for eight years before taking up his position in Singapore. Ridley had worked in the botanical department at the British Museum and had already developed a reputation as a scientist among the leadership of Kew after participating in collecting expeditions to Brazil. Before departure for Singapore, Ridley had even been sent for training at Cooper's Hill, the forestry school in England founded in 1885 on the principles of empire forestry in India. They both understood the importance of preserving and understanding the floral landscape, while also accepting that forests could provide some utilitarian value for society.[18]

While officials were employing global forestry management principles in Singapore, particularly in their planning and development, local ecological and cultural limitations soon came to influence their actual application, which quickly exposed tensions between the two approaches. This became clearer in the 1890s when Ridley attempted to transition the reserves from quick growth species to economically valuable species following larger empire-wide policies of exploitation. Ridley was particularly interested in Belian (*Rhizophora conjugata*) – 'probably the best timber in the Oriental region' – after he obtained seeds from Borneo. Few woodworkers in Singapore, however, were interested in working with such hardwoods as they took too long to mature and were difficult to process. As Ridley reported with some exasperation, 'the Chinese carpenter finds it more profitable to work with the softer and inferior woods, not only on account of their being easier to cut but also because they soon perish and have to be replaced'. This was most clearly seen in the sawmills of Singapore in which two species of local Dipterocarps (*Shorea macroptera* and *Shorea curtisii*) – a 'poor class of timber' – were the most popular trees to be hewn into planks. This eventually led Ridley to designate a few stands of non-native tropical hardwoods for use in the construction of prominent structures in Singapore, while more perishable native woods, which were more popular and easier to propagate, were planted in most of the designated forest reserves, where they proliferated. The efforts to create forest reserves in late nineteenth-century Singapore thus resulted in an emphasis on native species, due to ecological limitations when replanting began, as well as to the social and cultural needs of the society.[19]

Once these forest reserves had been established in a rudimentary manner to fulfil local needs relating to conservation, the matter of governing nature in the tropics also came into play. Ridley was an energetic botanist who is best known for the role he played in the development of rubber as a plantation crop. Although he had received preliminary training at Cooper's Hill, he remained an adherent of imperial botany, his primary administrative connection being to the Royal Botanic Gardens, Kew, not to officials in the Colonial Office. Based on his background, his main motivation was the development of healthy stands of forest to meet the needs of the surrounding society. The timber would not necessarily be cultivated for export and profit. Ridley was willing to work with what was ecologically most suitable for the forests under his purview, more in line with sustainable forest use policies that had been promoted throughout continental Europe earlier in the century under such luminaries as Alexander von Humboldt.[20] This would clash with a growing trend towards further exploitation of the forests.

The ascendance of forest exploitation also corresponded to changes in the administration of the British Empire, particularly under the tenure of Joseph Chamberlain, who was the Secretary of State for the Colonies from 1895 until 1903. Chamberlain emphasised the utility of science, a policy that created a formal bureaucracy of experts that shifted environmental administration away from personal networks of imperial and economic botany promoted from Kew. This resulted in funds and personnel being channelled towards research in tropical medicine, agriculture and forestry, as well as a range of new disciplines that would further integrate Britain's various possessions into the global economy. As Joseph Hodge has argued, this required new classes of colonial officials rooted in scientific and academic expertise to lead new departments throughout the British Empire, as well as a channelling of their energies towards supporting large, exploitive enterprises, including the management of forest resources.[21]

An early sign of how this transformation of administrative duties would affect forest policy occurred in 1895 when colonial officials transferred the Forest Department from the Singapore Botanic Gardens to the Land Revenue Office, which became 'responsible for the upkeep on the boundaries, protection and closure'.[22] Bureaucrats overseeing forest reserves in Singapore now simply issued permits to those who wished to harvest timber, a proper utilisation of the resource in alignment with imperial conservation policies. Ridley, who remained as Director of the Singapore Botanic Gardens, consistently opposed such an approach, as he wanted to nurture diverse and healthy stands of timber in limited access forest reserves in an attempt to establish more valuable, local tropical hardwoods in Singapore's interior. While Ridley was supportive of forest exploitation policies, he believed that simply viewing the forests of the region as quick revenue generators was not an economically sound policy, as well as being ecologically misguided. Under the management of the Land Revenue Office, Ridley adamantly argued that the forest would never be viable without a period of preservation to ensure the growth of healthy stands of timber. The entire episode was one in which, according to Ridley, officials in Singapore found him 'too damned scientific', leading them to make decisions that 'made the Straits Government stink in the nostrils of the naturalists'.[23] While this may have been true, by the end of the nineteenth century principles of rapid utilisation of resources had begun to take precedence in the forestry policy in Singapore and the British Empire.

The continual criticism Ridley directed at government officials concerning these policies led not only to the Land Revenue Office taking responsibility for forest management in Singapore, but also to the establishment of a Forest

Department in Kuala Lumpur, far from any influence that the contrarian botanist may have had. The basis for this new institution, which would replace the 'small Forest Department' and the Land Revenue Office, stemmed from a report the government commissioned in 1900 from H. C. Hill, a former official in India who also was influential in determining forestry policy in England at the time. The report that Hill submitted provided the template for applying empire-wide policies on governing nature in tropical South East Asia, and also reflected the ascendance of policies in which the primary role of the empire's forests was utilitarian.[24]

On arriving in Singapore, Hill visited the existing reserves with Ridley and W. L. Carter, the Collector of Land Revenue, thus reflecting the two choices that lay before him. In his report, Hill believed that the forest that existed after only fifteen years was sufficient to support a system of conservation. As he argued in the report, there was 'a belief in the minds of some people' that Singapore was deforested, although he did not find this to be true. 'There has been no actual denudation: for whenever land ceases to be cultivated a growth of grass, fern and soft woods quickly springs up on it. And, if this is not burnt over, the forest soon re-establishes itself.'[25]

Beyond his belief in tropical fecundity, and its ability to re-establish a forested Singapore, Hill argued that the cultivation of experimental plantations of tropical hardwoods, which Cantley and Ridley had both championed, was misguided due to their high costs and marginal benefits. Such species would take too long to mature for them to be profitable. As a solution, he proposed that all Crown Lands be more clearly differentiated. In reserve areas, there would be limited access and efforts would be made to encourage the growth of useful species, particularly those with industrial applications. Hill was particularly supportive of efforts to cultivate gutta percha (*Palaquium gutta*), which produced latex that was well suited for underwater telegraph cables. Mangrove forests, in contrast, produced 'a valuable supply of firewood' and their harvesting would allow 'these swamps' to 'be turned to any other purpose'.[26]

To allow for the implementation of these recommendations, Hill proposed that administration of the forests in Malaya and Singapore be simplified, following a system developed for India in which District Officers would assume responsibility for forests as part of their other responsibilities. It was 'the only possible starting point for any reasonably economical system of forest conservancy in the Straits Settlements'. This would entail a reduction in scientific officers and forest guards, whose employment was 'too expensive to be justifiable'. Instead, the District Officer and one trained forestry official – trained at the Dehra Dun Forest School in India, of course – would oversee the Crown Lands in each of the four Straits Settlements of Penang, Malacca, Singapore and Labuan. In Singapore, this official would 'arrange for the working of the mangrove tracts and weed gutta percha trees in reserves'. The Chief Forester

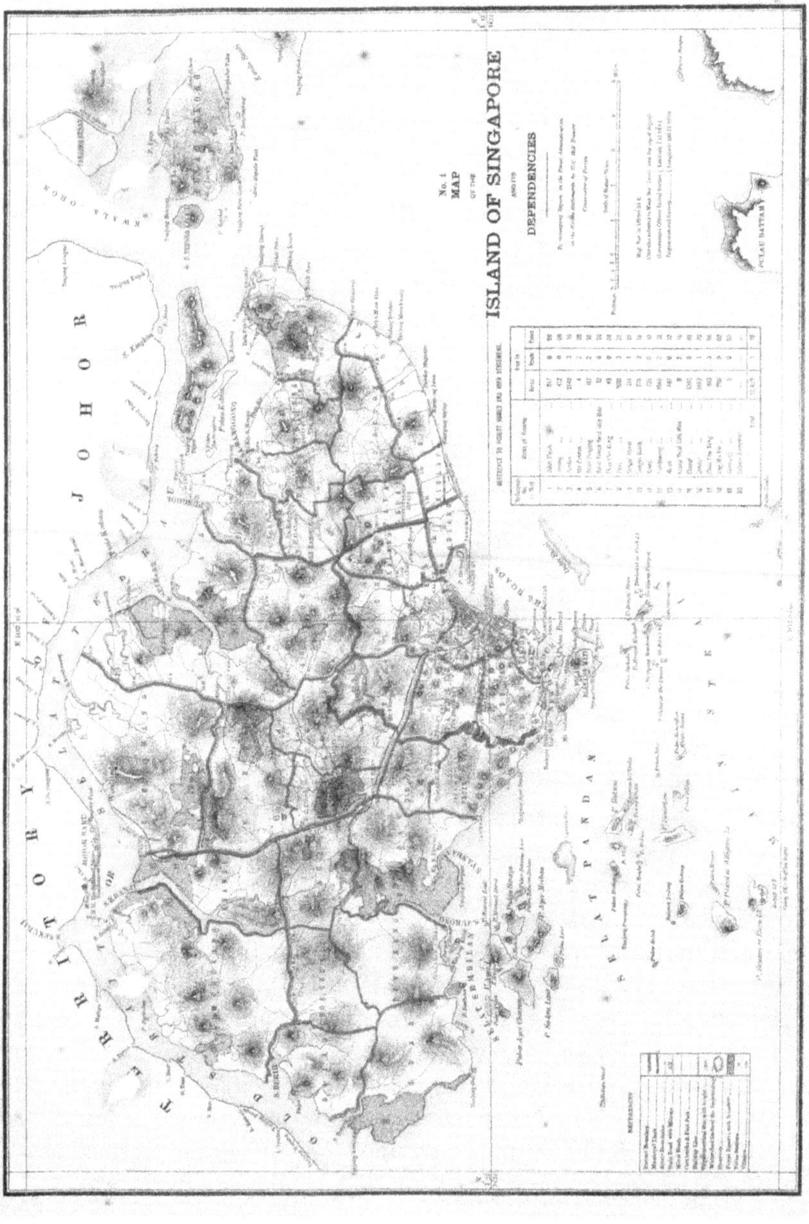

Map 3.1. Singapore's forest reserves, 1900. From H. C. Hill, *Report on the Present System of Forest Conservancy in the Straits Settlements, with Suggestions for Future Management* (Singapore: Government Printing Office, 1900). Photo: Singapore National University of Singapore, used with permission

Officer overseeing this entire system would be someone trained in Europe, and with experience working in Burma, which had been a timber mine under British imperialism. The government was uninterested in hiring anyone with local experience or knowledge to oversee the ecological health of the forests in Malaya and Singapore.[27]

The colonial government established the Forest Department for Malaya and the Straits Settlements in 1903. The forests of larger Malaya were now under the control of officials working out of Kuala Lumpur who were more aligned with a broader empire-wide forestry policy of exploitation of the resource, and had invariably trained in India. They were advocates of the doctrine of conservation and of utilising forest resources for the betterment of larger society, and pursued these goals throughout the Malay Peninsula, particularly after rubber plantations in the region began generating tremendous profits for British-owned firms. The Forest Department simply fulfilled policy ideas, while catering to the needs of economic profit for empire.[28]

The forest reserves of Singapore, with their limited space and lack of economically remunerative timber, quickly became an administrative afterthought. Much of this is summarised in a brief report on their status from 1901, when W. Langham-Carter, the Acting Collector of Land Revenue, described most of the Singaporean reserves as 'scrub' or 'poor jungle' that only merited being 'periodically visited both by the Collector of Land Revenue and the Forest Rangers'. For the next three decades, Forest Department reports ignored Singapore while silviculture remained under the purview of the Land Office, with officials exhibiting little concern over the ecological state of the wooded areas they oversaw. In the monotonous accounting of these reports, for example, A. M. Burn-Murdoch, the Conservator of Forests, simply wrote in 1905 that 'there is no available information' about the forests and 'merely the sums realised by the sale of passes to indicate the quantity' of timber harvested. Singapore had 50 square kilometres of forest reserves, constituting about 10 per cent of the land space, with minimal care or attention paid to them. Permits continued to be issued for timber harvesting in mangrove forests, while other areas were leased out to rubber tappers, as happened in a portion of the Sembawang Forest Reserve in 1909.[29]

This utilitarian approach to the forests of Singapore continued for decades, slowly devastating any ecological value they may have held. Although officials occasionally increased the areas designated as forest reserves, as happened in 1923 when the military returned an area of 450 hectares gazetted as the Mandi (today, Mandai) Reserve, there were many more instances in which the reserves and their boundaries were reduced or altered in numerous small ways. A large cemetery known as Bukit Brown developed on the edge of one forest reserve, for example, while the Singapore Golf Club and the Island Club carved out two different courses in the 1920s in the middle of the central water catchment area. While these developments were not technically inside the reserves, they did take place in strategically designated areas between alienated lands that the Land Office administered, leaving reserve lands as limited isolated biospheres on a small island.[30]

Matters grew even more dire within the reserves as squatter villages began to appear. Eventually, areas of the Ang Mo Kio and North Seletar reserves, 'which had no particular sylvicultural value', were 'let out to market gardeners' to support the cultivation of more vegetables. Much of this occurred through disinterest in any value the forest reserves might provide to society. At other times, these developments took place with direct government approval, through permits and rent-seeking on behalf of the Collector of Land Revenue, or through advice from the Singapore Botanic Gardens to various golf clubs to plant trees of the *Parkia* genus, which ultimately 'suffered incessantly from insect pests'.[31] Singapore had forests in the early twentieth century, although they were increasingly perceived as lacking any value, as it was difficult to utilise them for their timber. Officials believed it was more pragmatic to allow forests to transform into more suitable economic and social spaces.

From Forest Reserves to Nature Reserves

Policies that emphasised short-term economic benefits instead of long-term conservation were a standard global approach to forest management by the early twentieth century. This was true in the Malay Peninsula, with the forests of Singapore becoming an afterthought within these larger strategies due to their limited space and potential. Such an approach, however, came under scrutiny as scientific networks of foresters began to realise that massive exploitation was having long-term effects. Policies began to change. This resulted in officials designating new reserves throughout the empire, to be set aside for the ecological health of the forest.[32]

Singapore was at the forefront of this transformation as there was a growing acceptance that its forests would never be economically productive. As early as 1925 G. E. S. Cubitt, the Director of Forestry for Malaya, publicly argued that 'really effective management of the Singapore forests is possible only at a cost which the forests themselves do not seem to justify'. They were simply not worth the effort, and the forest reserves of Singapore had begun to fragment as they were increasingly modified for other purposes. Cubitt emphasised this shifting status when he noted that 'none of the reserves are of any great value, and considerable areas in them are already leased out on temporary occupation licenses to vegetable farmers', which was 'of greater importance to Singapore than the production of timber'.[33] Government officials had reached a point where the status of these reserves was being questioned openly.

Government perceptions of the worth of utilitarian conservation in the forest reserves of Singapore finally reached a turning point in 1930, when Bukit Timah Forest Reserve temporarily lost its protected status to accommodate the presence of a rifle range and squatter settlements. Officials eventually reinstated the status of the area as a reserve, but only after its boundaries were redrawn, resulting in a loss of 21 hectares. The reconstitution of this space as a forest reserve, however, signalled the formal end of imperial conservation policies in Singapore, as it was part of a larger proposal 'for the abandonment of all the reserves in Singapore, Bukit Timah excepted'. This was mainly due to squatters overrunning 'the remaining forests', which consisted 'of scattered areas of immature mangrove, which had been badly damaged by past overcutting'. Even if the authorities addressed these issues, J. P. Mead – the Director of Forestry – argued that 'it is doubtful whether they could ever be made to pay'.[34] Attempts to extract value from the timber of these forests had reached an end, and new approaches needed to be developed under these unique circumstances.[35]

In Malaya, concern about the 'natural diversity' of the forests began as early as 1915 when Cubitt – who also served as a forestry official in India at the same time – encouraged natural regeneration of the forests instead of cultivating desired species. The hiring of American botanist F. W. Foxworthy from the Philippine Bureau of Forestry to research how to utilise localised forest products contributed to these policies. Foxworthy and Cubitt began to differentiate the forest into that which could be utilised for profit and that which could be left untouched, or natural. In these efforts, Cubitt and Foxworthy received support from H. C. Robinson, the Director of the Federated Malay States Museum in Kuala Lumpur, who was also concerned with the environmental implications of utilitarian forestry management. Robinson proposed the decentralisation of policy and the creation of nature reserves, which would act as sanctuaries for protecting unique species in representative habitats in each sub-region.[36] Unique local ecological and political-social circumstances – including the diversity of forested areas, their increasing conversion into agricultural areas and the effects of the Great Depression – eventually forced officials to reconsider how natural spaces could be governed based on ideas that were gaining increasing acceptance globally.[37]

Early advocates for a more ecological approach to forest management in Malaya and Singapore placed their primary emphasis on protecting fauna, as 'the preservation of flora would in large measure be ensured' if legislation protected animals. They recommended a differentiation in areas, with forest reserves maintaining their utilitarian status while other designated spaces – nature reserves – would preserve ecological diversity.[38] These proposals reflected a decentralisation of forestry policy to fit local needs and conditions as 'there are small areas which it would be desirable to protect primarily on account of the interest of their flora', with the unique orchid species of Langkawi being one example. This change in policy in Malaya and Singapore was further altered after administrators contacted Richard Eric Holttum in 1936 for his thoughts on the state of forests in the region. As the Director of the Singapore Botanic Gardens, Holttum was a trained botanist with little interest in the utilitarian policies that had been promoted throughout the empire since the beginning of the twentieth century.[39]

Holttum argued that the limited colonial knowledge of the range of plant species throughout the Malay Peninsula made it more logical to make floral diversity the main purpose of these reserves. It was, after all, difficult to estimate how many faunal species had come under threat since the opening of 'the virgin territories of Malaya', which resulted in 'the forest-reserves [actually] exploited'. In line with larger changing policies, Holttum ultimately recommended the establishment of 'local reserves', particularly in Singapore, which would be accessible to 'all interested persons' seeking 'samples of the vegetation'. The ideal location for these smaller plots, according to Holttum, would be adjacent to forest reserves. For Singapore, he suggested that Bukit Timah Hill become such a 'botanical reserve', as it was 'the only area of varied natural forest in the island and is of considerable interest to botanists as a historical collecting ground'.[40]

Based on these various recommendations and shifting global understandings of the role of forests, the government created eight 'botanical reserves' throughout Malaya in 1937. Singapore contained three of these reserves, a reduction from the fifteen 'forest reserves' that had been conserved and exploited for forty years. While these areas continued to be designated as 'forest reserves', this was mainly 'for convenience of administration, not because they will be exploitable as forests'. At this time Holttum also became the 'Conservator of Forests' in Singapore, which was a further signal of the return of preservationists in overseeing greenery, in a move away from the remunerative policies of imperial forestry.[41]

In a report Holttum wrote after a decade in the position, he summarised the results that forty years of exploitative strategies had wrought on the forests of Singapore, and his position on being handed control over them in the late 1930s. 'The Forest Reserves', he succinctly recapped, were 'largely mangrove' and 'no longer worth maintaining for their production of timber'. Within these mangrove reserves, one was 'almost entirely cleared of large trees in 1937', while the other had been 'cleared of all but quite young trees'. This resulted in portions of three former forest reserves being 'handed over to the Botanic Gardens Department for maintenance as representative areas of natural vegetation'. Although they 'retained the status of Forest Reserves for purposes of administration', they were 'in effect Nature Reserves, primarily maintained for botanical interest'.[42]

The three botanic reserves in Singapore were Bukit Timah Forest Reserve, which was to be the primary nature reserve on the island, and two small mangrove areas, which were designated 'as samples of the natural vegetation'. Bukit Timah Forest Reserve consisted of 250 acres (101 hectares) and mainly covered the greater part of the hill of the same name. It contained 'a fine and varied stand of big trees and a steep valley – the Fern Valley – of botanical interest'. Bukit Timah Hill also offered 'magnificent views in every direction' and was home to a rich variety of 'mammalian fauna' including macaques, flying lemurs, wild boars and 'an occasional tiger', stated in reference to the earlier preference of zoologists to highlight the animal wealth of the region. The site, however, had other special considerations in its favour. It was also of 'historic importance' as it was the site of many early collectors, including Alfred Russel Wallace, and 'its perseveration' was 'more valuable still for the civic life of Singapore' as it offered 'a retreat wherein one can teach the rising generation to have a care for the wild things of wood and water'.[43]

The two mangrove reserves, in contrast, were in isolated areas. One was located at Kranji at the '14th mile on the road to Johor', in the northern reaches of the island, while the other was near the Pandan and Jurong rivers along the south-west coast. Both were devastated. 'The old trees' were gone, although there was 'much young growth' that 'adequately' preserved 'the character of mangrove', which Holttum hoped would allow 'local students of plant life' to become acquainted with its 'features'. Both these reserves, however, were poorly maintained and frequently suffered 'thefts of small timber'. To address this problem, a 'Chinese charcoal burner' was 'allowed to cut timber under strict supervision' in the second mangrove reserve, which was larger, in compensation for his help monitoring and patrolling the area.[44]

By 1939, the government expanded preservation of tropical biodiversity in Singapore when it designated these botanical reserves as 'nature reserves'. They were now 'representative plots' to be protected 'in perpetuity'. A Forest Ranger and six labourers were assigned to work Bukit Timah Forest Reserve. Their main tasks were preventing timber theft, cutting new paths to the summit, planting trees and providing identification labels to important trees, and finally the erection of small shelters at points 'where broad views could be obtained over Singapore town' and Malaya. While Bukit Timah Nature Reserve remained the primary natural spot on the island, regular patrols of the mangrove reserves continued, with occasional arrests of timber collectors. Botanists and coolies continued to collect specimens, identify and label trees and lay new paths. As one report summarised, 'new growth in areas formerly devastated continued satisfactorily'.[45]

Ten years later, with the Japanese occupation taking place in the interim, these reserves 'which had been devastated in 1936–37' had 'regenerated well' and contained 'dense thickets of young trees'. The mangrove reserves now contained 'all of the characteristic plants' expected of this tropical ecosystem and were 'big enough to give shelter to an interesting natural fauna'. In the meantime, the Bukit Timah Reserve now contained 'a fine sample of primitive forest, with a great variety of trees and smaller plants, some of which certainly [did] not occur elsewhere in the island'. Forests were no longer sites of economic utilisation but of burgeoning natural diversity.[46] The development of these areas was perceived as a positive example of environmental policy in the late British Empire.

The promotion of zones of 'natural diversity' in Singapore was further emphasised with the passage of the 'Nature Reserves Ordinance No 15 of 1951', which incorporated newly designated areas into the system, including a small seaside cliff on the southern side of the island at Labrador. In total, nature reserves now covered over 36 square kilometres. Most importantly, the ordinance now included the vast centre of the island, which operated as a water catchment area, thus leading to its current toponym 'the Central Catchment Area'. These nature reserves were 'dedicated, set aside and reserved for the purpose of the propagation, protection and preservation of indigenous fauna and flora of the Colony and for the preservation of objects and places of aesthetic, historical or scientific interest'.[47] In late colonial Singapore, its forests appeared to be preserved for their ecological value.

Yet the needs of a modernising, and decolonising, society continued to exert pressure on the nature reserves. The primary issue was fragmentation, as these spaces experienced pressure from the government seeking to promote economic progress. This was most clearly seen in the continued operation of three granite quarries bordering the Bukit Timah Reserve. These quarries threatened to 'seriously damage this final remaining piece of natural forest in the island'. Unless this reserve was allowed to 'be sufficiently large' to ensure it was 'self-regenerating', botanists at the Singapore Botanic Gardens argued, it would be difficult 'to maintain a humid internal climate in which the lesser forest plants can flourish'. Holttum went on to warn that 'ferns and other delicate plants' would disappear 'if the area is further restricted'.[48]

The continued operation of rock quarries near Bukit Timah Reserve reflected the continuation of utilitarian concerns in a modernising society that created pressure on the status of nature reserves. Such pressures existed in all the reserves on the island. Illegal timber cutting, for example, continued in the mangrove reserves, and although 'a number of arrests for thefts' were made, 'more than half the cases were discharged and in most of the others only nominal fines were imposed, an insufficient deterrent to future felons'. An even greater threat was a desire on the part of various government ministries to transform the land into economically productive areas. In the Pandan Nature Reserve, which featured stands of mangrove forests, for example, officials created over 36 hectares of prawn ponds in an attempt to promote aquacul-

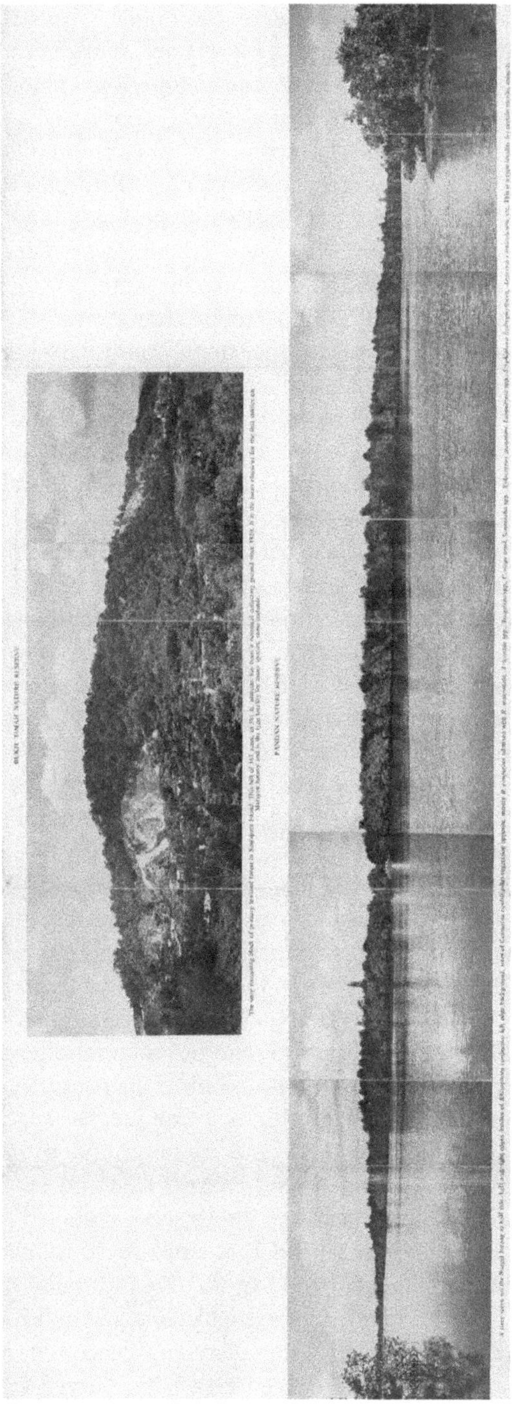

Figure 3.1. Bukit Timah Nature Reserve, 1957, featuring damage due to quarrying activities. From H. M. Burkill, *Annual Report of the Botanic Gardens Department for 1957* (Singapore: Government Printing Office, 1958), between pages 10 and 11. Photo: Singapore, National University of Singapore, reproduced with permission.

ture. At the northern Kranji Nature Reserve, the Public Works Department widened and realigned roads. Economic progress and modernisation clearly took precedence, despite these developments causing a loss of biodiversity. At the Kranji Nature Reserve, for example, there were reports that construction 'destroyed three of the four known specimens of *Cynometra ramiflora* on Singapore island', which remained a 'critically endangered' species in the early twenty-first century. These threats were further compounded by the continuation of work at the quarries bordering Bukit Timah Nature Reserve, which threatened to 'disfigure' the hill that gave the area its name.[49]

By 1957, as Malaya was heading into independence and Singapore into self-governance, the nature reserves on this small island in South East Asia had a doubtful colonial legacy. In December 1957 the Legislative Council reduced the Pandan Reserve from 555 to 150 hectares to accommodate the further expansion of prawn ponds. Despite arguments that efforts to promote aquaculture were doomed, as it would affect natural breeding and spawning grounds in the mangroves, the government perceived it as necessary to create employment and prosperity for a decolonising nation. Once these prawn ponds failed, as predicted, the area became a housing estate, replete with roads and other detritus of a modernising nation. At Bukit Timah, in the meantime, quarrying was 'a constant source of alarm', and there were instances in which blasting destroyed reserve boundary paths. When the Nature Reserve Board of Trustees applied to have the quarrying activities ceased and the various properties within the reserve transferred to the reserve, the lands were – instead – given to the Telecommunications Department for the construction of antennas and other devices.[50] Nature was to be cherished, but not as much as modernity and economic progress.

The forest reserves, and later nature reserves, of Singapore represent a model of how forested areas changed under imperial rule. The forests were created through imperial policies as a response to the devastating effects that spreading plantation capitalism had wrought upon the island, resulting in a denuded landscape. Through this reforestation programme, colonial officials selected the species that would cover the tropical island for properties they possessed, ranging from quick growth trees that would help conquer the space to highly valuable tropical hardwoods that would generate profit. The proliferation of any of these species, however, was dependent on social and environmental limitations. Once the forests had developed, they were exploited along principles of silviculture in the 'forestry creed' that emphasised utilitarian development, as enacted throughout the world in the late nineteenth century. In Singapore, however, the limited extent of available land and an expanding urban port resulted in severe limitations being placed on the conservation status of such forests. Eventually, a much-reduced area was reserved for botanical purposes in an attempt to preserve the biodiversity of the island. While these areas expanded in the late colonial era, the needs of the society continued to have precedence, raising the question of the limitations of all policies relating to nature conservation. While there may have been a desire to prioritise ecological diversity in the region, this was ultimately at the mercy of the continually modernising state, whether imperial or going through decolonisation, thus reflecting the pressures of governing nature in the tropics.

Timothy P. Barnard is Associate Professor in the Department of History at the National University of Singapore. His research interests include the environmental and cultural history of South Asian islands. Most recently, he has published *Nature's Colony* (2016) and *Imperial Creatures* (2019) with the University of Singapore Press.

Notes

1. Nathaniel Cantley, *Report on the Forests of the Straits Settlements* (Singapore: Government Printing Office, 1883).
2. S. Ravi Rajan, *Modernizing Nature: Forestry and Imperial Eco-Development, 1800–1950* (Oxford: Oxford University Press, 2006), 55; Cantley, *Report on the Forests of the Straits Settlements*; Timothy P. Barnard, *Nature's Colony: Empire, Nation and Environment in the Singapore Botanic Gardens* (Singapore: NUS Press, 2016), 50–65.
3. Fiona Williamson, 'Weathering the Empire: Meteorological Research in the Early British Straits Settlements', *British Journal for the History of Science* 48(3) (2015), 475–92; Charles Burton Buckley, *An Anecdotal History of Old Times in Singapore: From the Foundation of the Settlement under the Honourable East India Company on February 6th, 1819 to the Transfer to the Colonial Office as Part of the Colonial Possessions of the Crown on April 1st, 1867* (Singapore: Fraser and Neave, 1902), 735–41.
4. Cantley, *Report on the Forests of the Straits Settlements*, 10.
5. Ibid.
6. Rajan, *Modernizing Nature*, 6–9; Gregory A. Barton, *Empire Forestry and the Origins of Environmentalism* (Cambridge: Cambridge University Press, 2009); H. Colyear Dawkins and Michael S. Philip, *Tropical Moist Forest Silviculture and Management: A History of Success and Failure* (New York: CAB International, 1998); Richard Grove, *Green Imperialism: The Evolution of Environmentalist Ideas and Conservation Policies in Mauritius, the Caribbean and India, 1600–1860* (Cambridge: Cambridge University Press, 1995).
7. Benjamin Weil, 'Conservation, Exploitation, and Cultural Change in the Indian Forest Service, 1875–1927', *Environmental History* 11(2) (2006), 319–43; Rajan, *Modernizing Nature*; Barton, *Empire Forestry*.
8. Grove, *Green Imperialism*, 24–72; Barton, *Empire Forestry*, 31–32.
9. Miles Powell, 'Singapore's Lost Coast: Land Reclamation, National Development and the Erasure of Human and Ecological Communities, 1822–Present', *Environment and*

History 27(4) (2021), 635–63; Miles Powell, 'Harnessing the Great Acceleration: Connecting Local and Global Environmental History at the Port of Singapore', *Environmental History* 27(3) (2022), 441–66.

10. Cantley, *Report on the Forests of the Straits Settlements*, 19–20; Nathaniel Cantley, *Annual Report on the Forest Department, Straits Settlements, for the Year 1884* (Singapore: Government Printing Office, 1885), 1–2; Nathaniel Cantley, *Annual Report on the Forest Department, Straits Settlements, for the Year 1885* (Singapore: Government Printing Office, 1886), 6; Barnard, *Nature's Colony*, 64–65.
11. *Samanea saman*, due to its size, attracted the attention of Spanish explorers and colonists in the Americas. Alexander von Humboldt brought the tree to prominence following his travels in the early nineteenth century, when he described a particularly large specimen near Maracay, Venezuela. After botanists cultivated seeds in Europe, imperial botanical networks spread it throughout Asia and the Pacific as a shade tree. P. A. Durr, 'The Biology, Ecology and Agroforestry Potential of the Raintree, *Samanea saman* (Jacq.) Merr.', *Agroforestry Systems* 51(3) (2001), 223–37, here 227; Alexander von Humboldt, *Personal Narrative of a Journey to the Equinoctial Regions of the New Continent* (London: Penguin, 1995), 154–55.
12. Cantley, *Straits Settlements Report on the Forest Department, 1885*, 2.
13. Barton, *Empire Forestry*, 9–16; Nathaniel Cantley, *Annual Report of the Botanic Gardens Singapore, for the Year 1882* (Singapore: Government Printing Office, 1884), 1, 9, 15.
14. The contradictions in these approaches have led to much debate over whether the ultimate goal of these policies was preservation or exploitation. Weil, 'Conservation, Exploitation, and Cultural Change', 320; Grove, *Green Imperialism*; Ramachandra Guha, *The Unquiet Woods: Ecological Change and Peasant Resistance in the Himalaya* (Berkeley: University of California Press, 1990).
15. The pioneers of such policies in India were Berthold Ribbentrop, a German forester, and Dietrich Brandis, a German-British botanist, reflecting the global reach of such networks. Barton, *Empire Forestry*, 39–40; Donald Worster, *Nature's Economy: A History of Ecological Ideas* (Cambridge: Cambridge University Press, 1994).
16. Barnard, *Nature's Colony*, 64–65; Barton, *Empire Forestry*, 36–37.
17. Richard Drayton, *Nature's Government: Science, Imperial Britain, and the 'Improvement' of the World* (New Haven, CT: Yale University Press, 2000).
18. Barton, *Empire Forestry*, 70–72; Barnard, *Nature's Colony*, 44, 66.
19. Cantley, *Straits Settlements Report on the Forest Department, 1885*, 2; Nathaniel Cantley, *Report on the Forest Department, 1886*, 1–6; Henry Nicholas Ridley, *Annual Reports on the Forests Department, 1888* (Singapore: Government Printing Office, 1889), 2; Henry Nicholas Ridley, *Reports on the Gardens and Forest Departments, Straits Settlements* (Singapore: Government Printing Office, 1893), 8–9; Tony O'Dempsey, 'Singapore's Changing Landscape since c. 1800', in Timothy P. Barnard (ed.), *Nature Contained: Environmental Histories of Singapore* (Singapore: NUS Press, 2014), 33–35.
20. Rajan, *Modernizing Nature*, 28–30; Humboldt, *Personal Narrative*, 150–51; Eugene Cittadino, *Nature as Laboratory: Darwinian Plant Ecology in the German Empire* (Cambridge: Cambridge University Press, 1990).
21. This transformation was also occurring in France and Germany. As Rajan describes, authorities throughout Europe promoted 'new forestry' to 'improve the practice of

resource management to satisfy the goals of the new political economic doctrines'. Rajan, Mod*ernizing Nature*, 51, 34–51, 115–18; Joseph Morgan Hodge, *Triumph of the Expert: Agrarian Doctrines of Development and the Legacies of British Colonialism* (Athens: Ohio University Press, 2007), 8–9, 44–45; Drayton, *Nature's Government*, 255–60.

22. H. C. Hill, *Report on the Present System of Forest Conservancy in the Straits Settlements, with Suggestions for Future Management* (Singapore: Government Printing Office, 1900), 3; Barnard, *Nature's Colony*, 75–76.
23. Royal Botanic Gardens, Kew (RBGK): HNR/3/2/3: Henry Nicholas Ridley Notebooks, vol. 3, 1891–1896, 75; RBGK: DC/168/2: 'Letter from Ridley to Thiselton-Dyer', 1 October 1905, 94r; Barnard, *Nature's Colony*, 74–78; Timothy P. Barnard and Joanna Lee, 'A Spiteful Campaign: Forests, Agriculture, and Administering the Environment in Imperial Singapore and Malaya', *Environmental History* 27(3) (2022), 467–90.
24. Hill, *Report on the Present System of Forest Conservancy*, 9–10; Jeyamalar Kathirithamby-Wells, *Nature and Nation: Forests and Development in Peninsular Malaysia* (Singapore: NUS Press, 2005), 74–76; K. M. Wong, 'The Herbarium and Arboretum of the Forest Research Institute of Malaysia at Kepong – A Historical Perspective', *Gardens' Bulletin Singapore* 40(1) (1987), 15–30, here 15.
25. Rajan, *Modernizing Nature*, 112–13; Hill, *Report on the Present System of Forest Conservation*, 1–2.
26. This aspect of the report created a tremendous amount of tension, leading Ridley – in his personal correspondence and diaries – to decry this change in policy and to denigrate Hill for decades. HNR/3/2/4: Notebooks, vol. 4, 35–37; Hill, *Report on the Present System of Forest Conservation*, 5–9.
27. Barnard, *Nature's Colony*, 81–82; Hill, *Report on the Present System of Forest Conservation*, 9–10; Charles Lee Keeton, *King Thebaw and the Ecological Rape of Burma: The Political and Commercial Struggle between British India and French Indo-China in Burma, 1878–1886* (Delhi: Manohar Book Service, 1874).
28. Rajan, *Modernizing Nature*, 147–48; Kathirithamby-Wells, *Nature and Nation*, 74–76; Barnard and Lee, 'A Spiteful Campaign'.
29. W. Langham-Carter, *Report on the Forest Reserves of the Straits Settlements during the year 1901* (Singapore: Government Printing Office, 1902), 1–2; A. M. Burn-Murdoch, *Annual Report on the Forest Administration in the Straits Settlements for the Year 1905* (Singapore: Government Printing Office, 1906), 4; A. M. Burn-Murdoch, *Annual Report on the Forest Administration in the Straits Settlements for the Year 1909* (Singapore: Government Printing Office, 1910), 3–4.
30. Alex Josey, *Golf in Singapore* (Singapore: Asia Pacific Press, 1969); 'Field Firing at the Golf Course', *Singapore Free Press and Mercantile Advertiser*, 20 August 1924, 12.
31. E. Cubitt, *Annual Report on the Forest Administration in the Straits Settlements for the Year 1918* (Singapore: Government Printing Office, 1919), 1, 6; G. E. S. Cubitt, *Annual Report on the Forest Administration in the Straits Settlements for the Year 1923* (Singapore: Government Printing Office, 1924), 1; R. E. Holttum, *Annual Report of the Director of the Gardens, for the Year 1935* (Singapore: Government Printing Office, 1936), 20–21.
32. Weil, 'Conservation, Exploitation, and Cultural Exchange', 336–37; B. Ribbentrop, *Forestry in British India* (Calcutta: Central Printing Office, 1900).

33. G. E. S. Cubitt, *Annual Report on the Forest Administration in the Straits Settlements for the Year 1925* (Singapore: Government Printing Office, 1926), 6–7.
34. G. E. S. Cubitt, *Annual Report on the Forest Administration in the Straits Settlements for the Year 1927* (Singapore: Government Printing Office, 1928), 2; J. P. Mead, *Annual Report on the Forest Administration in the Straits Settlements for the Year 1929* (Singapore: Government Printing Office, 1930), 1; J. P. Mead, *Annual Report on the Forest Administration in the Straits Settlements for the Year 1930* (Singapore: Government Printing Office, 1931), 1.
35. During this period, when forestry officials were facing difficulties with land use, there was also a developing appreciation of the natural or ecological diversity of these areas, which would be termed 'biodiversity' today, corresponding to larger global trends. E. O. Wilson (ed.), *Biodiversity* (Washington, DC: National Academy Press, 1988).
36. Peter Szabo, 'Historical Ecology: Past, Present and Future', *Biological Reviews* 90(4) (2015), 997–1014; Kathirithamby-Wells, *Nature and Nation*, 170–75; J. G. Watson, *Mangrove Forests of the Peninsula* (Singapore: Fraser and Neave, 1929); K. D. Menon, 'A Brief History of Forestry Research in Malaya', *Commonwealth Forestry Review* 48(4) (1969), 309–16, here 312.
37. G. E. S. Cubitt, *Wood in the Federated Malay States: Its Use, Misuse and Future Provision* (Kuala Lumpur: Federated Malay States Government Printing, 1920); Kathirithamby-Wells, *Nature and Nation*; Menon, 'A Brief History of Forest Research in Malaya'; Ravan, *Modernizing Nature*, 146–54.
38. Singapore Botanic Gardens Library and Archives (SBG), Manuscript Box 1, Folder P, 'Letter to R.E. Holttum', 20 May 1936.
39. SBG, M1/P: Notes of Response on Back of 'Letter to R.E. Holttum', 20 May 1936; Menon, 'A Brief History of Forestry Research in Malaya', 312.
40. SBG, M1/P: Notes of Response on Back of 'Letter to R.E. Holttum', 20 May 1936; SBG: M1/P: 'Letter to R.E. Holttum from E.O. Shebbeare', 26 April 1939; E. J. H. Corner, *Annual Report of the Director of the Gardens for the Year 1937* (Singapore: Government Printing Office, 1938), 7–8; Lena Chan, 'Nature in the City', in Stephen Hamnett and Belinda Yuen (eds), *Planning Singapore: The Experimental City* (London: Routledge, 2019), 109–29, here 112.
41. R. E. Holttum, *Annual Report of the Director of the Gardens, for the Year 1938* (Singapore: Government Printing Office, 1939), 3; R. E. Holttum, *Botanic Gardens Department Annual Report for 1948* (Singapore: Government Printing Office, 1949), 6–7.
42. Holttum, *Botanic Gardens Department Annual Report for 1948*, 6.
43. Corner, *Annual Report, 1937*, 8.
44. Holttum, *Annual Report, 1938*, 3.
45. Corner, *Annual Report, 1937*, 8; R. E. Holttum, *Annual Report of the Director of the Gardens, for the Year 1939* (Singapore: Government Printing Office, 1940), 2–3; R. E. Holttum, 'Annual Report of the Director of Gardens, S.S. for the Year 1940', typed manuscript, Singapore Botanic Gardens Archives, 2.
46. Holttum, *Botanic Gardens Department Annual Report for 1948*, 7.
47. M. R. Henderson, *Botanic Gardens Department Annual Report for 1951* (Singapore: Government Printing Office, 1952), 6; J. W. Purseglove, *Annual Report of the Botanic Gardens Department for 1954* (Singapore: Government Printing Office, 1955), 8–9.
48. Holttum, *Botanic Gardens Department Annual Report for 1948*, 7.

49. J. W. Purseglove, *Annual Report of the Botanic Gardens Department for 1955* (Singapore: Government Printing Office, 1956), 13–14; J. W. Purseglove, *Annual Report of the Botanic Gardens Department for 1956* (Singapore: Government Printing Office, 1957), 14; '*Cynometra ramiflora* L.', Singapore National Parks website, https://www.nparks.gov.sg/florafaunaweb/flora/2/8/2834 (retrieved 29 August 2021).
50. H. M. Burkill, *Annual Report of the Botanic Gardens Department for 1957* (Singapore: Government Printing Office, 1958), 7–8; Barnard, *Nature's Colony*, 229–31.

Bibliography

Barnard, Timothy P. *Nature's Colony: Empire, Nation and Environment in the Singapore Botanic Gardens*. Singapore: NUS Press, 2016.

Barnard, Timothy P., and Joanna Lee. 'A Spiteful Campaign: Forests, Agriculture, and Administering the Environment in Imperial Singapore and Malaya'. *Environmental History* 27(3) (2022), 467–90.

Barton, Gregory A. *Empire Forestry and the Origins of Environmentalism*. Cambridge: Cambridge University Press, 2009.

Chan, Lena. 'Nature in the City', in Stephen Hamnett and Belinda Yuen (eds), *Planning Singapore: The Experimental City* (London: Routledge, 2019), 109–29.

Cittadino, Eugene. *Nature as Laboratory: Darwinian Plant Ecology in the German Empire*. Cambridge: Cambridge University Press, 1990.

Dawkins, H. Colyear, and Michael S. Philip. *Tropical Moist Forest Silviculture and Management: A History of Success and Failure*. New York: CAB International, 1998.

Drayton, Richard. *Nature's Government: Science, Imperial Britain, and the 'Improvement' of the World*. New Haven, CT: Yale University Press, 2000.

Durr, P. A. 'The Biology, Ecology and Agroforestry Potential of the Raintree, *Samanea saman* (Jacq.) Merr.' *Agroforestry Systems* 51(3) (2001), 223–37.

Grove, Richard. *Green Imperialism: The Evolution of Environmentalist Ideas and Conservation Policies in Mauritius, the Caribbean and India, 1600–1860*. Cambridge: Cambridge University Press, 1995.

Guha, Ramachandra. *The Unquiet Woods: Ecological Change and Peasant Resistance in the Himalaya*. Berkeley: University of California Press, 1990.

Hodge, Joseph Morgan. *Triumph of the Expert: Agrarian Doctrines of Development and the Legacies of British Colonialism*. Athens: Ohio University Press, 2007.

Kathirithamby-Wells, Jeyamalar. *Nature and Nation: Forests and Development in Peninsular Malaysia*. Singapore: NUS Press, 2005.

Menon, K. D. 'A Brief History of Forestry Research in Malaya'. *Commonwealth Forestry Review* 48(4) (1969), 309–16.

O'Dempsey, Tony. 'Singapore's Changing Landscape since c. 1800', in Timothy P. Barnard (ed.), *Nature Contained: Environmental Histories of Singapore* (Singapore: NUS Press, 2014), 17–48.

Powell, Miles. 'Singapore's Lost Coast: Land Reclamation, National Development and the Erasure of Human and Ecological Communities, 1822–Present'. *Environment and History* 27(4) (2021), 635–63.

———. 'Harnessing the Great Acceleration: Connecting Local and Global Environmental History at the Port of Singapore'. *Environmental History* 27(3) (2022), 441–66.

Rajan, S. Ravi. *Modernizing Nature: Forestry and Imperial Eco-Development, 1800–1950*. Oxford: Oxford University Press, 2006.
Szabo, Peter. 'Historical Ecology: Past, Present and Future'. *Biological Reviews* 90(4) (2015), 997–1014.
Weil, Benjamin. 'Conservation, Exploitation, and Cultural Change in the Indian Forest Service, 1875–1927'. *Environmental History* 11(2) (2006), 319–43.
Williamson, Fiona. 'Weathering the Empire: Meteorological Research in the Early British Straits Settlements'. *British Journal for the History of Science* 48(3) (2015), 475–92.
Wilson, E. O. (ed.). *Biodiversity*. Washington, DC: National Academy Press, 1988.
Wong, K. M. 'The Herbarium and Arboretum of the Forest Research Institute of Malaysia at Kepong – A Historical Perspective'. *Gardens' Bulletin Singapore* 40(1) (1987), 15–30.
Worster, Donald. *Nature's Economy: A History of Ecological Ideas*. Cambridge: Cambridge University Press, 1994.

CHAPTER 4

Rambouillet, Agricultural Stations and French Colonial Africa

Conserving and Improving Nature (1900–1930)

Raphaël Devred

The history of sheep and merinos in colonial Africa is one of zootechnicians, both in the literature and in the history itself. The publication of *Histoire de la recherche agricole française en Afrique tropicale francophone* by René Tourte – a former inspector-general at the French Office for Overseas Technical and Scientific Research (Office de la recherche scientifique et technique d'Outre-mer (ORSTOM)) – illustrates the mismatch between historians' and zootechnicians' knowledge.[1] Marshalling a vast range of protagonists, institutions, farms, plants and animals shaping colonial agriculture, this work acts as a reminder of the gap in the literature about the history of sheep, both in France and in its former colonies.[2] It invites historians of the environment to examine the subject of colonial sheep, in the wake of work by Diana Davis and William Beinart.[3] The Rambouillet imperial and national sheepfold seems to provide a coherent scale of study for examining the phenomenon of colonial sheep, both for its role in the emergence and development of zootechnics, and for the part it played in *merinisation*, a term these experts started using in the 1960s.[4]

This neologism was coined in 1964 when two Rambouillet zootechnicians – Raymond Laurans, the head of the establishment, and Jean Blanc, a former pupil – mounted an exhibition about the official beginnings of 'merinisation'. The exhibition covered the first phase when Spanish 'wool-producing animals' were imported and sent out around France in the late eighteenth and early nineteenth century.[5] This process was initiated by monarchs, aristocrats and men of learning such as Claude Carlier and Louis Jean-Marie Daubenton, in response to Colbert's and monarchs' plans to make France independent in sheep rearing for the industrial production of wool.[6] These trials were conducted using flocks of local French sheep, then Spanish sheep imported to France as of the 1770s.[7] These trials to acclimatise Spanish sheep and improve their wool, overseen by Daubenton at Montbard, convinced ministers and monarchs to acclimatise flocks in France. In 1786, Louis XVI had a flock imported to Rambouillet.

Merinisation then developed during the French Revolution, initially driven by crossing Spanish sheep (henceforth called merinos) with local French breeds, then by establishing breeding flocks.[8] The word 'merino' was a reminder of the African origins of the animal, as signalled in 1770, for it derives from the name of the Marinid Sultanate where the sheep originally came from.[9] In 1798, the French interior minister François de Neufchâteau turned Rambouillet into a sheep 'colony',[10] tasked with improving France's sheep bloodstock and with conserving the French breeding stock of sheep of Spanish-African origin.[11] Rambouillet was then used as the matrix for French merinisation.[12] It was a matter of starting from the pure Spanish breed now forming the flock and improving France's 'native' breeds – the term used by economists such as Jean-Baptiste Huzard and Abbé Tessier back in the 1780s to 1810s.[13]

This first phase in France, running from 1790 to 1840, was followed by a colonial phase, with the merinisation of the colonies' indigenous flocks using the French merinos. And it was especially Rambouillet that was behind the policy to colonise sheep breeds for the wool industry, driven by learned agronomists, naturalists, veterinarians and politicians. The Rambouillet sheepfold was thought of as a conservatory and a *nursery* for this improved and improvable breed of sheep, and viewed as a centre of expert knowledge.[14]

Zootechnics was built up as a discipline in the 1850s by zoologists such as Émile Baudement, the first professor to be appointed to the chair at the Institut national agronomique, in 1849. As livestock experts, zootechnicians led the project to acclimatise sheep in the colonies. Rambouillet became an international hub for the circulation of animals, knowledge and know-how. Additionally, by providing animals for experiments, Rambouillet supported the structuring of zootechnics as a science in the nineteenth century, and became a living laboratory for new husbandry practices. This role was consecrated when the Centre d'enseignement zootechnique was set up in 1955, followed by the Société d'ethnozootechnie in 1971.

And it was especially via its relations with Africa that Rambouillet scaled up its mission to 'improve' sheep. The case of the Rambouillet merinos in Africa provides a useful starting point for examining the history of the colonial sheep system. It is also a matter of enquiring into the role Rambouillet played in transforming African nature in general, and ovine nature in particular.

What the Figures Tell Us about Exports from Rambouillet to Africa

After having sold thousands of breeding sheep to various sheep breeders in France and Europe between 1793 and 1840, Rambouillet began its new mission to improve sheep and to provide colonies and foreign farmers with merinos. In 1852, a breeder from Cape Colony (1806–1910) became interested in Rambouillet breeding stock and acquired eighteen animals.[15] From 1868 to 1970, a total of 994 Rambouillet merinos were sent to Africa: 629 rams and 356 ewes.[16]

Table 4.1. Rambouillet merinos sent to Africa. Table data from: Bergerie nationale de Rambouillet, registers of serial numbers of rams and ewes (1858–1973). © Raphaël Devred.

Destination	Number	Dates
	Rams	
	629	1868–1970
French Equatorial Africa	3	1953
North Africa	2	1951
South Africa	277	1868–1934
Algeria	251	1868–1909; 1928; 1944–45
Angola	2	1931
French West Africa	11	1928, 1931, 1936
Ethiopia	5	1965
Réunion	4	1902
Madagascar	6	1896, 1912
Morocco	21	1922–29; 1963
Mauritania	1	1970
(British) Sudan	5	1899
Chad	2	1953
Tunisia	39	1896–1913; 1926; 1960
	Ewes	
	365	1868–1953
South Africa	164	1868–1928
Algeria	156	1868–1904; 1949–51
Angola	2	1930–31
Réunion	2	1902
Madagascar	15	1896
Morocco	5	1922
Sudan	3	1899
Chad	2	1953
Tunisia	16	1892–1924

The figures illustrate, first, the major problem facing the undertaking, which was seeking to merinise an entire continent with fewer than a thousand Rambouillets, especially as nearly half the exports were for breeders in South Africa (441 sheep). France's colonial sheep efforts were thus meant to issue from a drop of water, or rather a drop of blood, to be propagated in an ocean of African sheep via the system of experimental farm stations.[17]

From 1860 to 1880, most Rambouillets heading for Africa were sent to Cape Colony. It can be hard to ascertain where the animals departed from, since South African purchasers recruited sales agents in France, and the archives only indicate the names of these intermediaries. Two examples taken from the list of the 277 rams sent to Cape Town throw light on South African interest in Rambouillet sheep. The first indicates the reputation the establishment had acquired, for General Botha in person bought two rams in 1909 for his farm in the Transvaal;[18] as yet nothing further is known about Botha's farm.

However, it has been possible to move beyond the limits in the documentation in the second case, giving us a clearer idea of the types of exchange characterising these experimental farms.[19] The Rubidge family owned Wellwood Farm in Cape Colony, whose history is known thanks to the works of William Beinart.[20] On 25 March 1873, Charles Rubidge and his son Richard left Cape Town, setting sail from Table Bay aboard the *Asiatic* bound for England. On 10 May they were in France looking for the best breeding stock for their flock. Their search led them to Rambouillet sheepfold on 7 June 1873, where Charles bought two rams, which left Rambouillet on 13 June.[21] The objective of the South African breeders, who owned vast flocks of merinos, was to find improvers in France or Australia to ameliorate their animals' wool so as to provide industry with better material, and hence sell better products at better prices.

It is difficult to follow the trajectories of merinos in private South African farms: the archives have to be unearthed in the territories concerned, or descendants found who can provide this information – presupposing both that the farm still exists and that the documents have been conserved. But the French side of the merinisation endeavour is better documented, because it was the state and major landowners with national political influence who drove these acclimatisations, especially in Algeria. As of 1880 the colony became the second most common African destination for Rambouillet merinos, after South Africa. After 146 rams and 25 ewes were sent out to the Moudjbar national sheepfold in the north of the colony, private settlers in Algeria imported stock between 1885 and 1891: Omer Chevalier bought 21 rams and 15 ewes from Rambouillet between 1902 and 1911 for his private farm at Zdin, near Sétif. In the same period, Tunisia's Agricultural Services had 35 rams and 16 ewes sent out. Rambouillets were then tested in Morocco from 1922 to 1929: 19 rams and 5 ewes were sent at the request of the Casablanca Husbandry Service. Then in 1931, the Union ovine made a final shipment of 3 rams from France to Algeria, closing the great period of Rambouillets being sent to Africa, before a final flurry in the years 1945–55.[22]

Trials of Rambouillets in French Africa

Once the Rambouillets left the sheepfold outside Paris, the establishment lost trace of its sheep. The very extensive correspondence that has been conserved contains very few references to what happened to them.[23] Having said that, the Rambouillet sheepfold does hold issues of a review founded in 1929: *L'union ovine: Revue de l'élevage du mouton et du commerce de ses produits*. With the backing of the Ministry of Agriculture, Michel Lallour – at the time a sheep breeder in French North Africa – founded a syndicate in 1926, the Groupement national des éleveurs de moutons, or Union ovine de France, which produced this journal.[24] The Union ovine funded a network of schools in France, including Rambouillet. The syndicate was composed of three unions covering France's territory between them: the Union ovine de France, the Union ovine coloniale and the Union ovine d'Afrique du Nord. The purpose of the Union ovine de France was to 'familiarise both European and native breeders overseas with scientific and rational data about rearing'.[25] The journal's objective was thus to 'link up the breeder-producer with the industrial-consumer, scientists with practitioners, colonies with the metropole, and French sheep farmers with big breeding worldwide'.[26] In February 1931, in its third year of publication, the journal had a print run of five thousand copies, on sale for five francs in France and abroad.

The main readers were members of the three Unions ovines: rich landowners and sheep owners, wool merchants, sheepskin tanners, scientists (veterinarians, agronomists and zootechnicians) and industrialists, in both France and its colonies. They included figures such as Eugène Mathon, a wool merchant from Roubaix who was president of various bodies, such as the Syndicat des fabricants de tissus from 1912 to 1932, Roubaix commercial court from 1908 to 1911, the Comité central de la laine (which he founded in 1922) and the Union ovine as of 1926.[27] The Union's members wielded influence with the Ministry of Agriculture and state administrators, when they were not themselves officials. Further prosopographic study of the journal would reveal the Union ovine's networks, the institutions its members belonged to and any ties they may have had with the French parliamentary colonial party.[28]

On 16 June 1931, a dinner was held at the end of the 'colonial wool day' at the Exposition coloniale in Paris. The event illustrates the political support enjoyed by the syndicate, being attended by Paul Reynaud, the Minister for Colonies, Maréchal Lyautey, the general commissioner of the exhibition, and his delegate Marcel Ollivier, a former governor-general of Madagascar.[29] Also present was the regent of the Banque de France, Jean Balsan, who was also a member and administrator of the Union ovine, whose textile manufacturers made the French army's uniforms.[30] Then in the 1920s and 1930s, an industrial, political and scientific sheep lobby sprang up around the Union ovine, interested in expanding industrial wool sheep.

By reading the articles in the review from 1929 through to 1932, we may follow the traces of the Rambouillet merinos once they had left the farm.[31] The animals were shipped to colonial ports before joining their new flocks in Africa. Two shipments to Madagascar in 1896–1900 allow us to identify the role played by acclimatisers, together with the major issues they faced. On 19 November 1896, four rams and fifteen ewes were sent to the colony of Madagascar on order of the Ministry.[32] This shipment was described by the head of the Zootechnical Service, P. Geoffroy, in 1931:

> In 1896, at the request of Resident Laroche, the Minister of Agriculture sent the colony 4 rams and 15 ewes from the Rambouillet national sheepfold. For special reasons, the flock was initially held at Diego where an illness (probably heartwater) decimated it. 2 rams and 6 ewes were sent up to Imerina in 1897, and were initially held at Nanisana Agricultural Station, before being transferred to Botafe. The females remained sterile and, by 1900 nothing was left of the import. In 1901, General Galliéni renewed the experiment with 15 rams and 5 ewes from Sologne and 15 rams and 8 ewes from Algeria.[33]

Acclimatisation thus received support from central administrations, with ministers and colonial governors figuring among those involved in imports to Madagascar.[34] Shortly after the 1900 Exposition universelle in Paris, Madagascar sent a Madagascan ram and a young shepherd called Randremba to Rambouillet:[35] 'Randremba, who had looked after the live animals sent by the colony during the exhibition, was placed in the Rambouillet farm school to learn about sheep husbandry and particularly about crossing merinos with the Madagascan ram.'[36] Apparently Randremba was instructed in French techniques to become an intermediary for zootechnical knowledge, and thus contribute to the campaign to improve Madagascan breeds. He was to be sent back to Madagascar after one year.

These merino–Madagascan crosses are known thanks to a series of postcards demonstrating the short-lived success of this cross-breeding, an experiment that culminated in the flock being dispatched to the Rambouillet slaughterhouse after interest in the experiment waned.[37] Be that as it may, this exchange throws light on what happened to the Rambouillets: the animals and men were sent out to colonial experimental farms belonging to the government, in order to transmit 'proper' zootechnical knowledge and carry out the project to improve African sheep in accordance with French methods.

Improvements by the Stations: Zootechnical Methods

From 1830 to 1950, African sheep breeds – just like French breeds in the years 1780–1830 – were viewed as degenerate by zootechnicians, who sought to explain how this alteration had come about.[38] According to them, the deterioration in wool could be seen in the poor quality of African fleeces, in which there was more guard hair (thick fibres) than wool. As though mirroring the image of North Africa as the granary of Rome, zootechnicians thought that the Maghreb was the homeland of the original merinos whose golden fleece, turned into fine wool, would supply industry: this was in fact fanciful, for the breed had been created in Spain, then in Rambouillet, and the sheep were suited for meat and leather production, not just for wool as targeted by the Europeans. In *Les mérinos* in 1868, Émile Baudement,[39] one of the founders of zootechnics, explained that the merino breed was brought from Africa to Spain by Arabs,[40] but that in Algeria 'rearing practices have remained in an entirely primitive and stationary state that needs to be changed to improve wools'. 'It is precisely this change that the Spanish Arabs carried out; they were thereby superior to the Arabs of Africa, just as the sheep breed of Spain became superior to the African sheep, while being of the same blood.'[41]

This discussion about the African origins of the merino breed and the reasons for its degeneration continued as colonisation progressed. In an article in *L'union ovine* in 1930 examining the history of merinos in Africa, journalist and administrator Pierre Hamp noted about Algeria:[42]

> We arrived in the land of sheep where the merino breed probably came from, the golden fleece that had made Spain's fortune. This question of the origin of the merino sheep was going to become very important in the history of French textiles, as of the disembarkation at Sidi-Ferruch [marking the beginning of French rule]. If the merino was a Spanish breed brought to Africa when the Moors invaded Andalusia, the conquest of North Africa put us in contact with a bastardised breed we could purify. If, on the contrary, the merino was a Marinid animal, we were arriving on the continent that produced the finest wool fleece in the world. . . . Zootechnics hesitates. In the foothills of the Atlas Mountains in Morocco, there are sheep with dewlaps, with horns tight against the head, the sheep of the Beni-Ahssen from Kénifra, which has exactly the same form as the Rambouillet pure breeding stock, and which have not been crossed since their ancestors arrived from Segovia in 1786. If these animals originally came from Atlas to Segovia, why would the North

> African sheep, issuing from so fine a breed, be so corrupted? Why, whenever a merino ram is placed in what is meant to be its historic habitat, are so many precautions required to prevent its dying?[43]

As it was corrupted, the African sheep's nature had to be restored by improvement. What was needed was to use the descendant of the African merino, now a European breed, for 'it was the most perfect zootechnical product', 'a truly industrial breed',[44] which would provide a way of regenerating the blood of African sheep and thus turn the North African colonies into wool colonies, following the British model in South Africa and Australia. In addition to being central to what was said about sheep, the merino breed was at the heart of colonial rivalry between France and Britain, and was now to provide a way for France to close the gap in matters ovine.

However, this plan had to be implemented in a colonial context in which crop cultivation was prioritised over animal husbandry. Pierre Hamp noted that 'the legend of [North] Africa as the granary of Rome exerted a pull over the mind of many colonists, making them prefer cereal to flocks'.[45] Sheep zootechnicians set about dismantling this colonial myth to benefit their industry.[46] In 1931, for example, a veterinary inspector with the Algerian Husbandry Service, G. Trouette, wrote that 'we must no longer be hypnotised by this historical error, relayed and repeated thousands of times, [according to which] North Africa was the granary of Rome'.[47] For zootechnicians and some of the political elite in the metropoles, North Africa was sheep country, as illustrated by the Arab sheep civilisation they were trying to govern and improve.

While metropolitan governments were the initial driving force behind merinisation, with subsequent support in the colonies, metropolitan members of the Union ovine were lobbying the French government to conduct experimental trials in a series of stations in North Africa, Sudan and Madagascar. These experimental stations housed agricultural equipment and experts, with veterinarians, agronomists and zootechnicians working to disseminate French agricultural knowledge and techniques. The zootechnicians established breeding stock for plant strains (in nurseries and fields) and animals to first colonise, then improve African nature.

For sheep, the Husbandry Services established flocks of breeding stock – which they sometimes referred to as nurseries – in various sheep regions. The purpose was not just to improve native flocks, but also to sedentarise nomadic pastoralists and lead them progressively towards agriculture. These experimental farms were total projects, their purpose being to transform both 'native' agriculture and 'native' nature:

> The climate, remoteness and need to form an improved race make it almost impossible to establish prosperous private companies. Sudan, as the Colo-

ny's government has well understood, cannot be profitably exploited by the natives. What is thus needed is for the Peulh pastoralists to abandon their husbandry methods and become sedentary to farm a race created for them. . . . In this region, a sheep farm of 8,000 to 10,000 enclosed hectares divided into paddocks and provided with drinking water could be used, drawing on the experience and results obtained in official farms, to create and fix this race and to teach native farmers new husbandry methods.[48]

Husbandry Services were set up in each colony: Madagascar, French Equatorial Africa and French Western Africa. Zootechnicians and veterinarians established and looked after breeding stock, chosen from among the best breeds and merino flocks in France, Australia or Cape Colony.[49] All their care and husbandry technique was based on keeping the breeding stock 'pure', mating it each year with native sheep to be improved.[50] The scientists used two techniques already employed in France: building up from a pure merino flock, thanks to which the colonial services sold the lambs to native and colonial farmers; or else crossing merino sires with native flocks to produce crosses. This latter method was the one most used by the French administration in Africa. It was called the 'progressive' or 'indirect' method,[51] and resulted from crossing indigenous ewes with a merino ram for four generations, over three to four years.

Each year, when it was tupping season, the zootechnicians tried to convince native farmers to bring their flocks to the station for crossing. They separated each ram so as to control which individual reproduced and so establish the basis for animal selection in accordance with French methods.[52] To win them over, zootechnicians held full-size demonstrations. They acquired African flocks to be improved, which they placed in sheepfolds in stations, and tried to build up flocks that could illustrate the superiority of their husbandry techniques. As Pierre Hamp explains, the Arabs and North African nomads were to be convinced by demonstration and the uniformity of flocks:

There is such diversity, ranging from the White Fathers' black sheep in Thibar in Tunisia to the 'Petit Oranais' from the Berguent region south of Oujda. And it was a matter of giving uniformity to the colonial flock, of being an example for Arab flocks that contained a mix of everything, from astrakhans and merinos to sheep whose wool had no curls, like dog hair.[53]

An additional concern for the station was to try to impose a sedentary system on the (semi-)nomadic modes of husbandry, so as to sedentarise people and flocks. Many North African farmers were nomadic or transhumant pastoralists. The colonial authorities saw farm stations as a tool for this policy to pull in husbandry activities, that is, concentrate people and animals placed under the watchful eye of colonial husbandry experts.

The example of the Rambouillet merino allows us to sketch a map of these experimental establishments. In Algeria, there were sheepfold stations at Tadmit, Djelfa, Ben Chicao and Moudjbar; in Sudan, the Nioro farm and the Société d'élevage de Diré;[54] in Morocco there was Ain Kaddous at FezBab, Halima at Meknes, the central experimental station at Rabat for selection and trials, the autonomous experimental farm at Casablanca and the Menara at Marrakech.[55] In 1930, based on this model of government stations, the Union ovine d'Afrique du Nord set up five experimental centres in Morocco, at Khemisset, Oued Zem, Dar-Oulad-Zidouh, El Borouj and Sidi Larbi.[56] According to the note published by the Archives nationales pour les colonies, these agricultural establishments could even influence colonial demographic policy: 'a model sheepfold was established at Moudjbar in 1878, followed by a population centre in 1905–1908. It became the seat of the commune of M'Fatah, created by decree on 8 October 1956 in the department of Médéa.'[57]

Even though this description is false – the sheepfold disappeared in 1896 and the town was a *spahis* centre, not a colonial settlement – it encapsulates a component of the imperial imaginary, namely that sheep breeding and farming would civilise the country. But any such idea that the station would play a civilising and matrix role for a settlement zone in Moudjbar was likewise false: the nomads were already present, and the project only altered the region for a brief period. Still, it allows us to deconstruct the myth of colonial civilisation by French-style agricultural progress, which failed in the desert, as we will see. Being a propaganda journal, *L'union ovine* conveys the scientific and technical imaginary at work together with hopes for Algeria, rather than a grounded approach. It thus illustrates how zootechnicians projected these ideas onto African nature via the merinisation and station programmes. To acclimatise merinos in Africa, as had been done in Europe, the zootechnicians at the Husbandry Services established fixed sheepfolds that needed to be supplied with fodder,[58] entailing an alteration to the farms' environs. This led to an entire programme to transform nature around stations, which –zootechnicians believed and hoped, in accordance with the myth of progress – were to act as spearheads for agricultural colonisation.

Right across Africa, colonists said not only that the sheep were degenerate, but also that the 'natives' played a role in damaging the environment. Poor husbandry practices among the 'natives', especially Arabs and nomads, reputedly led to erosion; according for example to Jean Balsan: 'the native' puts too many sheep on his land, 'pasturage is destroyed, and animals die of hunger'.[59] To combat the overgrazing caused by oversized flocks, the colonial authorities sought to reduce the flock size of 'natives'.[60] Transhumance was strictly regulated, as was the use of burning pastureland.[61] Farming was now confronted with conservation programmes, particularly of forest: 'we have wisely safeguarded forests, often removing available pastureland there too'.[62]

In addition to general regulations, stations became places for experiments in restoring African agriculture. To regenerate Africa's degraded soils, scientists and colonists thought of using plants to try to fix the earth, and undertook large-scale plant trials. The purpose was to fix the earth using hedges and shrubs, while creating rich pastures for feeding improved breeds.

Botanist and agricultural engineer Guy Roberty (1907–71) provided the most complete study, explaining the method of experimenting plant by plant in government stations in Morocco in 1931, where Napier grass from central Africa was trialled, as was Kikuyu grass, Rhodes grass, Rescue grass, Natal grass, teff and crabgrass, along with legumes and a whole series of other grasses.[63] Roberty even provides a model plan for a station, made up of concentric rings lying around the farm, with a sheepfold, granaries, silos and storage buildings, followed by cereals and fodder produced by temporary meadows, with swathes of permanent meadows further out.[64]

Soil conservation was carried out by creating and restoring plant cover, mostly comprised of local plants and plants from other regions in Africa, though species from Europe or other colonies such as Australia were also used. Stations thus functioned as animal and plant conservatories whose purpose was to act as a hub for disseminating practices and species across native agricultural zones. They were to transform African nature by improving breeds, pastures, fields and the entire agricultural infrastructure. While this colonial agricultural programme seems to have partaken in generating what Alfred Crosby calls 'neo-Europes',[65] it raises more questions than it answers. What was the scale of environmental modification around stations? Did they affect the local economy? May we still detect traces of them today in these territories' modified architecture and ecosystems? If 'neo-European' sheep existed in Africa, what might this tell us about the existence of 'neo-Rambouillets'?

In assessing the programme of sending Rambouillet merinos to colonial stations, one observation stands out: all the state's attempts in the colonies seem to have failed within the first few years. One of the rare cases of temporary success helps us see the reasons for these failures. The Husbandry Service lent a Rambouillet ram raised in a sheepfold to a private farm, the Brogha sheepfold in Morocco, where it successfully bred with three Arles merinos. Using an unidentified workforce, the zootechnicians had green maize distributed from May to June, green or weathered alfalfa from June to September, silage from August to February, and 'each day the ram also received a litre of oat and maize grain'.[66] In this improved environment it was possible to cross the Rambouillet ram with other merinos, but only thanks to intensive care. In 1928, after four years there, he was replaced by another merino ram. But the new ram was not crossed with native races. All the other trials with Rambouillets mentioned in the journal were failures. The practices, techniques and knowledge of the French led the best merinos to almost certain and often unexplained death.

Rambouillets in the Desert: Misplaced Zootechnical Faith

L'union ovine provided several responses to the failure of trials to acclimatise knowledge, Rambouillets and agricultural techniques in the colonies.[67] Zootechnicians' confidence in their capacity to fashion the animal world was probably the main cause, with them forgetting or denying zootechnical realities – both of the Rambouillets and of Africa. They thought of the Rambouillet as the archetypal animal-machine: 'the Rambouillet merino is the prototypical sheep with fine dense wool. It is very robust and resistant, and capable of adapting to all regimes on all terrains and of acclimatising in all countries',[68] according to the establishment's advert in the *L'union ovine* in 1930.

Though heralded as the prototype of total acclimatisation, the Rambouillet was not what zootechnicians thought it was. Because they saw it as the direct descendant of Arabian merinos, the zootechnical experts took it out into the desert searching for its geographical origins. But by 1900 Rambouillets lived in sheepfolds, surrounded by damp meadows and temperate forests at the national sheepfold, and were abundantly nourished with cultivated fodder every day. Fear of losing the breed led the zootechnicians to keep the animals indoors, to prevent them from contracting foot rot or any other disease,[69] with complex ecological consequences for the sheep.[70] Merinos thus went from intensive standardised sedentary systems to a nomadic world of transhumant and pastoral systems in Northern and Southern Africa.

Zootechnicians such as Trouette – a Husbandry Service inspector, veterinarian, teacher at the Algerian School of Agriculture, editor of *L'éleveur nord-africain* and regular contributor to *L'union ovine* – were among the first to reach this conclusion: 'The merino, of whatever type, has not succeeded as an improvement in the South because we forgot that soil, climate and spontaneous vegetation command animal production, because the environment is the master, not the species.'[71] The main culprit for failure was nature itself. Experts had thus doubly forgotten nature – that of the merino, and that of African ecologies. In 1930, Trouette penned a history of zootechnical trials with Rambouillets in Algeria:

> In 1855, inspired by Bernis, the head of the Military Veterinary Service in Algeria, it was decided to create a sheepfold in the south of the country. It was located at Tadmit, and had 600 native sheep, and 15 merino rams were sent there from the Rambouillet imperial sheepfold in 1856. By the end of the year 11 rams had died. They had fertilised many ewes and 400 lambs were born. All without exception succumbed during winter 1856–1857. The region deemed to be unhealthy for Rambouillets, the sheepfold was transported to Birin and 60 Arles ewes were sent there together with 36 new sub-

Figure 4.1. Advertisement for the Rambouillet merino. From *L'union ovine*, Bergerie nationale de Rambouillet (1930). Photo: © BnF/Gallica, reproduced with permission.

jects from Rambouillet. In spring 1857, the latter, carried off by anthrax, went to join their forebears. Thus as of 1860 the sheepfold received almost solely Arles merinos. Finally, after having been successively transported to Djelfa, Ben Chicao, Berrouaghia and lastly Moudjbar, it was definitively shut down

> in 1900. Its influence left no trace in the regions in which it had been located. Not a single sheep, from Ben-Chicao to Laghouat, bears the slightest trace of any infusion of merino blood.[72]

Trouette uses the story to explain to his fellow agronomists, veterinarians, breeders and industrialists why the Rambouillet was not adapted to North Africa. Yet in the pages of the same journal, authors continued to vaunt the superiority of Rambouillet merinos over those from South Africa.[73] Or else they buried these reports: 'over the past twenty years they have conducted the most disparate, most extravagant and most inconclusive trials. The findings have been frequently denounced. They should not be overemphasised', opined veterinarian Dr H. Velu.[74] Other trials were conducted using other breeds, such as the Arles merinos from France and merinos from Australia and South Africa; further study of these sheep breeds could bring to light any improvements and their genetic consequences on 'indigenous' breeds before, during and after the colonial era.[75]

Veterinarians often relayed these accounts of colonial trials. P. Geoffroy (chief veterinarian with the Madagascan Zootechnical Service) and Durieux (the head of the Ambovombé station) thus related the history of the merino breed in Madagascar, blaming its failure on disease.[76] Zootechnicians identified African nature itself as the prime culprit – an invisible nature of parasites (fleas and ticks), viruses and bacteria, which killed merinos within a few months of their arriving in the stations. In Madagascar, diseases such as heartwater – transmitted by ticks – were the main cause of death for Rambouillets. According to the experts it was this, associated with other natural phenomena such as a drought and a cyclone in 1922, that explained why the Ambovombé sheepfold was shut down.[77] Geoffroy concludes: 'once again, success was insufficient, with no batch producing a snowball effect.'[78]

The case of the Rambouillet teaches us about the limits of experts' ecological and zootechnical knowledge: accounts of failures and of successful scientific experiments circulated but little. This resulted in a dual discourse in accounts about acclimatising Rambouillet merinos, thought of by some as an improvement machine, and by others as a breed that could not be adapted to Africa. In fact, what broke zootechnicians' scientific faith in conquering the deserts of African nature was not nature itself, but the problems of acclimatisation regimes for sheep and knowledge.

At the end of this exploratory enquiry, the history of the Rambouillet merino in Africa suggests that we should look more broadly at acclimatisation trajectories and regimes. In the eighteenth century, men of learning considered that the success of acclimatisation hinged on imitating the environmental conditions (soil, climate, pasturage) and husbandry techniques of the land whence the animals first came.[79] In 1786, the way the Spanish flock at Rambouillet was tended was planned to be a cross accommodating the husbandry of the Spanish-African flock and French soil.[80] But as soon as the flock arrived it was confronted with the freezing cold climate of northern France, the wet meadows of Rambouillet and sheep pox – not to mention that royal farm administrators had sold the fodder, for despite plans to leave the flock out in the fields, at the last moment it was taken indoors to protect it from the winter conditions.[81] The flock was carefully tended to, and by 1787 had been acclimatised to the Rambouillet climate and husbandry techniques.[82]

In the nineteenth century, zootechnics attributed merinos with Spanish-African origins, forgetting the difficult beginnings and total modification of the mode of life of animals now placed in sheepfolds and fed fodder from cultivated meadows – which developed in France in part due to the merinisation being conducted at Rambouillet.[83] Agronomic and zootechnical stations sought to transpose the intensive and productive systems used in France to the colonies. The (Rambouillet) merino was thus thought of as the central mechanism in an agrarian and civilising project, a means not only to fertilise the soil and produce industrial wool, but also to sedentarise African farmers and to effect their agronomic acculturation. The African model was devised on the regime developed at Rambouillet, at Arles and in the network of nineteenth-century national sheepfold stations, with merinos being crossed with native breeds. But it was local sheep farmers in the region of Arles, and in the Île-de-France and the Beauce near Rambouillet, who had adapted the cross-breed to their practices: the Arles cross-breed and the Beauce merino cross-breeds (which no longer exist) were the remnants of this successful acclimatisation regime.

The limitations of the French acclimatisation regime transpire when compared to the British regime. South African breeders managed to integrate merinos in a dual environmental and geotechnical context. Despite the destruction caused by colonial conquest, white sheep farmers adapted their farming by adopting the practices and knowledge of local peasants, whom they called 'Khoisans'.[84] Above all, they introduced (Rambouillet) merinos into well-established husbandry systems.[85] South African merino farms, likewise confronted with heartwater, provide proof that African nature was not strictly responsible for French failures. Sedentarisation progressively made the Rambouillet breed inadaptable unless kept in sheepfolds and lavished with care. It was Rambouillet acclimatisation and the work to conserve the breed in France that contributed to colonial failures.

Raphaël Devred is a PhD researcher in history and environmental humanities at the University of Versailles Saint-Quentin-en-Yvelines, at the Centre for the Cultural History of Contemporary Societies. Raphaël's dissertation focuses on the environmental history of the Rambouillet estate from the eighteenth to the twenty-first century. His research has been supported by the Fondation des sciences du patrimoine, in partnership with the National Archives and the National Domains of Chambord and Rambouillet.

Notes

1. René Tourte, *Histoire de la recherche agricole française en Afrique tropicale francophone et de son agriculture, de la Préhistoire aux temps modernes* (Paris: L'Harmattan, 2019), 4 vols.
2. One of the few recent history PhDs is by a veterinarian: Bernard Heude, *La Sologne: Des moutons, des landes et des hommes du xviiie siècle au Second Empire* (Rennes: PUR, 2012). Jean-Marc Moriceau's work on animal husbandry establishes a ranking in the study of animal species, with sheep rearing coming third, behind cows and pigs. Jean-Marc Moriceau, *L'élevage sous l'Ancien Régime: Les fondements agraires de la France moderne XVIe–XVIIIe siècles* (Paris: Sedes, 1999).
3. Diana K. Davis, *Resurrecting the Granary of Rome: Environmental History and French Colonial Expansion in North Africa* (Athens: Ohio University Press, 2007); William Beinart, *The Rise of Conservation in South Africa: Settlers, Livestock, and the Environment 1770–1950* (Oxford: Oxford University Press, 2008).
4. Jean Blanc and Raymond Laurans, *Les débuts de la mérinisation* (Rambouillet: Centre d'Enseignement Zootechnique, 1964).
5. In eighteenth-century France, sheep were sometimes referred to as 'bêtes à laines' (literally, wool-producing animals): Abbé Claude Carlier, *Traité des bêtes à laine ou méthode d'élever et de gouverner les troupeaux aux champs et à la bergerie* (Paris: Vallat la Chapelle, 1770), 2 vols; and for the first chronological phase, see Blanc and Laurans, *Les débuts de la mérinisation*.
6. Louis Jean-Marie Daubenton, *Instructions pour les bergers et les propriétaires de troupeaux* (Paris: Imprimerie de P.-D. Pierres, 1782).
7. One of the largest imports was organised by Minister Turgot in 1776, some of which went to Daubenton.
8. The entire archives of the Mesta, the livestock association that the Spanish sheep came from, amounting to two thousand documents since 1600, contain only a dozen references to merino wool, according to Julius Klein, *The Mesta: A Study in Spanish Economic History, 1273–1836* (Cambridge, MA: Harvard University Press, 1920), 5.
9. In 1770 Abbé Carlier noted this African origin, and mentioned the role played by Cardinal Jimenes (1436–17) in Spain's conquest of northern Africa and the export of African sheep to Spain in the sixteenth century: Carlier, *Traité des bêtes à laine*, 29–30. The Marinid were a Berber dynasty who ruled what is nowadays northern Morocco from the thirteenth to the fifteenth century. Louis and Jeanine Reveleau, 'La quête des bêtes à laine fine et en particulier en France au xviiie siècle, préparant la mérinisation', in Claude Guintard and Christine Mazzoli-Guintard (eds), *Élevage d'hier, élevage d'aujourd'hui: Mélanges d'ethnozootechnie offerts à Bernard Denis* (Rennes: PUR, 2004), 121–38.
10. François de Neufchateau, 'Avis aux cultivateurs et propriétaires de troupeaux, sur l'amélioration des laines', in *Recueil des lettres circulaires, instructions, programmes, dis-*

cours et autres actes publics, émanés du . . . Ministère de l'Intérieur (Paris: Imprimerie nationale, 21 December 1798), 256.

11. Charles-Philibert de Lasteyrie, *Traité sur les bêtes-à-laine d'Espagne; Leur éducation, leurs voyages, la tonte, le lavage et le commerce des laines, les causes qui donnent la finesse aux laines: Auquel on a ajout l'historique des voyages que font les moutons des Bouches-du-Rhône, et ceux du royaume de Naples; L'origine, les succès, l'état actuel du troupeau de Rambouillet, et les moyens de propager et de conserver la race espagnole dans toute sa pureté* (Paris: Imprimerie de la feuille du cultivateur, 1799), 11–12.
12. Éric Teyssier, 'La bergerie impériale d'Arles et l'introduction des mérinos en Provence', *Provence historique* 51(204) (2001), 193–212; Laurent Brassart, '"La ferme des animaux" ou l'invention d'une politique de l'animal utile sous le Consulat', *Annales historiques de la Révolution française* 377(3) (2014), 182–85.
13. Anon., 'Mémoire sur la question, quels seroient les moyens de perfectionner dans les Provinces Belgiques la laine des Moutons', in *Mémoires sur les questions proposées par l'Académie impériale et royale des sciences et belles-lettres de Bruxelles qui ont remportés les prix en 1776* (Bruxelles: J. L. de Boubers, 1776), 12; Augustin-François de Silvestre, *Essai sur les moyens de perfectionner les arts économiques en France* (Paris: Imprimerie de Mme Huzard, 1800), 30; Jean-Baptiste Huzard, *Instruction sur l'amélioration des chevaux en France, destinée principalement aux cultivateurs* (Paris: Imprimerie de Mme Huzard, 1802), 14; Abbé Alexandre-Henri Tessier, *Extrait de l'instruction de M. Tessier sur les bêtes à laine et particulièrement sur la races des mérinos* (Boulogne: Imprimerie Leroy-Berger, 1811 [1810]), 4.
14. Hélène Blais, *L'empire de la nature: Une histoire des jardins botaniques coloniaux (fin XVIII*e *siècle–années 1930)* (Ceyzérieu: Champ Vallon, 2023).
15. This figure comes from analysis of auction sale notebooks (1793–1853), held at the Archives départementales des Yvelines and the Archives nationales.
16. This figure comes from registers of all sheep at Rambouillet national sheepfold. Animals were documented in other ways on occasions, but this is the most rigorous document since it was used to trace the genealogy of sheep so as to manage the reproduction of the breed from 1858 to 1973.
17. Raf de Bont, *Stations in the Field: A History of Place-Based Animal Research, 1870–1930* (Chicago: University of Chicago Press, 2015); Raf de Bont, *Nature's Diplomats: Science, Internationalism and Preservation, 1920–1960* (Pittsburgh, PA: University of Pittsburgh Press, 2021); Blais, *L'empire de la nature*.
18. Archives de la Bergerie nationale de Rambouillet, registre-matricule béliers, year 1907. General Botha (1862–1919) was the founder and prime minister of the Union of South Africa (1910–19). He had distinguished himself in the Boer Wars (1899–1902) through various deeds, including capturing Winston Churchill on 15 November 1899.
19. I thank Robert Rubidge, Charles's grandson, who allowed me to consult his grandfather's diary, via the work Richard Rubidge, *The Merino on Wellwood: Four Generations* (n.p., n.d.).
20. William Beinart, *The Rise of Conservation in South Africa: Settlers, Livestock, and the Environment 1770–1950* (Oxford: Oxford University Press, 2008), 304–31.
21. Rubidge, *The Merino*, 14; Archives de la Bergerie nationale de Rambouillet, registre-matricule béliers, year 1871.
22. Archives de la Bergerie nationale de Rambouillet, registre-matricule béliers, year 1928.

23. Archives nationales (AN), 20160285/215-218.
24. Elian Judas Finbert, *La vie pastorale: Brebis et bergers* (Paris: René Julliard, 1942), 321.
25. Ibid., 322.
26. Michel Lallour, 'Pourquoi cette revue?', *L'union ovine*, November 1929, 7.
27. *L'union ovine*, 5 February 1931, 56.
28. Henri Brunschwig, 'Le parti colonial français', *Outre-Mers: Revue d'histoire* 162 (1959), 49–83; Marc Lagana, *Le parti colonial français: Éléments d'histoire* (Québec: Presses de l'Université de Québec, 1990), 40.
29. The Minister for Colonies Paul Reynaud figures on the cover of the 5 July 1931 issue of *L'union ovine*, 301, and Maréchal Lyautey on a signed photo printed inside.
30. He was additionally president of the Union des syndicats patronaux des industries textiles de France, of the Chambre syndicale des fabricants de drap militaire and of the Chambre de commerce et d'industrie de l'indre: Christine Mery-Barnabe, *De la manufacture royale de draps à l'usine Balsan, catalogue de l'exposition Châteauroux et les cités lainières d'Europe, Archives municipales de Châteauroux* (Châteauroux: Ville de Châteauroux archives municipales, 2010).
31. *L'union ovine* existed from 1929 to 1945, before being renamed *Le mouton* in 1946, then disappearing in 1957, against the backdrop of decolonisation and competition from new sheep rearing journals. *La revue ovine*, created in 1945 by former pupils and teachers at Rambouillet national sheepfold, was taken over by the new Fédération nationale ovine in 1946 and by *Pâtre* – created by Jean Blanc, formerly of Rambouillet, which as of 1951 was the regional supplement (for south-eastern France) of the *La revue ovine* – which still exists today.
32. Archives de la Bergerie nationale, registre-matricule béliers, year 1894; AN, 20160285/202.
33. P. Geoffroy, 'L'élevage du mouton à laine à Madagascar', *L'union ovine*, 5 April 1931, 184.
34. Joseph Gallieni, *Rapport d'ensemble sur la pacification, l'organisation et la colonisation de Madagascar (octobre 1896 à mars 1899)* (Paris: Henri Charles Lavauzelle, 1900), 343.
35. AN, 20160285/202: correspondence between Maurice and Jully, and especially the letters of 14 and 24 October 1900.
36. 'Les Pupilles malgaches du Comité de Madagascar à Paris', *Journal officiel de Madagascar et dépendances*, 6 April 1901, 5627.
37. Archives départementales des Yvelines, 3Fi202 67: A. Bourdier (ed.), 'Bergeries nationales de la Pommeraie. Troupeau expérimental–agneaux mérinos-malgaches', v. 1905; Christian Painvin, 'Bergerie nationale de Rambouillet: Les moutons de Madagascar (1896–1905)', *Mémoires et documents de la SHARY* (2021), 40.
38. Claude-Olivier Doron, *L'homme altéré: Races et dégénérescence (XVIIe–XIXe siècles)* (Seyssel: Champ Vallon, 2016).
39. Jean Marc Drouin, 'Baudement, Émile (1816–1863): Professeur de Zoologie appliquée à l'agriculture et à l'industrie (1852–1863)', in *Les professeurs du conservatoire national des arts et métiers: Dictionnaire biographique 1794–1955*, vol. 1 (Paris: Institut national de recherche pédagogique, 1994), 140–47.
40. This itinerary has been part of the history of sheep since the eighteenth century: Augustin-François de Silvestre, *Essai sur les moyens de perfectionner les arts économiques en France* (Paris: Imprimerie de Mme Huzard, 1800), 29.

41. Émile Baudement, *Les mérinos* (Paris: Librairie d'éducation et d'agriculture de Charles Delagrave et Cie, 1868), 96.
42. Pierre Hamp, the pen name of Henri Bourillon, was a journalist, politician and writer. He was invited to North Africa by Lyautey and André Gide in the 1920s, where he discovered colonial sheep farming. See: Pierre Hamp, *La laine, la peine des hommes* (Paris: Flammarion, 1931); Association pour l'étude de l'histoire de l'inspection du travail (AEHIT), *Henri Bourrillon dit Pierre Hamp: Inspecteur du travail et écrivain humaniste. Hommage de sept personnalités* (Paris: AEHIT, 1993).
43. Pierre Hamp, 'La laine d'Afrique', *L'union ovine*, 5 June 1930, 238.
44. Baudement, *Les mérinos*, 97.
45. Pierre Hamp, 'La foi dans le mouton en Afrique du Nord', *L'union ovine*, 5 July 1930, 280.
46. See especially Davis, *Resurrecting the Granary of Rome.*
47. G. Trouette, 'L'avenir de l'élevage ovin dans le Tell algérien', *L'union ovine*, 5 December 1931, 562–63.
48. G. Catelin, 'Les possibilités de développement de l'élevage du mouton à laine au Soudan', *L'union ovine*, 5 November 1931, 523.
49. For Madagascar: the Service vétérinaire des haras et de l'élevage created by Galliéni on 1 February 1903; for French West Africa: the Service zootechnique created by a decree dated 31 December 1904; and for Central Africa, which became French Equatorial Africa in 1910: the Service de l'élevage created in 1909. See Tourte, *Histoire de la recherche agricole*, vol. 4, 367–68.
50. G. Catelin, 'La création des centres types', *L'union ovine*, 5 November 1931, 519–20.
51. Geoffroy, 'L'élevage du mouton à laine à Madagascar', 187; V. M. Durieux, 'Le mouton à Madagascar', *L'union ovine*, 5 October 1932, 345.
52. J. Henrioud, 'L'élevage du mouton dans la petite et la moyenne colonisation', *L'union ovine*, 5 December 1931, 571.
53. Pierre Hamp, 'La foi dans le mouton en Afrique du Nord', *L'union ovine*, 5 July 1930, 282.
54. *L'union ovine*, 5 November 1931, 522.
55. Guy Roberty, 'À propos de l'affouragement du cheptel marocain: Essais entrepris dans les fermes et jardins du Protectorat et les Centres Union Ovine', *L'union ovine*, 5 March 1931, 132.
56. E. Miege (head of the Service des recherches agronomiques et de l'expérimentation in Morocco), 'Les centres Union ovine d'expérimentation fourragère au Maroc', *L'union ovine*, 5 November 1931, 516–18.
57. Archives nationales d'Aix en Provence (http://anom.archivesnationales.culture.gouv.fr/geo.php?lieu=Moudjebeur+(Alg%C3%A9rie, retrieved 22 March 2022); Marcel Lesne, 'Une expérience de déplacement de population: Les centres de regroupement en Algérie', *Annales de géographie* 388(71) (1962), 567–603.
58. E. Henrioud, 'Les améliorations par le mérinos et l'élevage du Châtillonais en Tunisie', *L'union ovine*, 5 May 1931, 232–33 (the author was the manager of the Douëmis estate).
59. Jean Balsan, *L'union ovine*, 5 April 1930, 154.
60. Roberty, 'À propos de l'affouragement du cheptel marocain'; G. Catelin (technical secretary of the Moroccan section of the Union ovine d'Afrique du Nord and task officer

in Sudan for the Union ovine coloniale), 'Les possibilités de développement de l'élevage du mouton à laine au Soudan', *L'union ovine*, 5 October 1931, 474–76.

61. *L'union ovine*, 5 November 1931, 522; Beinart, *The Rise of Conservation in South Africa*, 44–47.
62. Jean Balsan, *L'union ovine*, 5 April 1930, 154.
63. Roberty, 'À propos de l'affouragement du cheptel marocain', 133–34.
64. 'L'ensilage au Maroc: Foin fané ou foin ensilé? Silo fosse ou silo tour?' *L'union ovine*, 5 January 1932, 30–33; Roberty, 'À propos de l'affouragement du cheptel marocain', 131.
65. Alfred Crosby, *Ecological Imperialism: The Biological Expansion of Europe, 900–1900* (Cambridge: Cambridge University Press, 2004), 2.
66. P. Bezer, 'L'amélioration du mouton dans le bassin inférieur du Sebou: Une réalisation modèle: La bergerie du Brogha (domaine J. Rinieri)', *L'union ovine*, 5 March 1930, 106.
67. Colonial archives held in the countries of former stations could be used to complete this initial survey of the trajectory taken by the sheep.
68. Advert for Rambouillet merinos, published in *L'union ovine*, the colonial magazine, by the Rambouillet national sheepfold (1930).
69. Foot rot is a disease caused by bacteria in the hoof, causing animals to limp. It weakens sheep and may affect fertility.
70. On these aspects of ovine ecology and ethology, see: Michel Meuret, *Un savoir-faire de bergers* (Versailles/Dijon: Quae/Educgri, 2010); Vinciane Despret and Michel Meuret, *Composer avec les moutons: Lorsque les brebis apprennent à leurs bergers à leur apprendre* (Avignon: Cardère, 2016).
71. G. Trouette, 'Questions ovines algériennes: Croisement ou sélection?', *L'union ovine*, January 1930, 15. These reasons were identified by colonists more generally: *Les journées marocaines du mouton, Casablanca, 13, 14, 15 octobre 1934* (Casablanca, Imprimeries réunies, 1935), 29.
72. Trouette, 'Questions ovines algériennes', 13; G. Trouette, 'À la station d'élevage de Tadmit', *L'éleveur nord-africain: Organe de la Fédération des syndicats d'élevage et de la Société d'aviculture d'Algérie* 82 (20 May 1929), 1087–89.
73. G. Curasson, 'La conduite de l'élevage ovin au Soudan', *L'union ovine*, 5 April 1930, 160.
74. H. Velu, 'La production animale au Maroc et les grands problèmes de l'élevage marocain', *L'union ovine*, 5 October 1931, 465.
75. Ibrahim Belabdi et al., 'Genetic Homogenization of Indigenous Sheep Breeds in Northwest Africa', *Scientific Reports* 9(1) (2019), 1–13; Aliènor Bertrand, Anne Blondeau Da Silva and Dominique Taurisson-Mouret, *Liaisons pastorales: Coévolutions, ruptures, résistances* (Paris: Edisens, 2023).
76. Geoffroy, 'L'élevage du mouton à laine à Madagascar', 183–88.
77. It was created on 5 February 1921 and closed ten years later: Durieux, 'Le mouton à Madagascar', 348.
78. Geoffroy, 'L'élevage du mouton à laine à Madagascar', 185.
79. Louis Jean-Marie Daubenton, *Mémoire sur le premier drap de laine superfine du crû de la France: Lû à la rentrée publique de l'Académie royale des Sciences, le 21 avril 1784* (Paris: Imprimerie royale, 1784), 13.
80. Carlier, *Traité des bêtes à laine*, vol. 1, 31, 36.
81. 'Compte que rend au Roi en la personne de Monsieur le Comte d'Angiviller Gouverneur et administrateur général de son domaine privé de Rambouillet, François

Bourgeois régisseur de la ferme du Roi', Archives départementales des Yvelines (Montigny-le-Bretonneux), 60J 459 (3); 'Mémoire présenté à M. de Fleurieu, le 27 ventôse an 13 (mars 1805) par Bourgeois, économe de l'Établissement rural de Rambouillet', Archives départementales de l'Essonne (Chamarande), 52J 37.

82. 'Mémoire présenté à M. de Fleurieu, le 27 ventôse an 13 (mars 1805) par Bourgeois', Archives départementales de l'Essonne (Chamarande), 52J 37.
83. The practice of cultivating pastureland developed in the first half of the nineteenth century, especially in regions where merinos were adopted such as the Beauce and Seine-et-Oise. See Jean-Claude Farcy, *Les Paysans beaucerons au XIXe siècle* (Chartres: Société archéologique d'Eure-et-Loir, 1989), 64–65, 191–92 and 204.
84. Beinart, *The Rise of Conservation in South Africa*, 8.
85. Charles Philibert de Lasteyrie, *Histoire de l'introduction des moutons à laine fine d'Espagne* (Paris: Levrault Paris, 1802), 104 (in 1782, a breeder named Van Runen received four hundred pure-race animals from the Spanish king, which transited from Extremadura via Holland to Cape Colony where they joined a flock of a thousand sheep); Harold Burnell Carter, *His Majesty's Spanish Flock: Sir Joseph Banks and the au profit de Merinos of George III of England* (Sydney: Augus and Robertson, 1964), 278 and 438 (the British developed breeding using the royal flock at Kew and Windsor, overseen by Joseph Banks, following the French model).

Bibliography

Association pour l'étude de l'histoire de l'inspection du travail (AEHIT). *Henri Bourrillon dit Pierre Hamp: Inspecteur du travail et écrivain humaniste. Hommage de sept personnalités*. Paris: AEHIT, 1993.

Beinart, William. *The Rise of Conservation in South Africa: Settlers, Livestock, and the Environment 1770–1950*. Oxford: Oxford University Press, 2008.

Belabdi, Ibrahim, Abdessamad Ouhrouch, Badr Benjelloun, Mohamed Lafri et al. 'Genetic Homogenization of Indigenous Sheep Breeds in Northwest Africa'. *Scientific Reports* 9(1) (2019), 1–13.

Bertrand, Aliènor, Anne Blondeau Da Silva and Dominique Taurisson-Mouret. *Liaisons pastorales: Coévolutions, ruptures, résistances*. Paris: Edisens, 2023.

Blais, Hélène. *L'empire de la nature: Une histoire des jardins botaniques coloniaux (fin XVIIIe siècle–années 1930*. Ceyzérieu: Champ Vallon, 2023.

Blanc, Jean, and Raymond Laurans. *Les débuts de la mérinisation*. Rambouillet: Centre d'enseignement zootechnique, 1964.

Brassart, Laurent. '"La ferme des animaux" ou l'invention d'une politique de l'animal utile sous le Consulat'. *Annales historiques de la Révolution française* 377(3) (2014), 182–85.

Brunschwig, Henri. 'Le parti colonial français'. *Outre-Mers: Revue d'histoire* 162 (1959), 49–83.

Crosby, Alfred. *Ecological Imperialism: The Biological Expansion of Europe, 900–1900*. Cambridge: Cambridge University Press, 2004.

Davis, Diana K. *Resurrecting the Granary of Rome: Environmental History and French Colonial Expansion in North Africa*. Athens: Ohio University Press, 2007.

De Bont, Raf. *Stations in the Field: A History of Place-Based Animal Research, 1870–1930*. Chicago: University of Chicago Press, 2015.

———. *Nature's Diplomats: Science, Internationalism and Preservation, 1920–1960*. Pittsburgh, PA: University of Pittsburgh Press, 2021.

Despret, Vinciane, and Michel Meuret. *Composer avec les moutons: Lorsque les brebis apprennent à leurs bergers à leur apprendre*. Avignon: Cardère, 2016.

Doron, Claude-Olivier. *L'homme altéré: Races et dégénérescence (xviie–xixe siècles)*. Seyssel: Champ Vallon, 2016.

Drouin, Jean-Marc. 'Baudement, Émile (1816–1863): Professeur de Zoologie appliquée à l'agriculture et à l'industrie (1852–1863)', in *Les professeurs du conservatoire national des arts et métiers: Dictionnaire biographique 1794–1955*, vol. 1 (Paris: Institut national de recherche pédagogique, 1994), 140–47.

Farcy, Jean-Claude. *Les Paysans beaucerons au xixe siècle*. Chartres: Société archéologique d'Eure-et-Loir, 1989.

Heude, Bernard. *La Sologne: Des moutons, des landes et des hommes du xviiie siècle au Second Empire*. Rennes: PUR, 2012.

Klein, Julius. *The Mesta: A Study in Spanish Economic History, 1273–1836*. Cambridge, MA: Harvard University Press, 1920.

Lagana, Marc. *Le parti colonial français: Éléments d'histoire*. Québec: Presses de l'Université de Québec, 1990.

Lesne, Marcel. 'Une expérience de déplacement de population: Les centres de regroupement en Algérie'. *Annales de géographie* 388(71) (1962), 567–603.

Mery-Barnabe, Christine. *De la manufacture royale de draps à l'usine Balsan, catalogue de l'exposition Châteauroux et les cités lainières d'Europe, Archives municipales de Châteauroux*. Châteauroux: Ville de Châteauroux archives municipales, 2010.

Meuret, Michel. *Un savoir-faire de bergers*. Versailles/Dijon: Quae/Educgri, 2010.

Moriceau, Jean-Marc. *L'élevage sous l'Ancien Régime: Les fondements agraires de la France moderne XVIe–XVIIIe siècles*. Paris: Sedes, 1999.

Painvin, Christian. 'Bergerie nationale de Rambouillet: Les moutons de Madagascar (1896–1905)'. *Mémoires et documents de la SHARY* (2021), 40.

Reveleau, Jeanine, and Louis Reveleau. 'La quête des bêtes à laine fine et en particulier en France au xviiie siècle, préparant la mérinisation', in Claude Guintard and Christine Mazzoli-Guintard (eds), *Élevage d'hier, élevage d'aujourd'hui: Mélanges d'ethnozootechnie offerts à Bernard Denis* (Rennes: PUR, 2004), 121–38.

Teyssier, Éric. 'La bergerie impériale d'Arles et l'introduction des mérinos en Provence'. *Provence historique* 51(204) (2001), 193–212.

Tourte, René. *Histoire de la recherche agricole française en Afrique tropicale francophone et de son agriculture, de la Préhistoire aux temps modernes*. 4 vols. Paris: L'Harmattan, 2019. Retrieved 1 August 2024 from https://www.fao.org/3/a0217f/a0217f00.htm.

CHAPTER 5

Missing Conservation?

On the Puzzling Dearth of Nature Conservation in Mandate Syria and Lebanon

Diana K. Davis

The origins of nature conservation are most often located in efforts to protect beautiful woodland landscapes and to safeguard wildlife for hunting. Iconic examples frequently provided in the literature include the establishment of the Série de Fontainebleau, the first protected area in France in 1861, the foundation of Yellowstone National Park in 1872 in the USA, appropriated from Native Americans, and the creation of Parc National Albert in the Belgian Congo in 1925, commonly claimed to be the first in Africa. The earliest national park on the continent, though, was Les Cèdres de Téniet el-Haâd, established in French Algeria in 1923.

Similar efforts to protect forests and game were instituted in many colonial territories by most European imperial powers in the late nineteenth and early twentieth centuries.[1] France, for example, implemented similar forestry laws and policies in most of its colonial territories, as well as designating protected areas including several national parks. In the French Maghreb and in its sub-Saharan possessions, the early precedent of Algeria often informed later policies, and several nature reserves and national parks were in place before the end of the Second World War. In Syria and Lebanon, by contrast, any meaningful nature protection other than tentative forestry policies only began in the early 1990s. The dearth of meaningful nature conservation actions during the Mandate period in the Levant, when France directed the development of Syria and Lebanon, is thus rather puzzling.

This chapter analyses the historical development of nature conservation in Mandate Syria and Lebanon to tease out some answers to this puzzle, which appear to lie in domestic and international politics and economics, as well as in complex global knowledge networks. Based on a partial Mandate archive, it suggests that the politically unsettled and violent years of French rule complicated any substantial progress towards nature conservation, which occurred much earlier in several other parts of the Mediterranean Basin and in Africa.

The Mandate authorities' overwhelming focus on economic development further directed attention away from nature conservation other than utilitarian forestry. Being preoccupied with politics and economics, the Mandate authorities apparently overlooked indigenous forms of nature conservation such as the traditional Hema grazing reserve system. Overriding the colonial–postcolonial divide as commonly conceived, this chapter analyses the switch after independence from primarily French to predominantly British sources of knowledge, expertise and practice of nature conservation, concluding that these French and British traditions had in fact much more in common than is conventionally realised.

Modelling the Maghreb

French interests in the Levant were quite long-standing by the time France was named a Mandate power by the League of Nations in 1920 following the First World War and the dismantling of the Ottoman Empire. The French had maintained close ties with Lebanon since at least the 1860s, especially among the large Catholic community there, which France had intervened to protect. Indeed, the French had been dealing closely with the Ottoman Empire for decades, and had helped to write their Forest Code in 1870. As detailed by historian Edmund Burke III, French businesses held about 81 per cent of Ottoman internal debt in 1914, in addition to approximately 59 per cent of Ottoman external debt, and had a keen interest in the potential profits from oil in Mosul.[2] In Syria the French also had very substantial investments in infrastructure, various utilities and commodities such as silk and cotton. Although the Mandate was charged with helping the people of Syria and Lebanon work towards an independent government and successful self-rule, French policies appear to have been aimed more at using their Mandate to increase power and profits, and their administration was more 'colonial' than not.[3]

In its early years, the Levant Mandate was led by many French officials who arrived with significant experience in administering the French Protectorate of Morocco, hoping to emulate prior successes in the Maghreb. These were mostly military men thought to be capable of dealing with the prevailing hostility among the local population who were angry that France had dismantled the British-backed Arab government set in place by King Faysal.[4] On arriving in Syria, General Henri Gouraud, the first high commissioner, had decades of experience in Morocco. Gouraud, who ruled from 1919 to 1923, appointed many French colonial agents from Morocco who staffed key positions in the new administration.

Gouraud drew on Morocco for the great majority of his policies in Syria, including dividing up the territory into a number of smaller, semi-autonomous regions to enact a policy of 'divide and rule' along ethnic and religious lines, which had worked well for the French in Morocco.[5] To this end, and primarily to prevent a strong national identity developing that might lessen their power, in 1920 the French high commissioner and Mandate administration created 'Greater Lebanon' (with most of the Maronite Christians), the primarily Sunni Muslim state of Damascus (most of the southern half of today's Syria) and the state of Aleppo (most of the northern half of today's Syria).[6] In 1922 the Mandate administration created separate enclaves for minority groups concentrated in specific areas. The Shiite Alawites were given the Alawite state (Latakia) along the coast north of Lebanon. The Druze were given the state of Jebel Druze in the deep south. This had the intended effect of isolating the Christians, Druzes and Alawites from political life in most of Syria, which was dominated by conservative Sunni Muslims.

These administrative regions, though semi-autonomous in terms of economic responsibilities and with certain local laws and policies, were all ultimately under the final authority of the high commissioner for Syria and Lebanon. The history of these regions over two decades is complex, and also somewhat confusing because the administrative units changed and the nomenclature was not consistent. For example, in 1922 the short-lived Syrian Federation was created by merging the states of Damascus and Aleppo with the Alawite state. This was dissolved in 1924 when the states of Damascus and Aleppo were combined into a single large state, the state of Syria, comprising most of Syria with the addition of the heavily Turkish sanjak of Alexandretta along the north coast bordering Turkey. At this point, the Alawite state was once again separated out, leaving it isolated. The Bedouin territories in the eastern and southern parts of the Mandate were kept under a separate French military authority, the Contrôle Bedouin.

Although the French wrote a constitution for Syria in 1930 and allowed Syrians to hold elections, France retained absolute veto authority, meaning that political life in Syria was little more than a charade. For a brief three-year period from 1936 to 1939, all of Syria was united, including the Alawites and the Druze, but in 1939 nascent moves towards Syrian independence were scuttled when the French government suspended the constitution, and the Alawite and Druze states were separated out again. Also in 1939, the sanjak of Alexandretta was ceded to Turkey to appease Ataturk.[7] This administrative arrangement remained more or less stable throughout the Second World War, until the French pulled out in 1946. Yet their policy of 'divide and rule', entailing so many political-administrative upheavals for the local populations, was to have long-lived consequences: in the words of historian William Cleveland, 'the destructive political instability that came to characterise Syria after independence in 1946 must be traced, in part, to the institutionalised fragmentation practiced by the French Mandate authorities'.[8]

In Lebanon, although a constitution was approved in 1926, it did not provide authentic independence because no bilateral treaty was signed between Lebanon and France, and because the French high commissioner retained the power to dissolve parliament and even suspend the constitution – which he did in 1932, 1934 and 1939.[9] Moreover, until the end of the Mandate after the Second World War, France retained control over military affairs and foreign relations in Lebanon. This is why Mandate-wide legislation, such as the 1935 Forest Law (discussed below), applied to the entire Mandate including Lebanon, as well as to the Alawites and the Druze.

The successful experience in Morocco, whence 'divide and rule' was derived, was used as a model for other aspects of Mandate administration. One instance was the attempt to provide a wide array of social services to attract the local population to the French, in a bid to win their support and loyalty. Gouraud and subsequent high commissioners thus set about building roads, irrigation networks, experimental farms, schools, orphanages and railroads, together with agricultural development plans and health services for people and their livestock, most of which were inspired by previous experiences in the French Maghreb.[10] It was assumed that the 'politics of attraction' set in place by Gouraud and others would provide positive results in Mandate Syria, but political instability stymied any such hopes.

As discussed in more detail below, political upheaval plagued the French administration throughout the Mandate period, and after a serious rebellion in 1925–27 the Mandate authorities tried a different, more conciliatory strategy, often referred to as the 'politics of the notables'. Practising 'politics of the notables' entailed relying on well-to-do landed families to exert pro-French influence over a local client base. Part of this strategy further entailed pitting rural populations against urban populations to stave off nationalist organising. Fearing another revolt, the French also tempered some aspects of Mandate administration so as not to alienate the local populations.

Environmental Management without Conservation

Given the examples of Maghreb influence in so many sectors of Mandate administration, it is puzzling that so few aspects of nature conservation were developed during the Mandate. French colonial territories elsewhere had experienced a flurry of activity relating to various aspects of nature conservation, including many in forestry. The French colonial Forest Codes of Algeria (1903), Tunisia (1915) and Morocco (1917) were some of the earliest, with even earlier precedents found in the partial Algerian forest laws of the nineteenth century dating back to at least 1854.[11]

The comprehensive 1903 Algerian Forest Code served as a model and variously revised template for the Tunisian and Moroccan forest codes, as well as for the 1913 Forest Law in Madagascar, revised in 1930.[12] The 1930 Madagascar Forest Code in turn served as the model for the 1935 Forest Code for French West Africa.[13] French Equatorial Africa did not see a comprehensive Forest Code until 1946.[14] A Forest Law was drawn up for French Indochina in 1930, though, preceded by an April 1927 decree regulating hunting.[15]

The year 1927 also saw commercial hunting outlawed in French Equatorial Africa, as well as the creation of ten nature reserves in Madagascar where hunting had been regulated by law as early as 1907.[16] By 1926 a network of protected areas functioning as nature reserves had been created throughout French West and Equatorial Africa, officially totalling millions of hectares.[17] Algeria boasted five national parks by the time that Parc National Albert (later Virunga National Park) was established in the Belgian Congo in 1925.[18] Morocco created Toubkal National Park in 1942; Kruger National Park in South Africa was established in 1926. By the mid-1930s, dozens of game reserves and protected areas had also been created in the many territories of British Africa.

It is quite an anomaly, then, that hunting in Syria and Lebanon was not regulated until 1952, after the French Mandate.[19] Nor were official 'protected areas' created until the early 1990s.[20] Although protection of water and maritime resources attracted some legal attention early in the Mandate, with a 1921 law regulating coastal fishing and the protection of maritime fauna, practically the only natural resource to attract sustained administrative attention during the Mandate was forested areas, deemed by most French officials, foresters and scholars to be 'degraded'.

Prevalent concepts of 'degradation' stretched back centuries. Anglo-European interpretations of many arid and semi-arid Mediterranean and Middle Eastern landscapes as deforested, overgrazed and ruined by the 'natives' came to dominate over the course of the nineteenth century. This view was compounded by the widespread belief in desiccation theory, which claimed that deforestation caused the climate and environment to become desert-like, together with its corollary, that planting trees, even in arid areas, would 'bring back the rains' and create a 'proper climate and environment'. Over recent decades scientists and researchers have repeatedly shown this to be a false construct, yet it remains surprisingly persistent.[21]

During the Mandate period, many colonial powers, including most French and British agents, operated within this ideological framework, a kind of environmental orientalism that resulted in many policies and laws being devised to 'improve or repair' what was wrongly perceived as a 'strange, defective, and ruined/desertified' environment in many imperial territories, in comparison to Europe's 'normal and productive' environment.[22] Trees and forests were nearly always viewed within this framework, deforestation was assumed and the need for 'reforestation' was frequently proclaimed since it was believed that trees would increase rainfall and prevent erosion, thus improving agricultural conditions.

Although the French-written Ottoman Forest Law was officially in place in Mandate Syria and Lebanon, the Ottomans were frequently accused of not having implemented it, and ideas for reform were soon voiced. In 1921, for instance, a young French forester, Pierre Monnet, was sent to Syria for a year-long mission to study the best means to preserve and expand the forests. Monnet had graduated from the renowned French forestry school in Nancy in 1919.[23] In his report, he notes very poor forest conditions in most of the Mandate territory and blames 'centuries' of supposed deforestation, overgrazing and burning by the local populations, invoking the all-too-common French colonial environmental trope of long-standing degradation.[24]

During his mission, Monnet visited Cyprus to study its forests and forest policy, looking for ideas for the French Mandate territories. Cyprus had been under British administration since 1878. By 1921 they had mapped and demarcated all forests, after starting in 1879 to introduce a series of forest laws to preserve and extend the forests, which foresters and most observers perceived as degraded. Under the British, the orientalist environmental interpretation prevailed that Cyprus's landscape had been degraded by the Cypriots, a view that has also been proved incorrect.[25] But foresters on Cyprus had realised that beautiful forests would 'help to attract tourists and their gold ... *Forestry in Cyprus means money!*'[26]

Monnet was received and given tours by Cyprus's principal forest officer, Mr Unwin, a former forest officer in British India. Monnet was very impressed with what he saw in Cyprus, which he says partly inspired his recommendations for Syria.[27] Monnet's subsequent recommendations for the French Mandate included limiting and regulating grazing in forests (especially by goats), working to reduce forest fires and extending forested areas through 'reforestation'.[28] All these foci of Monnet's recommendations were also the foci of the Cyprus Forestry Laws, and both were strikingly similar to the Algerian and Moroccan Forest Codes. All these laws claimed most forest land as state land, and emphasised policies to prevent forest fires and to restrict or eliminate most grazing in forest areas. This is not surprising since the Cyprus Forest Law was based on the precedents of both the French-written Ottoman Forest Law and the Forest Law recently penned for British India.[29] Moreover, the first Forest Conservator for Cyprus had been the French forester Paul Madon.

Yet even before Monnet's 1921–22 mission, the Mandate authorities were passing local laws – which went above and beyond what was possible under the existing Ottoman Forest Code in Syria and Lebanon – to protect and extend forests in the territory. The 1870 Ottoman Forest Code had been written by French foresters invited as consultants by the Ottoman administration, and was informed by French forestry theories and French forest law and policy.[30] Probably the earliest local forest law in the French Mandate was decree no. 89 of 1 November 1920 on protecting the cedars of Lebanon, followed by a circular of 7 April 1922 governing the organisation of Lebanon's Forest Service.[31]

In 1922 another province, Latakia (in the Alawite state), passed local forest legislation in 1922 applicable only in Latakia, modelled directly on Algeria.[32] As part of this, Latakia created a local Forest Service that was run by a French forester from Algeria. By 1923 and 1924, other localised forestry regulations were promulgated, some of them dealing with forest taxes.[33] The great majority of these early efforts at legislation were in the most heavily forested provinces: Lebanon, Latakia and Alexandretta, all located along the Mediterranean coast, and all believed to be degraded and deforested to various degrees. They focused on delimiting and protecting local forests, regulating wood cutting, charcoal making and grazing, and seeking to eliminate fires. During the Mandate Latakia and Alexandretta were often lauded for taking the most and the earliest forestry actions.[34] Yet in these early years foresters often complained that there was no comprehensive and proper Forest Law for the Mandate as a whole.

Piecemeal efforts at forestry improvement continued throughout the 1920s and into the 1930s. In 1924, decree no. 2661 on reforesting Greater Lebanon was passed (on 9 August).[35] In addition to trying to prevent any destruction and most uses of trees, such policy instruments also regulated certain uses by requiring permits for actions such as chopping wood and charcoal making. Fines were issued when people were caught breaking these early forestry rules.

Forest departments were primarily self-funding and thus reliant on revenue from permits, taxes, fines and forest products, thus limiting their actions. Only when funds were available were modest attempts made at tree planting. But all these earlier forestry laws were abrogated in 1935 when a comprehensive Forest Law (Code Forestier) was passed for the entire Mandate territory (Syria and Lebanon). This was decree no. 226/LR of 7 October 1935.[36]

This comprehensive 1935 law was inspired by earlier French forestry laws, both domestic and colonial, and bears a striking resemblance to the Algerian and Moroccan Forest Codes. The Syrian Code contains 152 articles, in comparison to Algeria's 190 and Morocco's 84, but covers much of the same ground, including a shared emphasis on curtailing or eliminating grazing, on preventing forest fires (by various means including mandatory firefighting service), on reforestation of assumed deforested areas and on the expropriation of land that could be 'reforested' on grounds of public utility to protect water supplies, prevent erosion or stabilise sand dunes.[37] In 1938, it was requested that collective punishment be added to the new forest code in Latakia for cases of 'severe' forest fires, another striking similarity with the Maghreb forest codes.[38]

Forest Services also started to be set up in the 1920s, usually closely associated with agricultural departments. Throughout most of the Mandate period forestry was under the purview of the Agriculture Department, most probably due to the widespread belief, detailed above, that planting trees would increase rainfall, improve local hydrology and prevent droughts and erosion. Latakia, as mentioned above, had its own Forest Service in the early 1920s, directed by a French forester from Algeria, and by 1926 a committee for forest supervision had been created within the Syrian Department of Agriculture.[39] French Algerian forestry influence continued unabated, and in 1937 Latakia's Forest Service was being run by a brigadier forester from the Algerian Forest Service who had been in charge of Latakia's Forest Service for several years.[40]

In 1938, three years after the 1935 Forest Code was passed, a more substantial Forest Division was created within the Agricultural Department headed by Wajid Bey Malki, a graduate of the French forestry school, the École des barres. A separate Forestry Department, independent of the Agriculture Department, was not created until 1943, with Mr Malki being named its first director.[41] In Lebanon, the Forestry Department was still part of the Agriculture Department a decade after independence.[42]

Despite these laws and administrative developments, little was done in terms of forestry other than issuing permits for uses such as wood cutting and charcoal making, attempts to control grazing and fines for various infractions. Reforestation efforts, for instance, were quite limited outside a few circumscribed sites, usually in agricultural areas and the mountainous regions in Syria and Lebanon. In the early 1930s in Lebanon, for example, the number of reforested hectares was very modest, with 68 ha in 1932, 90 ha in 1933, 338 ha in 1934, 354 ha in 1935 and 180 ha in 1936.[43] Yet it was commonly claimed by scholars, administrators and others that about half a million hectares in Syria and Lebanon needed to be reforested.[44] So while the Mandate authorities developed many plans for forests and trees, few were in fact implemented.

Profits, Politics and Paper Conservation

The purpose of many of the forestry goals and activities discussed and framed during the Mandate period was to encourage and facilitate agriculture and, to a degree, tourism. It is widely acknowledged that the French sought to extract profits from their mandates in the Levant, yet to pacify the population they also needed a reliable domestic supply of basic foodstuffs such as grains. This was extremely difficult due to the region's limited, unpredictable and erratic rainfall. Most of the territory, excepting Lebanon and the coastal strip, has high rates of inter-annual precipitation variability. This causes non-equilibrium ecological dynamics stymying reliable predictable harvests and successful tree plantations.[45] The nearly continuous political problems in the French Mandate territory from 1920 into the Second World War, as detailed below, also caused many difficulties. As a result, what little nature conservation there was materialised primarily as 'paper conservation' – planned and written down in laws and policies, but infrequently and inconsistently executed on the ground.

As indicated earlier, the part of the natural world to receive the most attention under the Mandate in terms of 'nature conservation' was trees and forests rather than wildlife. The French had arrived to administer the Mandate with the widespread yet erroneous belief that the entire area of south-west Asia and the Near East was deforested, aridified and degraded, just as they wrongly believed most of North Africa had been ruined and turned to desert over the centuries.[46] Whereas in the Maghreb this declensionist environmental narrative had primarily been framed in lofty terms of saving civilisation (and furthering colonial agriculture) by resurrecting the allegedly more lush and forested former environment,[47] in Syria it was more often couched in calculating economic terms.

A report published in 1945, just as the French were preparing to leave the region, exemplifies the dominant attitudes and summarises much of the written analysis over the Mandate period together with the policies and projects undertaken up to that date. This report on reforestation and its agricultural and economic role in Syria and Lebanon was published by the prestigious French School of Engineers of Beirut (founded in 1913) under the patronage of the University of Lyon (France) and with the support of the Mandate Government.[48] It was written by Philippe Assouad, an engineer and the head of the Damascus city planning department. The first half of Assouad's report reprises the standard yet incomplete French and European view of the environmental, hydrological and economic role and value of forests, followed by an extended discussion of the 'vital' role of reforestation in the region, particularly for agriculture. Extrapolating from desiccation theory, it was believed that planting trees on a large scale would curtail droughts by increasing rainfall, ameliorate flooding, lead to better irrigation, decrease erosion and thus improve agriculture.[49]

Assouad's report echoes much of a related report produced at the beginning of the Mandate period. This report on agriculture in Syria and Palestine was written in 1922 for the Marseille Chamber of Commerce by Paul Parmentier, a professor in the science faculty at the University of Besançon and delegate on a French mission to Syria. It emphasises what was seen as forests' vital role in regularising rainfall, improving the climate and making agriculture highly productive. Parmentier repeated the by then standard but incorrect story that the region had been nearly completely deforested, and that reforestation was crucial to the success of agriculture by restoring the proper ('normal') hydrological regime, and thus to the success of the economy and thereby that of the

Mandate itself. He theorised that even hydro-electric power would be feasible if the mountains were 'reforested', since he believed that reforestation would increase and regularise the rains and create gushing rivers.[50]

This argument – that reforestation is a cheaper and faster way to improve the hydrologic situation (and irrigated agriculture) than building dams and reservoirs – dominated over the next two decades and lay at the heart of Assouad's 1945 report.[51] Reforestation was said to lead to improved agriculture and sustained profits, not only from agricultural products, but also from the sale of various forest products such as timber, turpentine, coal tar and charcoal. This was a utilitarian economic argument, not one to conserve nature for reasons such as biodiversity. Assouad concludes his report with the following exhortation:

> for Syria and Lebanon, mutilated and emptied by centuries of neglect and negligence, to become once again the 'land of milk and honey' that its [former] prosperity has made famous, it is indispensable that their forest armature, which protects the mountains, be restored . . . all this being a source of benefits and inexhaustible riches . . . which divine providence has for all times assigned to the tree in nature.[52]

He lauds the 1935 Forest Code for Syria and Lebanon for its focus on reforestation, usually to be financed by the state, and for its provisions allowing land to be expropriated for the 'public utility' of reforestation.

Whereas in Syria the justification for forestry focused primarily on improving irrigation conditions and agricultural production, in Lebanon tourism and its revenues were at least as important in driving forestry. The value of trees to create beautiful scenery and a cool, refreshing climate in the mountains of Lebanon was recognised early in the Mandate, especially for attracting people to spend their summer vacations. Some even hoped to make Lebanon 'the Switzerland of the Near East'.[53] By 1924 the Commission for Tourism had organised an annual Tree Celebration (Fête de l'arbre) involving schools to try to instil a love of trees among the people from an early age, believing that this would lead to less deforestation and more reforestation.[54] Much of the little reforestation work that actually took place in Lebanon was carried out alongside roads, to increase their appeal to visiting tourists as they travelled around.[55]

By the early 1930s the press was extolling Lebanon as 'the land of tourism'.[56] Around this time, the Friends of the Tree Society in Lebanon was set up, and one of its founding members and first director was Emile Edde, who was elected president of the nominally independent state in 1936.[57] Members of this society would sometimes help the Forest Service plant trees during its reforestation efforts.[58]

Although the elite, urban populations with the means to visit forests for pleasure were probably in favour of most of what the Forest Service was endeavouring to do, many of the poorer people in the French Mandate territories were not as happy with their activities or policies. As early as 1924, Mandate authorities were having to modify forest policies in order to quell unrest. The forest taxes were abolished in 1924, for instance, due to the population's perceiving them as 'difficult and vexatious'.[59] Yet because Forest Services had to be mostly self-financing, the number of fines levied on local people for forest law infractions and the resultant revenue soon began to increase, and went on increasing nearly every year throughout the Mandate. This probably contributed to increasingly negative feelings among Syrians and Lebanese towards French administration during the Mandate.

Such negative attitudes towards the French had been accumulating since the French army marched on Damascus in 1920 just a few months after the Mandate had been awarded. This event upset many people in the Mandate territories, and in several ways set the stage for a series of continual uprisings and rebellions over the next two decades of French administration. After riots broke out in the Jabal Druze region in 1922, these escalated to a fully fledged revolt by 1925, before becoming a national revolution backed by the growing nationalist movement opposing the French Mandate in both Syria and Lebanon,[60] which was crushed in 1926 by the French with more than fifteen thousand troops and days-long aerial bombardments.[61]

The attempted revolution did succeed however in forcing a change in French administration, with more autonomy granted to the Mandate territories. This is when the politics of the notables was adopted, though the French high commissioner still held vast powers, and French mandatory rule continued to be more like traditional colonial rule than that of a mandatory power preparing a new state for independence. The Great Depression then hit Syria particularly hard, and a series of droughts damaged crops and livestock, causing further immiseration. Such suffering was especially pronounced in the Badiya, the eastern steppe region of the Bedouin and pastoralists, kept under military control by Mandate authorities basically as a separate state.[62]

Never able to completely control these tribes, the French first tried to pacify the Bedouin with an indirect policy of appeasement and attraction, then later in the Mandate period with sedentarisation and direct control policies.[63] It mattered to the French to try to gain control of the Bedouin and to win them over as a bulwark against the growing nationalism in urban areas. In conjunction with the Mandate policy of trying to increase agricultural production in the rain-fed border regions of the steppe, traditional pastures and migration routes were slowly but increasingly circumscribed, meaning tensions and problems recurred over the course of the Mandate.

Instability in the Mandate territories was mirrored in (or a reflection of) significant instability in France, where there was a succession of thirty-three different governments between 1920 and 1940.[64] This instability had profound effects on Mandate Syria. In 1940, France fell to Nazi Germany and Mandate administration was aligned with the Vichy government. Vichy Syria was thus invaded in 1941 by the British in Palestine and Iraq, followed by Free French forces. French control continued for several more years and the final withdrawal of French troops did not take place until April 1946, after the end of the Second World War. With attention and resources focused on politics, revolts and then war during these turbulent years, nature conservation was a low priority, and little was accomplished even in forestry.

The British Middle East Office, the UN FAO and International Networks in Perspective

By 1947, a year after the French finally pulled out of Syria, foresters from the British Middle East Office (BMEO) headquartered in Cairo were being invited by the Syrian and Lebanese governments to report and advise on their forests. The newly independent countries were concerned not only about their forests and forest resources, but also about agricultural production, as the French had been during the Mandate. Most BMEO forest advisors had either trained in India or else had served in the Indian Forest Service for many years. A few also had experience in forestry in Cyprus. The first British report on Lebanon makes clear the prevailing scientific attitudes: 'forestry is the shield of agriculture, particularly in the Mediterranean environment . . . . The restoration of a balanced agricultural economy is impossible without the restoration of forests.'[65] This is identical to what the French believed and had sporadically attempted to put into practice over the previous quarter century.

Although the switch from French to British (and anglophone) expertise in forestry and other environmental issues may seem abrupt, it was understandable given Syrian nationalists' antipathy to French Mandate rule, regional tensions and the international politics of the time. Yet this apparent change in expertise was not as complete as it may seem, and the continuity of environmental thought behind the very similar policy recommendations for forests and the environment is more complex, international and deeply rooted than hitherto recognised. This is one reason that environmental policy in the early years of independence in the former mandates was very similar to what had preceded it, belying any clear or distinctive 'colonial–postcolonial' divide.

On the one hand, long-standing Anglo-French rivalry in the region, the early rumblings of the Cold War and Britain's wish to create an 'informal empire' in the Middle East (with access to petroleum) were all instrumental in

British interest in helping to 'develop and modernise' Syria and Lebanon so as to gain influence, resources and markets.[66] On the other hand, Syria and Lebanon were mostly eager to cut ties with France given the violence and grave problems encountered during the Mandate, as explained above. This enmity was obvious in an early statement on forest lands by the new Syrian government in 1955, which made the exaggerated claim that 'the intentional negligence of forestry by the French Mandatory authorities, who did not protect forest trees . . . left the [eastern] region barren and treeless'.[67]

Many of Britain's early development efforts in the greater Middle East region, including agriculture and forestry initiatives, were conducted through the British Middle East Office (BMEO). The BMEO was created in 1945 out of the facilities, networks and many of the personnel of the Second World War British Middle East Supply Centre (MESC, 1941–45).[68] The Americans displayed a very similar interest in the region for nearly identical reasons, and were also invited to provide expert guidance in these sectors.[69]

Syria and Lebanon became members of the United Nations Food and Agriculture Organization (FAO) in 1945, the year the FAO was created and the same year that France became a member. Within a few years FAO staff began publishing reports on forestry in the former mandates.[70] As a result of the 1951 Supplemental Agreement (TA-230/s/2) signed between the Syrian government and the FAO, one of the FAO's European forestry advisors, Mr van Bottenburg, was sent to assess forests and to advise the government. The background material of his report and other early FAO forestry reports appears to have come mainly from the BMEO forestry reports.

During van Bottenburg's two-year stay in Syria (1952–54), a new Forest Law was drawn up and promulgated in 1953 (Qanun-il-Hiraj No. 66). He explains that the earlier 1935 French Forest Law 'had excellent qualities for that period' but had produced little effect due to 'the absence of an adequate organisation'. He further clarifies that the new 1953 Syrian Forest Law 'incorporated the best elements of the 1935 [French] Act, adjusting them to the current situation'.[71] Despite the government's antipathy to the French Mandate and the apparent dominance of British foresters in the early years of Syrian independence, there continued to be significant French influence, a fact not recognised in the literature.[72] For example, a graduate of the French forestry school (Nancy) was the principal forester in Syria in 1955.[73] By 1961, five (out of six) of the senior foresters working in Syria had trained at Nancy, with the sixth having trained in the USA.[74] At approximately the same time, in 1957, the seven most senior foresters in Lebanon had trained in France at the Nancy Forestry School.[75]

The 1961 FAO report on Syria by Manohar Das Chaturvedi illustrates the international character of scientific forestry at this time. Chaturvedi had been the Inspector-General of Forests in India during the 1950s before becoming an FAO forestry advisor. Like BMEO forester Maitland before him, he highlights the percentage of total land area forested, known as the rate of forest cover (or rate of woodedness), which he believes to be quite important for forest policy. He lists the forest cover rates for all the primary regions of the world, noting a 27.6 per cent global average, citing, not surprisingly, the 1952 National Forest Policy of India as his source. This led Chaturvedi to conclude that the current 2.4 per cent forest cover rate in Syria was inadequate and needed to be raised to at least 20 per cent for most of the country except the eastern steppe zone (the Badiya). Chaturvedi also cited as an inspiration the 1894 National Forest Policy of India, which 'has stood the test of time'.[76] It might seem from his report, then, that much of the influence on forestry in early independent Syria derives primarily from the British and their experiences in India and Cyprus.

Although the connection between the importance accorded to forestry – an idea of partly German origin – and British colonial influence in the Middle East and elsewhere has been highlighted in the existing literature, the reality is more complex.[77] The notion of the rate of woodedness (percentage of forest cover) discussed by Chaturvedi, for instance, originated in French forestry, more precisely in French colonial Algeria, and was then exported around much of the world through its forestry schools and writings published on the topic.[78] Known as the *taux de boisement* in French, it became a very influential and widely applied 'normal' rate of woodedness with Euro-centric origins, usually about 30–33 per cent. In arid and semi-arid lands it is essentially meaningless, and when used as a policy guideline can be ecologically harmful. The *taux de boisement* represents a colonial environmental ideology that has had a particularly long and significant impact in India, although much scholarship assumes that a distinctively German influence prevailed in British Indian forestry.

As I have detailed elsewhere, despite the fact that German techniques of sylviculture were particularly influential in many parts of the world, the idea of desiccation – that deforestation causes the climate to aridify and that planting trees will 'bring back and regularise the rains' – is of French origin, and infused most French writing and practice in forestry and in nature conservation more generally.[79] Although not recognised widely in current literature, the influential German forester Franz Heske explained over eighty years ago that 'to the French belongs the credit for having first recognised and safeguarded by law some of these effects of the forest [on rain, hydrology and climate . . .] towards the end of the 18th century'.[80] From this date on, education, writings and professional relationships established the problematic idea of desiccation – particularly that deserts and arid lands are anthropogenic, caused by deforestation and overgrazing, necessitating massive reforestation – which came to dominate nearly all environmental thinking in Europe, North America and many other parts of the world.[81] From the early nineteenth century onward it was frequently proclaimed that nearly the entire land surface of the globe had once been – and should again be – forested. In the words of the late Richard Grove, by the late 1880s 'the typologies of anti-desiccation forest policy in the French and British colonial states were so closely inter-related that they can be said to have constituted a single ruling philosophy rather than two separate traditions'.[82]

A substantial reason for this is that from 1867 to 1893, 88 per cent of British foresters sent to India were trained in France, and for the period 1875 to 1884 all such British foresters were trained in France and no longer in German schools.[83] Another reason is the emergence of a truly international forestry and natural science literature that was widely and closely read by a great many influential people in multiple languages, as was common until the mid-twentieth century. Thus though not widely recognised, the French influence on British forestry in India – and later Cyprus – has been significant and widespread.

After the Second World War, British and FAO consultants often held up Cyprus – only about 60 miles away from the Syrian and Lebanese coastline and with a similar ecology and climate – as an example to follow for environmental conservation in Syria and Lebanon. French influence was once again at work in Cyprus. The earliest reports on the forests were conducted by French foresters including G. de Montrichard in 1873, L. Bricogne in 1877 and Madon after the British took control of the island in 1878.[84] When de Montrichard visited Cyprus in 1873, the island was nominally under the French-written Ottoman Forest Law, but the report he wrote resulted in regulations for a fee-based system of tree-cutting.[85] He described forests as ruined by 'the pastoral customs of the Orient', and recommended the 'extinction' of goats, 'reforestation', pasture management, fire suppression and restrictions on the movements of forest users.[86] These recommendations – which were strikingly similar to existing forest policy in French Algeria – laid down policy guidelines that were largely followed over the next half a century, and which substantially influenced subsequent Forest Laws and policies on the island.

The first forester to be appointed Principal Forest Officer for Cyprus by the British was a Frenchman, Paul Madon, who arrived in 1880 with experience in French Algeria.[87] Madon served for just over two years, during which time he stimulated reforestation and made several proposals, nearly identical to de Montrichard's, which were influential for years to come. A later British forest consultant credits Madon with having started plantations with 'Australian trees' (eucalyptus), describing Madon's work as very influential, 'distinguished by excellence' and 'of the utmost value to the student of Forestry today'.[88] This forest advisor, D. Hutchins, notes that the main goals for Cyprus forestry as laid down by Madon, namely planting many trees and reducing or eliminating shifting cultivation, forest fires and goat grazing, were on the way to being accomplished, with tackling goat grazing the last to be achieved. Cyprus's forest laws further reflect this mixed influence, being based on both the British-Indian Forest Laws and the Franco-Ottoman Forest Law.[89]

Environmental thinking and policy, therefore, were more consistent than they may first appear when analysing the shift from French influence to British influence at the end of the Second World War in Syria and Lebanon. A strong focus on using afforestation to try to improve overall environmental conditions and so increase agricultural production was carried over from the French Mandate to early independence policy. Other than regulating hunting in 1952, no significant nature conservation efforts were made in Syria or Lebanon – other than forestry initiatives and a few rangeland development projects – until the early 1990s, and this despite calls for the creation of national parks and protected areas as early as 1952.[90]

In fact, the lack of significant protected areas and absence of any national parks in the Middle East was bemoaned at the 1954 UNESCO Symposium on the Protection and Conservation of Nature in the Near East.[91] This major conference in Beirut was attended by over fifty international participants, and organised and sponsored by UNESCO in cooperation with Lebanon's Friends of the Tree Society. But it was not until nearly four decades later, in 1991, that the first protected area was established in the former French Mandates. This long delay was caused by many complex factors driven primarily by a combination of widespread poverty, national economic development priorities, and political instability in Lebanon (the long civil war) and Syria (multiple uprisings) as well as across the region more generally, including ongoing tensions with Israel.[92]

Conservation Hidden in Plain Sight?

Despite all the attention focused on forests and trees during the French years and early independence period, the first protected area in the former Mandate territories was established in 1991 in eastern Badiya, the arid Syrian steppe where few trees grow. This is the Al Talila Reserve for wildlife preservation and the reintroduction of the oryx and sand gazelle, created with IUCN and FAO funding.[93] Lying only 30 kilometres or so east/south-east of Palmyra (Tadmor), this 22,000 ha. biodiversity reserve is centred on an area long used as a critical grazing reserve for camels raised by the Bedouin.[94] The core Al Talila biodiversity protection area is surrounded by a larger rangeland rehabilitation project of approximately 130,000 ha. organised into grazing cooperatives.

A second, larger protected area was created in 1993, Jebel Abdul Aziz nature reserve in the semi-arid north-east of Syria. It is more than double the size of Al Talila, comprising 49,000 ha.[95] In Lebanon, the first protected area was the Horsh Ehden forested nature reserve, established in 1992.[96]

Figure 5.1. Camels of the Sba' tribe grazing in the Al Talila Protected Area in 2007, a rare rainstorm in the background. This tribe is no longer allowed to graze their camels on their customary pastures in Al Talila. Courtesy of the photographer: Professor Dawn Chatty.

It is likely that the Al Talila Reserve was once a traditional 'Hema' pasture reserve. The Hema (Hima) system is a grazing reserve system based on communal property, and founded on tribal structure and customary law.[97] It is not an open-access system but a well-regulated system for reserving pastures for grazing during hard times such as drought periods. The system was common in many parts of the Middle East during the Ottoman period and earlier, and probably dates back much further, being partly based on the Koran and related writings, though it may in fact predate Islam.[98]

On being 'rediscovered' by the Syrian government in the 1970s, a simplified version of pastoral cooperatives was deployed in the country's steppe region in a bid to raise livestock production, seeking to ameliorate what was assumed to be range deterioration from overgrazing.[99] Recent research has demonstrated, however, that environmental degradation in the Syrian steppe is primarily due to agricultural expansion in marginal lands, not to 'overgrazing'.[100]

The indigenous Hema system, however, had been mostly suppressed during the 1950s and 1960s due to the independent Syrian government's anti-nomad sentiment, leading it to abolish customary law among the nomad tribes, and to nationalise their tribal lands in 1958, allowing for substantial agricultural encroachment in rangelands. The Ba'ath party, which came to power in 1963, accentuated this negative trend, writing into the Ba'thist constitution that 'nomadism is a primitive social state . . . The party strives for the sedentarisation of nomads [and] for the abolition of tribal custom'.[101]

It appears from the available archival records that the French Mandate authorities did not recognise Hema for what it was, given their focus on pacifying and sedentarising the Bedouin for much of the Mandate period. The steppe and desert regions of Syria did not come under French control until the early 1930s. Even then, large parts of the steppe lands remained outside their control, and it is very likely that the Hema system, controlled by the Bedouin and not the state, continued during the Mandate period and into the 1940s.[102] Nature conservation, in the form of Hema, was thus probably hiding in plain sight during the Mandate period, practised by the Bedouin tribes in Syria. This indigenous form of nature protection – like local rangeland conservation systems more generally – is rarely recognised in the literature on nature protection.

It is noteworthy that Al Talila is only about 15 miles from Palmyra, a key node of control ever since the Mandate period. Palmyra, historically a prosperous oasis trade centre and a key desert caravan station, was where the Contrôle Bédouin was founded in 1921.[103] By 1931 the first 'Assistance médicale au desert' clinic was operational in Palmyra and was sending out mobile units to better reach and pacify small villages and the nomads.[104] Palmyra was also the headquarters of the earliest Méhariste Camel Company (one of only three of these specialised military units formed in the early 1920s), and by the early 1930s was home to an Air Force squadron and a Foreign Legion outpost. Not coincidentally, the pipelines constructed in 1931 to carry oil from Iraq to the Mediterranean port of Tripoli passed right next to Palmyra.[105] The archival record, however, does not reveal any French attempts at nature conservation in this region during the Mandate period.

Today Palmyra is the site of extremely important natural gas fields and is at the nexus of several pipelines. In 2015, civil war fighting at Palmyra's petroleum installations cut off about 45 per cent of Syria's gas and electricity sources.[106] The year 2015 also saw the civil war hit the Al Talila reserve, and fighting has continued in the area until recently. The location of the first nature reserve in Syria, at the long-strategic location of Palmyra, thus raises questions for future research about how nature conservation, war and control of nomads may intersect – a topic that deserves greater scrutiny.[107]

Acknowledgements

For all their help, I would like to thank Professor Dawn Chatty, the staff at the Centre des archives diplomatiques de Nantes (CADN), the staff at the Oxford

University Plant Sciences Library, Gregory Barton, Jennifer Harbster, Robert Blecher, Daniel Neep, Ghalia Martini, Mohamed Ali Mekouar and Ibrahim Nahal.

Diana K. Davis, a geographer and veterinarian by training, is now a Professor of History and Geography at the University of California, Davis. Her publications include the award-winning 2007 book *Resurrecting the Granary of Rome: Environmental History and French Colonial Expansion in North Africa* (Ohio University Press) and *The Arid Lands: History, Power, Knowledge* (MIT Press, 2016). She also co-edited *Environmental Imaginaries of the Middle East and North Africa* (2011, with Edmund Burke III). Her current research focuses on the environmental history of San Francisco.

Notes

1. André d'Aubréville, A. Barbey and E. Barclay (eds), *Contribution à l'étude des réserves naturelles et des parcs nationaux* (Paris: Paul Lechevalier, 1937).
2. Edmund Burke, 'A Comparative View of French Native Policy in Morocco and Syria, 1912–1925', *Middle Eastern Studies* 9(2) (1973), 175–86.
3. Peter Sluglett, 'An Improvement on Colonialism? The "A" Mandates and Their Legacy in the Middle East', *International Affairs* 90(2) (2014), 413–27.
4. Daniel Neep, *Occupying Syria under the French Mandate* (Cambridge: Cambridge University Press, 2012).
5. David W. Lesch, *Syria* (Cambridge: Polity Press, 2019), 47–55.
6. William Cleveland and Martin Bunton, *A History of the Modern Middle East* (Boulder, CO: Westview Press, 2009), 217–25.
7. Philip S. Khoury, *Syria and the French Mandate: The Politics of Arab Nationalism, 1920–1945* (Princeton, NJ: Princeton University Press, 1987), 58–59.
8. Cleveland and Bunton, *A History*, 232.
9. Ibid., 225–28.
10. There are many references in archival materials to successful French precedents in North Africa that proved to be helpful in the new Mandate. For one influential example, citing Augustin Bernard's work on nomads in Algeria, see p. 14, Service des renseignements, *Les tribus arabes de Syrie, edition provisoire* (Confidential, not to be reproduced), Beirut: Haut-Commissariat de la République française en Syrie et au Liban, 1922, Centre des Archives Diplomatiques de Nantes (CADN).
11. Diana K. Davis, *Resurrecting the Granary of Rome: Environmental History and French Colonial Expansion in North Africa* (Athens: Ohio University Press, 2007).
12. Gouvernement général de Madagascar, 'Arrêté promulguant dans la colonie de Madagascar et Dépendances le décret du 28 août 1913, établissant le régime forestier applicable à Madagascar', *Journal officiel de Madagascar et dépendances* 18 (October 1913), 1202–9; Pierre Montagne and Alain Bertrand, 'Histoire des politiques forestières au Niger, au Mali et à Madagascar', in Alain Bertrand (ed.), *L'État et la gestion locale durable des forêts en Afrique francophone* (Paris: CIRAD, 2006), 54–83.

13. Gouvernement général de l'Afrique occidentale française (GGAOF), 'Arrêté promulguant en Afrique occidentale française le décret du 4 juillet 1935, sur le régime forestier de l'Afrique occidentale française', *Journal officiel du Sénégal* 24 (July 1935), 599–606.
14. Association des ingénieurs du génie rural des eaux et des forêts (AIGREF), *Des officiers royaux aux ingénieurs d'État dans la France rurale (1219–1965)* (Paris: Éditions Technique et Documentation, 2001), 207.
15. Francis Evrard, 'Biogéographie indochinoise et réserves naturelles', in André d'Aubréville (ed.), *Contribution à l'étude des réserves naturelles et des parcs nationaux* (Paris: Paul Lechevalier, 1937), 257–61.
16. G. Petit, 'Les "Réserves naturelles" de Madagascar', in André d'Aubréville (ed.), *Contribution à l'étude des réserves naturelles et des parcs nationaux* (Paris: Paul Lechevalier, 1937), 229–39.
17. AIGREF, *Des officiers royaux*, 213.
18. Diana K. Davis, 'Enclosing Nature in North Africa: National Parks and the Politics of Environmental History', in Alan Mikhail (ed.), *Water on Sand: Environmental Histories of the Middle East and North Africa* (Oxford: Oxford University Press, 2012), 159–79, here 159.
19. Arreph El-Khoury, 'Game Preservation in Syria', in UNESCO (ed.), *Symposium on the Protection and Conservation of Nature in the Near East* (Cairo: Middle East Science Cooperation Office, 1954), 55–56.
20. Mohammad Cheikho, 'Syrian Arab Republic', in C. T. S. Nair and H. A. Nour (eds), *Public Sector Forestry Institutions in the Near East* (Cairo: FAO, 2011), 124–32.
21. Roy H. Behnke and Mike Mortimore (eds), *The End of Desertification? Disputing Environmental Change in the Drylands* (Dordrecht: Springer, 2016); Diana K. Davis, *The Arid Lands: History, Power, Knowledge* (Cambridge, MA: MIT Press, 2016).
22. Diana K. Davis, 'Imperialism, Orientalism, and the Environment in the Middle East: History, Policy, Power, and Practice', in Diana K. Davis and Edmund Burke III (eds), *Environmental Imaginaries of the Middle East and North Africa* (Athens: Ohio University Press, 2011), 1–22.
23. R. J., 'Nécrologie: Pierre Monnet (1893–1967)', *Revue forestière française* 11(1) (1967), 723–24.
24. Pierre Monnet, 'Une mission forestière en Syrie', *Revue des eaux et forêts* 62 (1924), 445–54. This ubiquitous narrative of deforestation in the region has been undermined by recent paleo-ecological data that shows some deforestation in some areas but no deforestation in many others. The reality of environmental change over time is much more complicated than was recognised during the colonial/Mandate period. Lara Hajar, '*Cedrus libani* (A. Rich) Distribution in Lebanon: Past, Present and Future', *Comptes rendus: Biologies* 333(18) (2010), 622–30.
25. A. T. (Dick) Grove and Oliver Rackham, *The Nature of Mediterranean Europe: An Ecological History* (New Haven, CT: Yale University Press, 2001); Oliver L. Pescott et al., 'The Forest on the Peninsula: Impacts, Uses and Perceptions of a Colonial Legacy in Cyprus', in Ana Isabel Queiroz and Simon Pooley (eds), *Histories of Bioinvasions in the Mediterranean* (New York: Springer, 2018), 195–217.
26. David E. Hutchins, *Report on Cyprus Forestry* (Nicosia: Government of Cyprus, 1901), 48, emphasis in the original.
27. Monnet, 'Une mission forestière', 453.

28. It is important to note that the term 'reforestation' usually signifies planting trees in areas that were previously deforested. The term 'afforestation' is reserved for planting trees where they have not grown within historical memory. The term reforestation, though, is often used ideologically to try to claim non-forested land as forest land that has been 'deforested', even when there is no evidence of deforestation.
29. Roza I. El-Eini, *Mandated Landscape: British Imperial Rule in Palestine, 1929–1948* (London: Routledge, 2006), 191.
30. Diana K. Davis, 'From the Divine to the Desertified: The Foundational Case of Deserts in the Middle East', *Global Environment* 12(1) (2019), 56–83; Selçuk Dursun, 'Forest and the State: History of Forestry and Forest Administration in the Ottoman Empire' (PhD diss., Sabanci University, 2007).
31. M. Lagarde, 'Code Forestier', in *Recueil des actes administratifs du Haut-Commissariat de la République française en Syrie et au Liban* (Beirut: Imprimerie Jeanne d'Arc, 1935), 426–69.
32. Ministère des affaires étrangères de la République française (MAE), *Rapport sur la situation de la Syrie et du Liban (année 1933)* (Paris: Imprimerie nationale, 1934), 128.
33. Lagarde, 'Code Forestier'.
34. Anon., *La Syrie et le Liban sous l'occupation et le Mandat français 1919–1927* (Nancy: Berger-Levrault, 1927); MAE, *Rapport sur la situation*, 27.
35. Lagarde, 'Code Forestier', 469.
36. Ibid.
37. Ibid., 432–33, 449–51 and 457. The Syrian Forest Code appears to have been drafted, in part, by Antoine Salha, who had studied law in Montpellier and went on to became the Chief of the Forest Service in Lebanon in the 1930s. Victor Maitland, *Report on Forestry and Soil Conservation in Syria* (Cairo: British Middle East Office (BMEO), 1947).
38. MAE, *Rapport sur la situation de la Syrie et du Liban (année 1938)* (Paris: Imprimerie nationale, 1939), 132–33. It does not appear, though, that serious fires were frequent, nor that lighting forest fires as a form of protest was as significant or widespread in the Mandate territories as it was in French Algeria.
39. Cheikho, 'Syrian Arab Republic', 126.
40. MAE, *Rapport sur la situation de la Syrie et du Liban (année 1937)* (Paris: Imprimerie nationale, 1938), 127.
41. Maitland, *Report on Forestry*, 9.
42. Herbert F. Mooney, *Some Observations on Forestry in Lebanon* (Beirut: BMEO, 1957), 31.
43. MAE, *Rapport sur la situation de la Syrie et du Liban (année 1936)* (Paris: Imprimerie nationale, 1937), 157.
44. Philippe Assouad, *Le reboisement: Son rôle agricole et économique en Syrie et au Liban* (Beirut: École française d'ingénieurs, 1945), 42.
45. Davis, *The Arid Lands*.
46. Ibid.
47. Davis, *Resurrecting the Granary of Rome*.
48. Assouad, *Le reboisement*.
49. While forests in humid continental ecosystems like parts of France and Europe sometimes do function in this way, they do not in most drylands around the globe.

50. Paul Parmentier, *L'agriculture en Syrie & en Palestine* (Le Mans: Imprimerie Monnoyer, 1922), 36–37. Planting trees in dryland environments like most of Syria, in fact, has been shown to cause numerous problems including aridifying soils, lowering water tables and disrupting soil ecology and nutrient flows. Joseph W. Veldman et al., 'Comment on "The Global Tree Restoration Potential"', *Science* 366(6463) (2019), 7976.
51. For other iterations of these ideas, see Anon., *La Syrie et le Liban sous l'occupation*, and Mohammed Sarrage, 'La nécessité d'une réforme agraire en Syrie' (PhD diss., University of Toulouse, 1935).
52. Assouad, *Le reboisement*, 46.
53. Maitland, *Report on Forestry*, 4.
54. MAE, *Rapport sur la situation de la Syrie et du Liban (année 1924)* (Geneva: Société des nations, 1925), 55.
55. MAE, *Rapport sur la situation de la Syrie et du Liban (année 1933)* (Paris: Imprimerie nationale, 1934), 151.
56. I. Maklouf, 'Le Liban: Pays du tourisme et de la villégiature', *La revue du Liban et de l'Orient méditerranéen* 6(41) (1934), 12–13.
57. Karim Makdisi, 'The Rise and Decline of Environmentalism in Lebanon', in Alan Mikhail (ed.), *Water on Sand: Environmental Histories of the Middle East and North Africa* (Oxford: Oxford University Press, 2012), 207–29.
58. MAE, *Rapport sur la situation de la Syrie et du Liban (année 1937)* (Paris: Imprimerie nationale, 1938), 156.
59. MAE, *Rapport sur la situation de la Syrie et du Liban (année 1924)* (Geneva: Société des nations, 1925), 80.
60. The origins of this revolt appear to lie in local discontent over changes in land tenure and local politics enforced by the French authorities, not over forestry.
61. Robert Aldrich, *Greater France: A History of French Overseas Expansion* (London: MacMillan, 1996).
62. Christian Velud, 'French Mandate Policy in the Syrian Steppe', in Martha Mundy and Basim Musallam (eds), *The Transformation of Nomadic Society in the Arab East* (Cambridge: Cambridge University Press, 2000), 63–81.
63. Dawn Chatty, 'Environmentalism in the Syrian Badia', in David G. Anderson and Eeva Berglund (eds), *Ethnographies of Conservation: Environmentalism and the Distribution of Privilege* (New York: Berghahn, 2003), 87–99.
64. Carol Hakim, 'The French Mandate in Lebanon', *American Historical Review* 124(5) (2019), 1689–93.
65. Maitland, *Report on a Visit*, 1.
66. Roger Louis, *Ends of Imperialism: The Scramble for Empire, Suez and Decolonization* (London: I. B. Tauris, 2006); Edward Spears, 'The Middle East and Our Policy There', *Journal of the Royal Central Asian Society* 32(1) (1945), 156–65; Board of Trade, *Report of the British Goodwill Trade Mission to Iraq, Syria, The Lebanon and Cyprus* (London: His Majesty's Stationery Office, 1946).
67. Directorate-General of Information Syria, *Syria. Vol. II: Economy and Finance* (Damas: Government of Syria, 1955), 48 and 52. This eastern region is 'naturally' an ecological steppe zone with very few trees.
68. Gregory A. Barton, 'Environmentalism, Development and British Policy in the Middle East, 1945–65', *Journal of Imperial and Commonwealth History* 38(4) (2010), 619–39.

69. International Agricultural Collaboration (IAC), *Report of the United States–Lebanon Agricultural Mission*, vol. 7 (Washington: Office of Foreign Agricultural Relations, 1948).
70. J. Rolley, 'Forest Conditions in Syria and Lebanon', *Unasylva* 2(2) (1948), 1–6; FAO, 'Forestry in the Middle East', *Unasylva* 6(3) (1952), 1–24.
71. Maarten van Bottenburg, *Report to the Government of Syria on Forestry Development* (Rome: FAO, 1955), 30.
72. Barton, 'Environmentalism'.
73. Van Bottenburg, *Report*, 31.
74. Mahonar Das Chaturvedi, *Report to the Government of the United Arab Republic on a National Forest Policy for Syria* (Rome: FAO, 1961), 23.
75. Mooney, *Some Observations*, 11.
76. Chaturvedi, *Report*, 10, ii.
77. Barton, 'Environmentalism'; S. Ravi Rajan, *Modernizing Nature: Forestry and Imperial Eco-Development 1800–1950* (Oxford: Clarendon Press, 2006).
78. Diana K. Davis and Paul Robbins, 'Ecologies of the Colonial Present: Pathological Forestry from the *taux de boisement* to Civilized Plantations', *Environment and Planning E: Nature and Space* 1(4) (2018), 447–69; Joanny Guillard, *Au service des forêts tropicales: Histoire des services forestiers français d'outre-mer, 1896–1960* (Nancy: AgroParisTech, 2014 [1999]).
79. Davis, *The Arid Lands.*
80. Franz Heske, *German Forestry* (New Haven, CT: Yale University Press, 1938), 173.
81. Richard Grove, *Green Imperialism: Colonial Expansion, Tropical Island Edens and the Origins of Environmentalism, 1600–1860* (Cambridge: Cambridge University Press, 1995).
82. Richard Grove, *Ecology, Climate and Empire* (Cambridge: White Horse Press, 1998), 30.
83. Charles Guyot, *L'Enseignement forestier en France: L'école de Nancy* (Nancy: Crépin-Leblond, 1898).
84. Robert S. Troup, *Colonial Forest Administration* (Oxford: Oxford University Press, 1940), 397–98.
85. A. Unwin, 'The Cyprus Forests and Forest Department', *Empire Forestry Journal* 10(2) (1931), 218–24.
86. Gérard de Montrichard, 'L'Île de Chypre', *Revue des eaux et forêts* 13 (February 1874), 37.
87. Gary S. Dunbar, 'The Forests of Cyprus under British Stewardship', *Scottish Geographical Magazine* 99(2) (1983), 111–20.
88. Hutchins, *Report on Cyprus*, 11–13.
89. El-Eini, *Mandated Landscape*, 191.
90. Hebert F. Mooney, *A Further Note on Forestry in Syria* (Beirut: BMEO, 1952).
91. Hebert F. Mooney, 'Forest Conservancy in the Near and Middle East', in UNESCO (ed.), *Symposium on the Protection and Conservation of Nature in the Near East* (Cairo: Middle East Science Cooperation Office, 1954), 99–105.
92. Géraldine Chatelard, 'ONG environmentales arabes et gestion des aires protégées: Des acteurs entre histoires nationales et paradigmes mondialisés', in Caroline Abu-Sada and Benoît Challand (eds), *Le développement, une affaire d'ONG? États et bailleurs dans le monde arabe* (Paris: Karthala-IREMAM, 2011), 191–225; Makdisi, 'The Rise and Decline', 207–29.

93. Cheikho, 'Syrian Arab Republic', 124–32; Gianluca Serra, Douglas Williamson and Caterina Batello, *From Indifference to Awareness: Encountering Biodiversity in the Semi-Arid Rangelands of the Syrian Arab Republic* (Rome: FAO, 2003).
94. Chatty, 'Environmentalism'.
95. Ministry of State for Environmental Affairs, *The Fourth National Report on Biodiversity in the Syrian Arab Republic* (Damas: Syrian Arab Republic, 2009).
96. Brandon Anthony and Diane Matar, 'Protected Areas in Selected Arab Countries of the Levant Region (Syria, Lebanon & Jordan): An Evaluation of Management and Recommendations for Improvement', in Tony Povilitis (ed.), *Topics in Conservation Biology* (Rijeka: InTech, 2012), 1–26.
97. Lisa Triulzi, 'The Bedouin between Development and the State: A Syrian Case Study', *Arab World Geographer* 5(2) (2002), 85–101.
98. Lutfallah Gari, 'A History of the Hima Conservation System', *Environment and History* 12(2) (2006), 213–28.
99. Dawn Chatty, 'Assumptions of Degradation and Misuse: The Bedouin in the Syrian Badiya', in Dawn Chatty (ed.), *Nomadic Societies in the Middle East and North Africa Entering the 21st Century* (Leiden: Brill, 2006), 737–54.
100. Françoise Debaine and Ronald Jaubert, 'The Degradation of the Steppe, Hypotheses and Realities', *Arab World Geographer* 5(2) (2002), 124–40; Jon Rae, 'Pastoralists, Present and Future: Rangeland Development in West Asia and North Africa', in Stephen G. Reynolds and John Frame (eds), *Grasslands: Developments, Opportunities, Perspectives* (Enfield: Science Publishers/FAO, 2005), 383–99.
101. As quoted in Jonathan Rae et al., *Tribes, State, and Technology Adoption in Arid Land Management, Syria* (Washington, DC: International Food Policy Research Institute, 2001), 16.
102. Personal communication from Professor Dawn Chatty, 18 December 2021.
103. Neep, *Occupying Syria*.
104. Robert Blecher, 'Desert Medicine, Ethnography, and the Colonial Encounter in Mandatory Syria', in Nadine Méouchy and Peter Sluglett (eds), *The British and French Mandates in Comparative Perspective* (Leiden: Brill, 2004), 248–68; Cyrus, 'L'Armée française du Levant: Son œuvre civilisatrice', *Le monde colonial illustré*, June 1937, 150–51. Medical assistance was provided to some Bedouins by small plane deliveries as early as 1921–22. Anon. 'L'Aviation militaire française au Levant', *Le Figaro*, 1922, 6.
105. Christina P. Grant, *The Syrian Desert: Caravans, Travel and Exploration* (New York: MacMillan, 1938), 294–95.
106. Yezid Sayigh, 'The War over Syria's Gas Fields', *Diwan: Middle East Insights from Carnegie*, 2015, http://carnegie-mec.org/diwan/60316, retrieved 13 April 2022.
107. Guillaume Blanc, *The Invention of Green Colonialism* (Cambridge: Polity, 2022); Bridget L. Guascari, 'The National Park: Reviving Eden in Iraq's Marshes', *Arab Studies Journal* 23(1) (2015), 128–53.

Bibliography

Aldrich, Robert. *Greater France: A History of French Overseas Expansion*. London: MacMillan, 1996.

Anthony, Brandon, and Diane Matar. 'Protected Areas in Selected Arab Countries of the Levant Region (Syria, Lebanon & Jordan): An Evaluation of Management and Recommendations for Improvement', in Tony Povilitis (ed.), *Topics in Conservation Biology* (Rijeka: InTech, 2012), 1–26.

Association des ingénieurs du génie rural des eaux et des forêts (AIGREF). *Des officiers royaux aux ingénieurs d'État dans la France rurale (1219–1965)*. Paris: Éditions technique et documentation, 2001.

Barton, Gregory A. 'Environmentalism, Development and British Policy in the Middle East, 1945–65'. *Journal of Imperial and Commonwealth History* 38(4) (2010), 619–39.

Behnke, Roy H., and Mike Mortimore (eds). *The End of Desertification? Disputing Environmental Change in the Drylands*. Dordrecht: Springer, 2016.

Blanc, Guillaume. *The Invention of Green Colonialism*. Cambridge: Polity, 2022.

Blecher, Robert. 'Desert Medicine, Ethnography, and the Colonial Encounter in Mandatory Syria', in Nadine Méouchy and Peter Sluglett (eds), *The British and French Mandates in Comparative Perspective* (Leiden: Brill, 2004), 248–68.

Burke, Edmund. 'A Comparative View of French Native Policy in Morocco and Syria, 1912–1925'. *Middle Eastern Studies* 9(2) (1973), 175–86.

Chatelard, Géraldine. 'ONG environmentales arabes et gestion des aires protégées: Des acteurs entre histoires nationales et paradigmes mondialisés', in Caroline Abu-Sada and Benoît Challand (eds), *Le développement, une affaire d'ONG? États et bailleurs dans le monde arabe* (Paris: Karthala-IREMAM, 2011), 191–225.

Chatty, Dawn. 'Environmentalism in the Syrian Badia', in David G. Anderson and Eeva Berglund (eds), *Ethnographies of Conservation: Environmentalism and the Distribution of Privilege* (New York: Berghahn, 2003), 87–99.

———. 'Assumptions of Degradation and Misuse: The Bedouin in the Syrian Badiya', in Dawn Chatty (ed.), *Nomadic Societies in the Middle East and North Africa Entering the 21st Century* (Leiden: Brill, 2006), 737–54.

Cheikho, Mohammad. 'Syrian Arab Republic', in C. T. S. Nair and H. A. Nour (eds), *Public Sector Forestry Institutions in the Near East* (Cairo: FAO, 2011), 124–32.

Cleveland, William, and Martin Bunton. *A History of the Modern Middle East*. Boulder, CO: Westview Press, 2009.

Davis, Diana K. *Resurrecting the Granary of Rome: Environmental History and French Colonial Expansion in North Africa*. Athens: Ohio University Press, 2007.

———. 'Imperialism, Orientalism, and the Environment in the Middle East: History, Policy, Power, and Practice', in Diana K. Davis and Edmund Burke III (eds), *Environmental Imaginaries of the Middle East and North Africa* (Athens: Ohio University Press, 2011), 1–22.

———. 'Enclosing Nature in North Africa: National Parks and the Politics of Environmental History', in Alan Mikhail (ed.), *Water on Sand: Environmental Histories of the Middle East and North Africa* (Oxford: Oxford University Press, 2012), 159–79.

———. *The Arid Lands: History, Power, Knowledge*. Cambridge, MA: MIT Press, 2016.

———. 'From the Divine to the Desertified: The Foundational Case of Deserts in the Middle East'. *Global Environment* 12(1) (2019), 56–83.

Davis, Diana K., and Paul Robbins. 'Ecologies of the Colonial Present: Pathological Forestry from the *taux de boisement* to Civilized Plantations'. *Environment and Planning E: Nature and Space* 1(4) (2018), 447–69.

Debaine, Françoise, and Ronald Jaubert. 'The Degradation of the Steppe, Hypotheses and Realities'. *Arab World Geographer* 5(2) (2002), 124–40.

Dunbar, Gary S. 'The Forests of Cyprus under British Stewardship'. *Scottish Geographical Magazine* 99(2) (1983), 111–20.

Dursun, Selçuk. 'Forest and the State: History of Forestry and Forest Administration in the Ottoman Empire'. PhD diss., Sabanci University, 2007.

El-Eini, Roza I. *Mandated Landscape: British Imperial Rule in Palestine, 1929–1948*. London: Routledge, 2006.

El-Khoury, Arreph. 'Game Preservation in Syria', in UNESCO (ed.), *Symposium on the Protection and Conservation of Nature in the Near East* (Cairo: Middle East Science Cooperation Office, 1954), 55–56.

Gari, Lutfallah. 'A History of the Hima Conservation System'. *Environment and History* 12(2) (2006), 213–28.

Guascari, Bridget L. 'The National Park: Reviving Eden in Iraq's Marshes'. *Arab Studies Journal* 23(1) (2015), 128–153.

Grove, Richard. *Green Imperialism: Colonial Expansion, Tropical Island Edens and the Origins of Environmentalism, 1600–1860*. Cambridge: Cambridge University Press, 1995.

———. *Ecology, Climate and Empire*. Cambridge: White Horse Press, 1998.

Grove, A. T. (Dick), and Oliver Rackham. *The Nature of Mediterranean Europe: An Ecological History*. New Haven, CT: Yale University Press, 2001.

Guillard, Joanny. *Au service des forêts tropicales: Histoire des services forestiers français d'outre-mer, 1896–1960*. Nancy: AgroParisTech, 2014 [1999].

Hajar, Lara. '*Cedrus libani* (A. Rich) Distribution in Lebanon: Past, Present and Future'. *Comptes rendus: Biologies* 333(18) (2010), 622–30.

Hakim, Carol. 'The French Mandate in Lebanon'. *American Historical Review* 124(5) (2019), 1689–93.

Khoury, Philip S. *Syria and the French Mandate: The Politics of Arab Nationalism, 1920–1945*. Princeton, NJ: Princeton University Press, 1987.

Lesch, David W. *Syria*. Cambridge: Polity Press, 2019.

Louis, Roger. *Ends of Imperialism: The Scramble for Empire, Suez and Decolonization*. London: I. B. Tauris, 2006.

Makdisi, Karim. 'The Rise and Decline of Environmentalism in Lebanon', in Alan Mikhail (ed.), *Water on Sand: Environmental Histories of the Middle East and North Africa* (Oxford: Oxford University Press, 2013), 207–29.

Montagne, Pierre, and Alain Bertrand. 'Histoire des politiques forestières au Niger, au Mali et à Madagascar', in Alain Bertrand (ed.), *L'État et la gestion locale durable des forêts en Afrique francophone* (Paris: CIRAD, 2006), 54–83.

Neep, Daniel. *Occupying Syria under the French Mandate*. Cambridge: Cambridge University Press, 2012.

Pescott, Oliver L., et al. 'The Forest on the Peninsula: Impacts, Uses and Perceptions of a Colonial Legacy in Cyprus', in Ana Isabel Queiroz and Simon Pooley (eds), *Histories of Bioinvasions in the Mediterranean* (New York: Springer, 2018), 195–217.

Rae, Jonathan. 'Pastoralists, Present and Future: Rangeland Development in West Asia and North Africa', in Stephen G. Reynolds and John Frame (eds), *Grasslands: Developments, Opportunities, Perspectives* (Enfield: Science Publishers/FAO, 2005), 383–99.

Rae, Jonathan, et al. *Tribes, State, and Technology Adoption in Arid Land Management, Syria*. Washington, DC: International Food Policy Research Institute, 2001.

Rajan, S. Ravi. *Modernizing Nature: Forestry and Imperial Eco-Development 1800–1950*. Oxford: Clarendon Press, 2006.

Sayigh, Yezid. 'The War over Syria's Gas Fields'. *Diwan: Middle East Insights from Carnegie*, 2015. Retrieved 1 August 2024 from http://carnegie-mec.org/diwan/60316.

Sluglett, Peter. 'An Improvement on Colonialism? The "A" Mandates and Their Legacy in the Middle East'. *International Affairs* 90(2) (2014), 413–27.

Triulzi, Lisa. 'The Bedouin between Development and the State: A Syrian Case Study'. *Arab World Geographer* 5(2) (2002), 85–101.

Veldman, Joseph W., et al. 'Comment on "The Global Tree Restoration Potential"'. *Science* 366(6463) (2019), 7976.

Velud, Christian. 'French Mandate Policy in the Syrian Steppe', in Martha Mundy and Basim Musallam, *The Transformation of Nomadic Society in the Arab East* (Cambridge: Cambridge University Press, 2000), 63–81.

CHAPTER 6

Between Empire and Development

The Ubiquitous Life and Career of Arthur Hugh Bunting

Joseph M. Hodge

In 1976, an article with the intriguing title 'Ubiquitous Agronomist' appeared in the UK-based international science magazine *New Scientist*. The article chronicled the career of a university professor by the name of Arthur Hugh Bunting, who had 'devoted his life to improving third world agriculture' and seemed to be everywhere at the same time in his determination to do so.[1] At the time, Bunting was chair of Agricultural Development Overseas at the University of Reading (in England), a position created specifically for him and funded by the UK's Overseas Development Administration (ODA). He would hold this unique position – as the article noted, there was no comparable chair anywhere in the world – until his retirement from academic life in 1982.[2] Much of his time and energy as chair was devoted to extending the benefits of the so-called green revolution 'in some way or other', but especially through setting up various international research institutes and initiatives under the umbrella of the Consultative Group on International Agricultural Research (CGIAR).[3]

The author of 'Ubiquitous Agronomist' provides her readers with some insight into the background and early years of the highly respected Professor Bunting. She notes, for example, that as a young man Bunting had been 'profoundly affected' by a pamphlet written by his father on the future of the African continent; that he was selected to be the Rhodes Scholar for the Transvaal in 1938; earned a DPhil in plant physiology at Oxford in 1941; worked as the Chief Scientific Officer of the infamous East African Groundnut Scheme in Colonial Tanganyika in 1947–51; and moved from there to British-controlled Sudan to set up the Central Rainlands Research Station at Jebel Tozi, before accepting the position of chair of agricultural botany at Reading in 1956.

Absent from the piece, however, is any mention of the surprisingly rich history of the Bunting family and their roots in Wesleyan Methodism and nineteenth-century radical Liberalism, or of the fact that Hugh's father, Sidney Percival Bunting, played a leading role in the founding of the Communist Party of South Africa (CPSA) and that Hugh himself was for a time a registered member of the Communist Party of Great Britain (CPGB). In point of fact, he was dismissed from the Groundnut Scheme after being accused of orchestrating a communist plot among its staff, and was consequently debarred from further employment in the British colonial empire. Such anecdotes may seem ill-fitted to a narrative about the life and illustrious career of a world-renowned agricultural scientist, someone appointed a Companion of the Order of St Michael and St George (CMG) for his services to international agriculture. Hugh himself seems to have contributed to the censoring and glossing over of his youthful digressions and his family's subversive past.

Yet as out of character as these episodes may at first appear, re-situating them within the storyline is essential, not only for understanding the complexity and ambiguity of Bunting as an individual, but more importantly, I would argue, for understanding the complexity and ambiguity of the history of international development, to which so much of his life and career was dedicated. This chapter is about the colonial and postcolonial career of Arthur Hugh Bunting, and the transnational connections of several generations of the Bunting family, which help place his career within a historically deeper and spatially broader context.

But more than this, to put a twist on a line from David Lambert and Alan Lester, this study 'is about some of the ways that individual people made the history of [development], and some of the ways that [development] made them'.[4] It is also a prime example of why studying 'development' is essential for understanding environmental change and conservation in the 'tropics'. The scale of natural resource extraction and management involved in postwar initiatives like the Groundnut Scheme meant, as Tom Robertson observes, that wherever 'development' took place, it transformed nature and remade landscapes radically.[5]

Bunting lived a colourful and unusual life, there is no doubt, but it was by no means singular or unique. He was one of many former colonial experts and technical advisers who subsequently went on to have 'second' or 'postcolonial' careers working in international development or other related fields overseas. In the case of British colonialism, many officers who were hired after the Second World War transitioned to working for international organisations, or else for UK-based overseas aid agencies, after they retired. Many also continued serving in former colonial territories for a number of years following independence.

Bunting's fellow scientist E. B. Worthington is a prime example. Worthington is best known for his groundbreaking study, *Science in Africa*, published in 1938 as the companion volume to Lord Hailey's *African Research Survey*, in which he conceived of ecology as the base science that connected all other fields of knowledge about Africa.[6] After the war, Worthington worked as a scientific and development adviser in sub-Saharan Africa, commissioned, for example, to prepare Uganda's first five-year development plan in 1946, and then appointed the first Secretary-General of the intercolonial Scientific Council for Africa South of the Sahara (CSA) from 1950 to 1955. In 1958, with the assistance of the CSA and the Commission for Technical Cooperation in Africa South of the Sahara (CCTA), Worthington produced a revised version of his earlier compendium, titled *Science in the Development of Africa*. As the updated name suggests, emphasis was placed on how research contributed to development, stating firmly that 'there will be no sound development without a thorough application of scientific methods and long-term basic research'.[7] Significantly, Worthington went on to become a leading international ecologist, first as scientific director of the UK's Nature Conservancy from 1957 to 1965, and then scientific director of the International Biological Program (IBP) from 1964 to 1974. Bunting and Worthington had been colleagues during their time in East Africa, and they remained in touch on returning to the UK. Bunting, Worthington and other international actors were part of a 'diaspora of sciences' (as described by Guillaume Blanc in the introduction) at the heart of the transnational circulations between Europe, Africa and Asia, and that came to define the history of nature management and environmental intervention in the twentieth century. It also reminds us that the history of conservation and development cannot be separated, as the same professionals often interacted, participating in overlapping networks of expertise and working to construct policies that sought to reconcile the imperatives of both.

The jump from late colonial to postcolonial development expert was particularly pronounced among professional personnel in departments such as agriculture, forestry, surveying, veterinary science and conservation. Their level of specialised knowledge and years of experience in the field made them highly qualified to join the burgeoning network of intergovernmental organisations such as the United Nations and its subsidiary bodies and specialised agencies, or to be hired as advisers by international companies and consultancy firms.

In other words, these individuals' careers were not simply imperial or even transnational, but also transhistorical. Their career patterns and life histories establish connections between different places and times that conventionally have been thought of as separate and distinct geographical spaces or historical temporalities. Most crucially, their careers and work experiences reveal an important human (and not simply ideological or institutional) thread of continuity across the seemingly fundamental divide of decolonisation and the end of empire.

In plotting the lifepath and career trajectory of Bunting, this chapter argues that former British colonial personnel continued to be involved in international affairs in varied ways. Drawing on recent approaches in historical geography and the 'new imperial history', this study examines the family, life history and career of Arthur Hugh Bunting, a man whose personal and professional experiences and connections spanned the late colonial–postcolonial divide and linked transcontinental spaces throughout Africa with Britain, in ways both expected and not so expected, for more than eighty years.

In this way, I wish to show that Bunting's background and past experiences were brought to bear on new sites, but also how his engagement with those sites helped to change his perspective on African rural development in important ways. In particular, his experience in tropical Africa imparted to him a lasting appreciation for studying ecology and for having sufficient time to experiment and build up knowledge through pilot schemes in such difficult environments before jumping headlong into large-scale projects and top-down interventions.[8]

The Colonial and Postcolonial 'Careering' of Arthur Hugh Bunting

Writing a broader and more deeply rooted history of development requires not just the inclusion of a more heterogeneous range of experiences and perspectives into our narratives. It will require researchers, as Corinna Unger suggests, to challenge the historical methodology further by thinking more profoundly about how to write histories of development that adopt a transnational and comparative framework of analysis; histories that do not leave states out, but enable us to see through and even beyond them in ways that will allow us to overcome what she describes as the prevailing 'methodological nationalism' of many studies.[9] Examining the circulations of technical experts, scientists, project managers and engineers, which as Suzanne Moon notes became increasingly global in scope as they moved across countries and between states and various commercial and intergovernmental organisations, is one way of opening up new and more critical lines of inquiry about the wider history of development.[10]

Retracing the careering of Arthur Hugh Bunting, together with the multiplicity of family connections and circulations that helped form the many worlds between and through which he travelled (both spatially and imaginatively), provides fertile ground for exploring this transnational history. 'A.H. Bunting still gives the impression of a man in a hurry', wrote the author of 'Ubiquitous Agronomist' in the 1970s, 'a single-handed yachtsman on a very large and lumpy sea'.[11] Like the image of a man who appears to be everywhere at the same time, that of a 'single-handed sailor' performing multiple duties and navigating his vessel through treacherous seas is an apt analogy for the lifepaths that colonial experts-turned-development practitioners like Bunting were able to chart and steer.[12] In reconstructing the multiple, interlocking family, religious, professional and scientific networks and circuits through which generations of Buntings journeyed and at times sojourned, my analysis identifies three 'hubs' or 'nodes' that were crucial in the making of Arthur Hugh Bunting's colonial and postcolonial career:

(1) The first is a text written by his father in 1933, when Hugh was 17 years old and in his first year of study at the University of Witwatersrand, titled *An African Prospect and Appeal to Young Africa, East, West, Central, South*. Hugh claimed on several occasions that this was the most influential piece of writing on his life. He believed it was addressed partly to him, which profoundly affected his psyche after his father's premature death in 1936. It convinced him of the importance of being a 'working scientist' dedicated to the practical development of Africa, with a sense of responsibility for the continent. I see this text as a sort of moment in time that functions as a site or node through which all the circulations, connections and relations of generations of the Bunting family flowed.

(2) The second 'node' is the ill-fated East African Groundnut Scheme in Tanganyika of 1947–51. This was one of the largest and best-known of the late colonial development schemes implemented by the British, through the Overseas Food Corporation, which was set up just after the war to rapidly increase the supply of colonial products for export. It was the first 'development project' that Hugh Bunting was involved in directly, serving as the Chief Scientific Officer for the scheme. Again, we can think of the scheme as a 'hub' through which various networks of individuals, practices and ideas intersected and overlapped.

(3) Our third moment occurs during Bunting's postcolonial tenure as Chair and Dean of Agriculture at the University of Reading. In 1968, he organised the first of two international conferences at Reading on change in agriculture, bringing together the world's leading agricultural scientists and technologists, together with colleagues in the social sciences, to discuss the challenges of rural development in the 'less-developed world'. 'It is unlikely' Bunting noted at the time, 'that so authoritative an interdisciplinary group has ever previously been assembled'.[13] Through these conferences – as well as his advising and consultancy work for various institutions and research centres, and his leadership in training a whole generation of students from the UK, Africa and other 'developing regions' – Bunting forged his own personal and professional networks of 'working scientists' who reshaped agricultural and environmental science and development in Africa and Asia.

Family Background and Early Years in South Africa

In trying to come to terms with Arthur Hugh Bunting and who he was, I have found myself having to peel back several generations of the Buntings' family history and the ideas associated with them. This is because the name of Bunting has long been associated with reform and even revolution. The Buntings were originally from Manchester, England. William Bunting, a tailor and 'thorough Radical', was born in Monyash, Derbyshire, and later settled in Manchester, where he married Mary Redfern in 1778. Originally from Over Haddon near Bakewell, Mary was the first in her family to be 'awakened' and converted to Methodism. Their only son, Jabez Bunting, would become the most prominent Wesleyan Methodist minister of his generation, elected to serve as president of the annual Wesleyan Methodist Conference four times between 1820 and 1851 when he retired.[14]

Under Bunting's leadership, with his firm belief in church order and discipline, Wesleyan Methodism was institutionalised, and revivalist challenges put down. Yet he began his career as an itinerant preacher and is remembered as the founder of the Wesleyan Methodist Missionary Society. After many years on the circuit, he was appointed secretary of foreign missions and settled in London in 1833. He became president of the Theological Institution for seminary training a year later.[15] Though he never ventured beyond Britain's shores, his name and influence was felt across early nineteenth-century Methodist missionary circuits and sites. It is, for example, in his honour that the mission station of Buntingville (or Buntingdale) in what is today the Eastern Cape of South Africa was founded by W. B. Boyce in 1830.[16]

Wesleyan Methodism remains an important thread connecting the larger Bunting family. The same can be said for Victorian liberal social reform. Hugh's grandfather, Percy William Bunting, was editor of the *Contemporary Review* and a prominent Victorian liberal and lawyer involved in social and political reform, and the welfare of modern Methodism. He devoted himself to the Forward Movement, which endeavoured to organise nonconformity as a national religious force. He was active in the National Vigilance Association and the moral purity campaign, as was his wife, Mary Hyett Lidgett, who opposed 'white slavery' (sex trafficking) and was an early member of the National Union of Women Workers. His sister, Sarah McLardie Amos, was also active in many of these public causes.

And through Hugh's grandmother's side of the family, he was related to Rev. John Scott Lidgett, president of the United Methodist Conference of 1932 and founder of the Bermondsey Settlement. Through the Lidgetts, the family had other South African connections: Lidgett & Sons was a shipping company, which set up the Lidgett Land Company in Natal to bring Wesleyan Methodist settlers to South Africa to settle on plots. The scheme failed but the enterprise was rebranded as the Lidgett Land Company and Wattle Plantation. Another cousin, Frank Lidgett McDougall, spent two years in South Africa helping run the wattle plantation until he emigrated to Australia in 1909. He served as economic adviser to Lord Stanley Bruce (Australian prime minister, 1923–

29), and was one of the driving forces, along with Lord John Boyd Orr, in the founding of the Food and Agriculture Organization in 1945.

One way to write the story of Arthur Hugh Bunting is to see his life and career as the culmination of a long family history and tradition of Methodist reform including missionary civilising impulses and liberal social reform causes, leading logically to twentieth-century international humanitarianism and development. But there is a twist to this story. Hugh's father, Sidney Percival Bunting, was one of the founding members of the Communist Party of South Africa (CPSA). He was educated at Magdalen College, Oxford, reading in 'Greats' (classical literature, philosophy and *literae humaniores*) and graduating in 1894. He articled with a firm of solicitors, Freshfields, and was admitted to the Supreme Court in 1900.

But rather than follow the conventional path for a man of his position, Sidney chose to become an officer in the British forces and was deployed to South Africa during the Anglo-Boer War. He developed an interest in languages, learning Afrikaans as well as Zulu. After the war, Sidney decided to remain in South Africa, not returning to England again until 1922 on his way to Moscow for the Fourth World Congress of the Communist International. He went to South Africa as a conservative imperialist, Hugh writes, but while there he began to take on 'the labour point of view'.[17] He was admitted as a solicitor of the Supreme Court of the Transvaal in 1902, and became associated with the South African Labour Party (ISL) in Johannesburg. He also sat as a Labour member in the Legislative Assembly of the Transvaal. His socialist politics led him to declare against the war in 1914.

In 1916, Sidney married Hugh's mother, Rebecca Notlowitz, a Jewish émigré from Lithuania, and a year later Hugh was born in Johannesburg. Sidney believed in humanist universalism, self-improvement and the missionary 'family of man'.[18] Yet, interestingly, Hugh recalls that there was no religion in the home while he was growing up. In the context of early twentieth-century Johannesburg, it seems that international Communism had replaced Wesleyan Methodism as a guiding life force.

In 1919, the ISL became the Communist Party of South Africa and developed links with the Comintern. Hugh's father was active in the 1922 Rand Strike in which white miners protested the acquisition of skills by black workers for work underground. He worked as a dispatch motorcycle rider for the strikers and was arrested. It was around this time that Hugh first went to school. It was also around this time that the family went back to England, and Hugh and his brother, Brian, stayed with their aunt while their parents went to Moscow. On their return, life began to change. The movement towards African workers and African trade unions became more significant. His father was active in meetings and as a lawyer representing African clients. Fewer and fewer white friends visited the family in the 1920s, but African visitors increased in number. Hugh remembers visiting African homes and playing with African children, as the family became more and more socially isolated from white society.

In 1927, Sidney stood for election to Parliament in Transkei but did not win; after the campaign the family's house on Regent Street was sold, and they went to live in a poorer area in Bezuidenhout Valley. In 1928, they returned again to Europe for the Sixth Comintern Congress. The party line for South Africa was now in favour of a 'native republic'. Sidney accepted this position, but he also defended the place of white South Africans, whom he felt played an important part and could not simply go somewhere else.

His initial opposition to the black republic thesis left Sidney a marked man in the Comintern's eyes, and on the family's return to South Africa he met with increasing hostility in the CPSA. He continued to speak publicly, mostly to black audiences, despite threats of attack by nationalists. There were bitter divisions and recriminations within the Party. A new generation of Africans were rising to leadership roles, men like Moses Kotane, J. B. Marks and Edwin Thabo Mofutsanyana, who targeted Sidney, who was secretary of the Party. In 1931 he was expelled from the Party, leaving him a broken man whose health began to steadily deteriorate. He suffered a series of strokes and died of a cerebral haemorrhage in May 1936.

As Allison Drew writes: 'Bunting's life demonstrates that individual members of the imperial diaspora could play a radicalising role.'[19] His identity, as the title of Drew's biography of Sidney Bunting reminds us, was forged somewhere between and through empire and revolution. I am struck by A. H. Bunting's family background, so deeply connected with the provincial, middle-class Englishness about which Catherine Hall has written so authoritatively, as well as by the rich imperial connections of the Bunting family, from Jabez to Sidney and Sidney's brother Sheldon (a civil engineer in India), and to Hugh himself; theirs was a family of empire. The evangelical Christian universalism of his great-great-grandfather, as well as his grandfather's liberal social reformism, seem to have shaped Hugh's father's imperial and international humanitarian principles. As Drew puts it, 'Once again, he turned to the words he had heard in his childhood: this time it was "the bitter cry of outcast Africa" of which he wrote.'[20] Only for Sidney it was law and politics rather than religion that served as the vehicles for his activism. As Hugh observed, his father 'saw poverty as something to be cured by social change, not passively endured'.[21]

What is interesting for the purposes of this study is how Sidney Bunting's legacy impacted and helped shape the life and career of his eldest son, Hugh. One of the last things Sidney did in life was write a short pamphlet in 1933, *An African Prospect and Appeal to Young Africa*,[22] which, as noted earlier, greatly influenced Hugh. This piece is a diatribe against the foreign capitalist exploitation of the assets of Africa and the serfdom and dispossession of the continent's indigenous inhabitants of the soil, and the resulting economic and ecological degradation. The shareholders are parasites. It is also a critique of liberal imperialism, which he says has left the native Africans penned in remote and congested reserves, and their own development restricted. It is not enough, he writes. He calls instead for a 'New Order', one that will give all people of the continent and the different peoples of the world a concern in the world's wealth and will exclude all privileges of class or race. The Soviet Union is the example that is breaking down the citadel of imperialist ideas and empire worship.

Bunting then envisions the possibility of a Soviet Africa, and what this might mean: A Union of African Workers and a Peasants' Republic. He cautions that this needs to be a realistic application of their lessons and principles to African conditions; and he criticises the liberal imperialist's passion for 'Five-Year Plans for Africa'. What is needed is a planned administration for Africa with all its assets by and for the direct benefit of Africa, black and white alike, on a pan-African scale. Every African patriot is to be called upon to do their part, to overthrow imperialist rule and set up the new order. He imagines a pan-African State Planning Commission and Supreme Economic Council, which would secure the best technical advice and scientific 'patriotism'; the sincere services of specialists, engineers, organisers, managers and auditors of the whole continent; research to maximise existing production; control of the banks, transport, factories and workshops; and an agricultural and hydrographic survey and irrigation plans on a continental scale. He sees large-scale agricultural industry at the highest technical level, and collective farming throughout vast open spaces that will revolutionise rural life and standards beyond recognition. This will include universal education, housing, public health, medical services, sanitation and social welfare. As Sidney concludes:

> The demand of a nation of a hundred and twenty million human beings for release from their chains must someday sweep the length and breadth of the whole African continent like a tidal wave: the mightier its sweep, the more irresistible its power and the speedier and more bloodless its victory . . . May it come quickly![23]

These are the powerful words that inspired Hugh Bunting to make a career as an international scientific researcher, teacher and advocate for African development. With these words and the bitterness of his father's death still fresh in his mind, he decided to leave South Africa behind and go into self-imposed exile in England. As Hugh concludes his memoir of his father:

> For humane principle . . . he had given up, stage by stage, all that he had been brought up to expect and to achieve. For his own reasons, he had virtually

> abandoned his family ties. His hopes for the white workers of South Africa were entombed in the Colour Bar Act passed by the Nationalist/Labour coalition in 1928 . . . Most of his former white friends had recoiled from his political ideals and marriage. In the end he was cast out . . . [even] by some of the Africans to whose service he had devoted his last years.[24]

Hugh Bunting's Early Career and Involvement in the East Africa Groundnut Scheme

If Hugh's father's career was one shaped by socialist politics, the law and trade unionism in South Africa, Hugh took a very different path, though one influenced, as his memoir to his father suggests, by the same broad humanitarian principles. Hugh would not follow his father into politics, but instead was interested in a career in science. Although he began his career inspired by the same socialist and communist influences that had shaped his father's life and activism, as time went by Hugh's ideology and philosophical approach to the world were increasingly framed by the notion of international scientific cooperation – an idea that attracted many scientists of Bunting's generation, from John Boyd Orr and E. J. Russell to Julian Huxley and E. B. Worthington.

It was in the final years of his father's life that Hugh went to university and decided to devote himself to applying natural science to improving the lives and agriculture of the peoples of Africa. He earned his BSc and MSc at the University of Witwatersrand in botany and physiology between 1934 and 1938, with an emphasis on fieldwork under the direction of the renowned South African ecologist J. F. V. Phillips. Bunting credited Phillips with turning him into 'a botanist by profession and conviction'.[25]

Then, in 1938, he was awarded a Rhodes Scholarship for the Transvaal, and entered Oriel College, University of Oxford, where he studied plant biochemistry under W. O. James. He worked on the metabolism of carbohydrates under anaerobic conditions in seedling barley, earning his DPhil in 1941. At Oxford, he also met his future wife, Elsie Muriel Reynard, through their common membership in the Communist Party. Muriel was a student at Lady Margaret Hall, and they married in 1941 and had three sons.

Perhaps the most significant appointment in terms of his future career was as an Assistant Chemist from 1941 to 1945 at the Rothamsted Experimental Station in Harpenden, England, working under E. M. Crowther. At Rothamsted, he investigated organic manuring by looking at the agronomic effects of sewage sludge. Although he was an agronomist, the influence of ecology and field-based studies on his thinking can be seen in the way he viewed agricultural science as the pinnacle of sciences because it depended on the integration of ideas from many different disciplines.[26]

It is interesting to note that Bunting and his wife Muriel were being watched by the Secret Services at this time, who were aware that he had been in contact with the Communist Party since 1939 and was known to be a member since 1942.[27] In 1942, for example, as a member of the Harpenden branch of the CP, he organised a public meeting at the Friends Meeting House in Harpenden to mark the one-year anniversary of the USSR entering the war.[28] The meeting was sponsored by the Left-Wing and Labour Club. Muriel, Bunting's wife, was noted for holding strong left-wing political views.

In 1943 he was elected to the Executive Committee of the Association of Scientific Workers, and in 1945 he was elected as secretary of the CP's National Scientific Advisory Bureau, remaining active in Party scientific affairs until the end of 1946, and making frequent visits to Party HQs. At the annual meeting of the Association of Scientific Workers in 1945, Bunting stressed the need to develop the earth's resources to meet the peoples' needs, claiming that, if properly used, science could create wealth for the people, rather than allowing 'monopolists' interests', as in the past, to 'pervert and frustrate' science by the control of production.[29]

After the war, Hugh felt the need to return to Africa, although not to South Africa where he realised it would have been impossible for him to avoid politics. He was selected to be a member of the Medical Research Council's human nutrition research unit from 1945 to 1947, where he worked on agricultural aspects of nutrition in the tropics, and visited Gambia twice.[30] It was at this time that he began referring to himself as 'Hugh' rather than 'Arthur'. Then, in 1947, at the height of the 'second colonial occupation' and postwar Fabian development drive, he was appointed Chief Scientific Officer for the Overseas Food Corporation's (OFC) infamous East African Groundnut Scheme, in Tanganyika.

The Communist Party did not want him to take the job because they felt he would only be serving imperialist interests, but Hugh felt that it was an opportunity to put his belief in the power of scientific production into practice and thus might do some good. One can imagine how the calling of his father to all African patriots must have influenced his thinking here. He was, as his father advocated, going to help introduce scientific, large-scale agricultural industry at the highest technical level 'throughout vast open spaces' of Tanganyika, and in the process revolutionise African rural life and standards beyond recognition.

It is interesting to note that Hugh's old mentor at Witwatersrand, Professor John Phillips, was also involved in the scheme, hired at the end of 1948 to be the Agricultural General Manager, and later promoted to joint General Manager.[31] The scheme failed, according to Bunting, 'for the same sorts of reasons which would prevent a baby who could scarcely crawl from winning the 200 metres hurdles'.[32] In other words, very little research had been done on the feasibility of the scheme prior to its implementation. In June 1951 Bunting was dismissed, in part because he had openly criticised the Minister of Food, John Strachey, for not being honest with the British public about the scheme. He was also barred from further official employment with the British Colonial Office and service.

At least that is the story on the face of it, but it is worth probing deeper into the circumstances surrounding Bunting's dismissal and exclusion from British colonial Africa. In December 1949, the Minister of Food, John Strachey, made an unannounced visit to East Africa amid mounting criticism and crisis of the Groundnut Scheme. There he investigated an alleged communist plot to raise trouble among the scheme's employees. Both the district commissioner and Strachey believed that Bunting was the person behind the 'criminal activities', as they described them in a cable, which involved among other things discussing Marxism and dialectical materialism with Tanganyikan 'natives'.[33]

Strachey's visit to Kongwa was in response to the threatened resignation of several heads of OFC departments and their demand for the resignation of the chairman, Leslie 'Dick' Plummer. The rebellion, according to the deputy chairman, had been deliberately engineered by Bunting. Along with Bunting, the minister implicated Barbara Hill, secretary to Alan Woods, in the plot. Strachey felt the situation was serious enough to discuss the matter with the prime minister as well as with security services. The Colonial Office decided they did not want Bunting in Tanganyika, and they encouraged Special Branch to find 'continuing Communist associations on Bunting's part' in order to strengthen their hand.[34] In response to Strachey's request, security services began an investigation of Bunting's activities and whereabouts, which lasted for nearly a decade. They kept a close watch on Bunting's travels and contacts, although in the end they concluded he did not constitute a threat. There was no evidence he was any longer in contact with the CP, and the charges against him by Strachey were found to be baseless.[35]

We might see this event as a smear campaign against Bunting, fabricated by Strachey and others who were trying to deflect criticism for the failure of the scheme away from themselves, by blaming it on some sort of communist agitation. Or possibly, Bunting really was working to undermine the scheme by radicalising the African workers involved. It is possible that he was working within this imperial project on one level, but also working against it on another – an example of working within and against imperialist designs, by sowing the seeds of their demise.

Whether or not the accusations brought against Bunting were true, this episode effectively put an end to his colonial career. What is also interesting is that the scheme did not shake his faith in large-scale technical development, although it did inject a strong dose of humility in him.[36] All his life he continued to hold an optimistic view of the potential of science and development to improve the condition of humanity. As the article cited at the beginning of the chapter noted: 'In spite of many errors . . . he stoutly defended the idea of such a project – in the right circumstances.'[37]

Following the 'Groundnut Affair',[38] Bunting's services were sought by Dr Kenneth Mellanby, the first principal of Ibadan University, Nigeria, on the recommendation of none other than Professor Phillips, who had left the scheme in 1951 to join the staff of the University of the Gold Coast. It seems, however, that the offer was dropped because the West African Security Office objected to accepting persons of communist sympathies for employment in the teaching profession. Despite the ban on employing him within the colonial empire, Hugh was appointed through the Foreign Office as a Senior Research Officer working for the Ministry of Agriculture in Sudan from 1951 to 1956, where he set up and headed the Central Rainlands Research Station at Jebel Tozi. Secret Services concluded that although he still kept in touch with communist scientists in the UK, he was not under the direction of the CP and it was unlikely he would lend himself to communist interests in the Sudan.

What I find interesting about the second 'hub' or moment in Hugh Bunting's career is the potentially subversive nature of his ideas and activities. Hugh was never really a 'colonial scientist' per se so much as a scientist who worked in the colonies and used the imperial networks through which he travelled to further his goal of using science in the service of development. I also find the lineage of humanitarianism that runs throughout Bunting's family history – from Victorian civilising mission to socialist internationalism to international scientific cooperation – intriguing.

Bunting's Time as Chair and Dean of Agriculture at the University of Reading

The final and arguably most influential 'hub' in Hugh's life history began in 1956, when he left the 'tropics' to take up the position of Chair of Agricultural Botany at the University of Reading. He would spend the rest of his career at Reading, helping to establish the university as an important centre for international agricultural research and education. Through his connections and travels, Bunting helped build the global professional networks of international agricultural science and rural development.

By the 1960s he had become a highly sought-after consultant, serving for example on UNESCO's commission to the University of the West Indies to advise on the future of the former Imperial College of Tropical Agriculture. He was instrumental in the founding of the International Institute of Tropical Agriculture (IITA) at Ibadan, Nigeria in 1967, and later served as chairman of its board of trustees. Bunting was also a member of the council of the new Ahmadu Bello University in Nigeria until 1968, and in 1967 was part of a commission on agricultural education in Kenya, which led to the creation of the faculty of agriculture at the University of Nairobi.

Hugh saw himself as an 'applied scientist'. Many of his publications throughout his career utilised an ecological framework or displayed an ecological bent.[39] He was joint founding editor of the *Journal of Applied Ecology* from 1964 to 1968. His most significant contribution during the Reading years was the organising of two important international seminars on agricultural change and rural development: 'Change in Agriculture' in 1968, followed by 'Policy and Practice in Rural Development' in 1974.[40] Both conferences contributed to debates over increasing biological and agricultural productivity through the so-called 'Green Revolution' and its impact on the environment and farmers in developing countries. In particular, many of the participants at these seminars helped establish what became known as Farming Systems Research (FSR), including Bede N. Okigbo, David W. Norman and Mike P. Collinson.[41]

Apart from cultivating connections among like-minded scientists and practitioners through hosting professional forums, Bunting exerted a tremendous influence on the last generation of colonial agricultural scientists. For a brief period in the 1960s, colonial probationary officers were required to take a course at Reading as part of the agricultural studentships sponsored by the UK's Ministry of Overseas Development, and many of them studied under Bunting. Many of his students also had careers overseas working for international agencies such as the World Bank, the FAO and the UNDP. Bunting also served two terms as dean of the Faculty of Agriculture and Food at Reading from 1965 to 1971. As dean, he was instrumental in building up a corps of faculty who, like him, had substantial scientific field experience working overseas in tropical regions of the former colonial empire.

Bunting stepped down as Chair of Agricultural Botany in 1974, taking on the mantle of a new role as Professor of Agricultural Development Overseas, a position funded by the UK's Overseas Development Administration (ODA) and specially created for him. Bunting held this position until he retired in 1982. It was during this final tenure that Hugh became particularly influential in the work and activities of the Consultative Group on International Agricultural Research (CGIAR), arguing that the world could produce enough food to feed the projected global population increases, but only if a political solution could be found to distribute the food from productive to unavoidably unproductive regions. In this sense, he continued to follow in the footsteps of Lord John Boyd Orr, the first director general and visionary of the FAO. Even in retirement, Bunting remained active as a consultant on agricultural research and development until his death in 2002 at the age of 85.

The life and career of Arthur Hugh Bunting is a particularly rich example of the links between colonialism and development, and of the remobilisation of former colonial personnel by international organisations and institutions in the afterlife of empire. Of course, not all former colonial recruits came from the same kind of background as Bunting. What the case of Hugh Bunting reminds us is that not all former colonial technical officers and scientific practitioners shared the same upbringing and early life experiences. One can detect certain patterns, such as the large number of personnel whose families had strong ties and connections to the empire, but one also finds a great deal of variation and many unexpected twists and turns. While it would be misleading to suggest that the radical roots and potentially subversive views held by Bunting were common, there were others who were motivated by similar desires to help improve the lives of colonial peoples.

What is also interesting about Bunting is how he reimagined his father's past, as well as how others remembered his own past. For example, in his 'Memorials of My Father', Hugh writes that his father, Sidney Bunting, was more of a realist than an idealist, and was not in any intellectual sense a 'Marxist'.[42] Hugh believed that his father was more of a 'pragmatic nineteenth-century English liberal' who, in the context of South Africa in the early twentieth century, found that his views placed him on the extreme left of the political spectrum where the only people who would listen and engage with him were communists and socialists. In this sense, Hugh was not at all surprised that in the end his father was expelled from the CPSA.

However, a close reading of the sources makes it clear that this was simply not the case, as Hugh knew full well. It is clear from *An African Prospect* as well as from letters Sidney wrote to his wife, Rebecca, that both he and Rebecca were committed communists, who read and knew Marx's writings very well.[43] It is also true that Hugh's brother, Brian, and his mother remained committed communists throughout their lives. It seems Hugh was not being entirely honest about his past. It is interesting in this sense that by the 1970s and 1980s his communist past and strong socialist views about development that he had held in his earlier years were being erased from his life's narrative, as if the origins of development were being rewritten and sanitised.

I think it is important that we not forget the entangled connections between the colonial and early postcolonial eras, nor the intersection of imperialist and anti-imperialist and socialist visions of development, both of which shared a technological faith in science and the state. In this sense, I would concur with Uma Kothari that: 'To produce a postcolonial development that critically re-evaluates development theory and practice and disconnects it from what has variously been termed its neo-colonial, re-colonial and imperial context and articulations requires, as a starting point, a historical analysis that identifies the particularities, and varied expressions, of the continuities between colonialism and development.'[44] The research presented here, it is hoped, offers a more critical understanding of these continuities, and of the overlapping histories of science, decolonisation, development and conservation in sub-Saharan Africa.

Joseph Morgan Hodge is Associate Professor of History at West Virginia University in Morgantown, WV. He is the author of *Triumph of the Expert: Agrarian Doctrines of Development and the Legacies of British Colonialism* (Ohio University Press, 2007). As a specialist in the transnational history of development and science, he also co-edited *Science and Empire* (Palgrave Macmillan, 2011, with Brett Bennett) and *Developing Africa* (Manchester University Press, 2014, with Gerald Hödl and Martina Kopf).

Notes

1. Angela Croome, 'In Person: Ubiquitous Agronomist', *New Scientist*, 14 October 1976, 90.
2. Ibid., 92.
3. The CGIAR is a worldwide network of agricultural research centres, founded in 1971, and originally funded by the Rockefeller Foundation, the World Bank, the FAO and the UNDP. It originally supported four centres: the International Maize and Wheat Improvement Center (CIMMYT) in Mexico, the International Rice Research Institute (IRRI) in the Philippines, the International Center for Tropical Agriculture (CIAT) in Italy and the International Institute of Tropical Agriculture (IITA) in Nigeria.
4. David Lambert and Alan Lester, 'Introduction: Imperial Spaces, Imperial Subjects', in David Lambert and Alan Lester (eds), *Colonial Lives across the British Empire: Imperial Careering in the Long Nineteenth Century* (Cambridge: Cambridge University Press, 2006), 1–31, here 1.
5. Thomas Robertson, 'Cold War Landscapes: Towards an Environmental History of US Development Programmes in the 1950s and 1960s', *Cold War History* 16(4) (2015), 417–41, here 423.
6. E. B. Worthington, *Science in Africa: A Review of Scientific Research Relating to Tropical and Southern Africa* (London: Oxford University Press, 1938). See also Peder Anker, *Imperial Ecology* (Cambridge, MA: Harvard University Press, 1997), 214–18.
7. E. B. Worthington, *Science in the Development of Africa: A Review of the Contribution of Physical and Biological Knowledge South of the Sahara* (London: Commission for Technical Co-Operation in Africa South of the Sahara, 1958), x.
8. Nicholas Westcott, *Imperialism and Development: The East African Groundnut Scheme and Its Legacy* (Woodbridge: James Currey, 2020), 192.
9. Corinna R. Unger, 'Comment on Joseph Hodge, On the Historiography of Development (Part I and II)', *Humanity*, 28 April 2016, retrieved 1 August 2024 from http://humanityjournal.org/joseph-hodge-roundtable/.
10. Suzanne Moon, 'Global and Usable Histories of Development', *Humanity*, 2 May 2016, retrieved 1 August 2024 from http://humanityjournal.org/joseph-hodge-roundtable/.

11. Croome, 'In Person', 92.
12. On the notion of the lifepath see: Stephen Daniels and Catherine Nash, 'Lifepaths: Geography and Biography', *Journal of Historical Geography* 30(3) (2004), 449–58.
13. Arthur H. Bunting, *Change in Agriculture* (London: Duckworth, 1970), 6.
14. Thomas Percival Bunting, *The Life of Jabez Bunting with Notices of Contemporary Persons and Events* (London: Longmans & Roberts, 1859), 1–2, 8 and 10–11; W. R. Ward, 'Bunting, Jabez (1779–1858)', in *Oxford Dictionary of National Biography* (Oxford: Oxford University Press, 2007), retrieved 1 August 2024 from http://www.oxforddnb.com/view/article/3947.
15. W. R. Ward, *Early Victorian Methodism: The Correspondence of Jabez Bunting, 1830–1858* (Oxford: Oxford University Press, 1976), xi; Ward, 'Bunting, Jabez'.
16. Reverend Joseph Whiteside, *History of the Wesleyan Methodist Church of South Africa* (London: Elliot Stock, 1906), 193–97.
17. Arthur H. Bunting, 'Memorials of My Father', in Edward Roux, *S. P. Bunting: A Political Biography*, ed. Brian Bunting (Bellville: Mayibuye, 1993), 34–49, here 39.
18. Catherine Hall, *Civilising Subjects: Metropole and Colony in the English Imagination 1830–1867* (Chicago: University of Chicago Press, 2002), 5 and 98.
19. Allison Drew, *Between Empire and Revolution: A Life of Sidney Bunting, 1873–1936* (London: Pickering & Chatto, 2007).
20. Ibid., 135.
21. Bunting, 'Memorials of My Father', 35.
22. Sidney P. Bunting, *An African Prospect and Appeal to Young Africa: East, West, Central, South* (Johannesburg, 1933).
23. Ibid., 30.
24. Bunting, 'Memorials of My Father', 38.
25. Barbara Pickersgill, 'Bunting, Arthur Hugh (1917–2002), Crop Scientist', in *Oxford Dictionary of National Biography* (Oxford: Oxford University Press, 2006), https://www.oxforddnb.com/view/10.1093/ref:odnb/9780198614128.001.0001/odnb-9780198614128-e-77029, retrieved 24 March 2022.
26. This is similar, as noted earlier, to the position taken by E. B. Worthington.
27. 'Minute by C. Grose Hodge', B.l.a., 17.2.50, Secret Services – Arthur Hugh Bunting, Elsie Muriel Bunting, UK National Archives [TNA], KV 2/3250.
28. May 1942, E. J. Russell Collection, MERL Special Collections, University of Reading.
29. 'Extract from Special Branches Report', National Archives, United Kingdom (NAUK), KV 2/3250, 26.4.45, 'Secret Services, Arthur Hugh Bunting, Elsie Muriel Bunting'.
30. Here he was following in the footsteps of John Boyd Orr, who had also studied nutrition in the tropics for the Empire Marketing Board, and then became involved in the movement for a World Food Plan working with Hugh's cousin Frank MacDougall and the League of Nation's Mixed Committee on Nutrition.
31. Westcott, *Imperialism and Development*, 134.
32. Pickersgill, 'Bunting, Arthur Hugh'.
33. 'Note for C.C. 208, 664 – Arthur Hugh Bunting, by G.R. Mitchell' and 'Note by G.R. Mitchell', NAUK, KV 2/3250, B.1. 21.12.49 and B.1. 6.1.50, 'Secret Services, Arthur Hugh Bunting, Elsie Muriel Bunting'.
34. 'Item 65a Note by G.R. Mitchell', NAUK, KV 2/3250, B.1. 14.2.50, 'Secret Services, Arthur Hugh Bunting, Elsie Muriel Bunting'.

35. 'Letter from Maurice Webb, to Ministry of Food', 17 May1950, NAUK, KV 2/3251 'Arthur Hugh Bunting, Muriel Elsie Bunting'.
36. Andrew Bennett, Jeremy Elston and Amir Kassam, 'An Appreciation of Hugh Bunting', *Tropical Agricultural Association Newsletter* (September 2006), 4.
37. Croome, 'In Person', 91.
38. The term 'Groundnut Affair' comes from Alan Wood's seminal account. See Alan Wood, *The Groundnut Affair* (London: Bodley Head, 1950).
39. Bennett, Elston and Kassam, 'An Appreciation', 4.
40. Arthur H. Bunting (ed.), *Change in Agriculture* (London: Duckworth, 1970); Guy Hunter, Arthur H. Bunting and Anthony Bottrall (eds), *Policy and Practice in Rural Development* (New York: Osmun, 1976).
41. See Michael P. Collinson (ed.), *A History of Farming Systems Research* (Wallingford/Rome: CABI/FAO, 2000).
42. Bunting, 'Memorials of My Father', 39.
43. Brian Bunting (ed.), *Letters to Rebecca: South African Communist Leader S.P. Bunting to His Wife, 1917–1934* (Bellville: Mayibuye, 1996).
44. Uma Kothari, 'From Colonial Administration to Development Studies: A Post-Colonial Critique of the History of Development Studies', in Uma Kothari (ed.), *A Radical History of Development Studies* (London: Zed, 2005), 47–66, here 63.

Bibliography

Anker, Peder. *Imperial Ecology*. Cambridge, MA: Harvard University Press, 1997.

Bennett, Andrew, Jeremy Elston and Amir Kassam. 'An Appreciation of Hugh Bunting'. *Tropical Agricultural Association Newsletter*, September 2006.

Collinson, Michael P. (ed.). *A History of Farming Systems Research*. Wallingford/Rome: CABI/FAO, 2000.

Daniels, Stephen, and Catherine Nash. 'Lifepaths: Geography and Biography'. *Journal of Historical Geography* 30(3) (2004), 449–58.

Drew, Allison. *Between Empire and Revolution: A Life of Sidney Bunting, 1873–1936*. London: Pickering & Chatto, 2007.

Hall, Catherine. *Civilising Subjects: Metropole and Colony in the English Imagination 1830–1867*. Chicago: University of Chicago Press, 2002.

Hunter, Guy, Arthur H. Bunting and Anthony Bottrall (eds). *Policy and Practice in Rural Development*. New York: Osmun, 1976.

Kothari, Uma. 'From Colonial Administration to Development Studies: A Post-Colonial Critique of the History of Development Studies', in Uma Kothari (ed.), *A Radical History of Development Studies* (London: Zed, 2005), 47-66.

Lambert, David, and Alan Lester. 'Introduction: Imperial Spaces, Imperial Subjects', in David Lambert and Alan Lester (eds), *Colonial Lives across the British Empire: Imperial Careering in the Long Nineteenth Century* (Cambridge: Cambridge University Press, 2006), 1–31.

Moon, Suzanne. 'Global and Usable Histories of Development'. *Humanity*, 2 May 2016. Retrieved 2 August 2024 from http://humanityjournal.org/joseph-hodge-roundtable/.

Pickersgill, Barbara. 'Bunting, Arthur Hugh (1917–2002), Crop Scientist', in *Oxford Dictionary of National Biography* (Oxford: Oxford University Press, 2006). Retrieved 2

August 2024 from https://www.oxforddnb.com/view/10.1093/ref:odnb/9780198614128.001.0001/odnb-9780198614128-e-77029.

Robertson, Thomas. 'Cold War Landscapes: Towards an Environmental History of US Development Programmes in the 1950s and 1960s'. *Cold War History* 16(4) (2015), 417–41.

Unger, Corinna R. 'Comment on Joseph Hodge, On the Historiography of Development (Part I and II)'. *Humanity*, 28 April 2016. Retrieved 2 August 2024 from http://humanityjournal.org/joseph-hodge-roundtable/.

Ward, W. R. *Early Victorian Methodism: The Correspondence of Jabez Bunting, 1830–1858*. Oxford: Oxford University Press, 1976.

———. 'Bunting, Jabez (1779–1858)', in *Oxford Dictionary of National Biography* (Oxford: Oxford University Press, 2007). Retrieved 2 August 2024 from http://www.oxforddnb.com/view/article/3947.

Westcott, Nicholas. *Imperialism and Development: The East African Groundnut Scheme & Its Legacy*. Woodbridge: James Currey, 2020.

Whiteside, Reverend Joseph. *History of the Wesleyan Methodist Church of South Africa*. London: Elliot Stock, 1906.

Wood, Alan. *The Groundnut Affair*. London: Bodley Head, 1950.

CHAPTER 7

The Adamsons, *Born Free* and the Late Colonial Era

Images That Helped to Change the Animal World

William Beinart

Wildlife literature, photography and film on Africa have long been popular in Western countries and beyond and have been closely connected with wildlife conservation.[1] This chapter focuses on a specific moment in Kenya, during the late 1950s and early 1960s, when popular media projected significant new developments in thinking about wildlife. Joy and George Adamson were protagonists of a change in human relationships with wild animals, particularly lions, whose lives became imbricated with visual media. Their ideas and practices reached a wide audience through illustrated books, the main focus here, and later through film. The material here draws on a broader project, with Katie McKeown, that explores popular writers, film-makers and photographers operating largely in East Africa from 1945 to 1980.[2]

The Adamsons advocated – and to an extent practised, though not without contradictions – a new relationship with wild animals that aimed to build trust, affection and empathy, as well as protection. Their writing and photographs, as well as the films made about them, resonated with popular audiences and helped to reshape attitudes and approaches to animals. I will suggest that these media representations were a significant – and neglected – element of modern environmentalism that took different routes from those that have largely been written into the mainstream of environmental history.

This chapter focuses on the photographic images in three books by Joy Adamson – *Born Free* (1960), *Living Free* (1961) and *Forever Free* (1962) – that sold millions of copies.[3] Joy Adamson was a prolific artist and photographer. Roughly fourteen thousand mostly black-and-white negatives survive, covering the period 1956 to 1962, of which about 330 (2.3 per cent) were published in the three original hardbacks. The early paperbacks had fewer images. Some colour photos and slides were also taken from 1960 and there are an additional *c.* two thousand photos covering the making of the film *Born Free* (1964–65), as well as further sequences from Joy's work with cheetahs and leopards. The negatives were kept in albums, usually with date and place recorded and occasionally with captions. From 1958, when the idea of publication arose, some contact prints and postcard-size enlargements were made so that it was easier to review the collection.[4]

The Adamsons' ideas and images were generated on the peripheries of a fading empire, rather than in the heart of Western society and institutions. George was committed to his role as a colonial official. He led a small patrol of loyalists during the Mau Mau rebellion, and claims to have turned some poachers into game keepers.[5] Unlike some of the settler writing at the time, Joy's texts barely touch on Mau Mau and the myths about it.[6] Nevertheless, the books have been seen as essentially colonial texts and she does use some common stereotypes. I suggest that these are more multidimensional texts, written at a moment of social change. Colonialism – especially at this time – was not a uniform set of practices, and the Adamsons were in some respects on the edges of white colonial society. Their context was specifically African, and the images were generally celebratory. They contributed to generating a new aesthetic, arising largely from a Western vision, but powerfully influenced by locations, animals and people in Africa.

Joy Adamson as Artist

Joy Adamson's adult life was profoundly shaped by living in a British colony and her audience was initially British and American. But she was born Friederike Gessner in Austria in 1910 and spoke English with a strong accent till the end of her life. She studied crafts and art in Vienna, and also worked with a photographer. The name Joy was given by her second husband, botanist Peter Bally, whom she met on a ship to Kenya in 1937; she adopted it in a British colonial world.[7] Over the next six years, Joy focused on botanical painting and travelled to different part of Kenya, often separately from her husband. She had great powers of observation, and her many paintings are precise and colourful. She later recalled, 'I was entranced not only by living under canvas and so close to nature, but even more by having the opportunity to observe . . . plant life'.[8]

In 1942 she met game warden George Adamson (b. 1906), stationed at Isiolo, north of Mt Kenya, who was responsible for the huge, arid Northern Frontier District (NFD).[9] They married in 1944; it was a troubled relationship but provided both with valuable support, ideas and opportunities. Immediately after the war, Joy started sketching African people and, during a visit to London, studied briefly at the Slade School of Fine Art.[10] She was commissioned by Louis Leakey, director of the Coryndon Museum in Nairobi, to paint the indigenous people of Kenya. A limited number of these paintings were done on safaris with George to visit pastoralist communities for his work, but for the most part she again travelled independently.

Part of her task was to record traditional dress. However, for Joy, this was not simply an exercise in documenting vanishing cultures; it was an artistic enterprise. She wrote: 'apart from making an historical record, I was mainly interested in catching the character of my sitters so that I should not only produce an ethnographical dossier but also a collection of living people'.[11] Joy could not communicate easily with most of her subjects. She depended, in her own mind, not simply on observation, but on 'holding the attention of my sitters with my eyes . . . . [I] thus automatically, but unconsciously, . . . was able to convey the spirit of my models'.[12] It is important for this chapter that Joy aimed to give her subjects agency in the way the images were constructed. The roughly six hundred portraits are now mostly divided between State House and the National Museum in Nairobi.

Joy's systematic photography also began with this project, for which about six thousand negatives covering 1947 to 1956 survive in the Nairobi museum. The majority of these are rather formal black-and-white portraits of African people, designed to inform her colour paintings. However, she had a sufficiently alert eye to take photographs of many individuals and groups in more informal settings, such as ngomas (dances). A picture of Samburu women on the move, their donkeys laden with their shelters, records traditional life but also seems to express the confidence of African people at home in their environment. She was using a Leica with limited depth of focus and the closer-up images of African people, especially in strong light, are generally more effective than the landscapes and crowd scenes.

The Early Lion Photographs

Part of George Adamson's duty as a game warden was to deal with wildlife considered dangerous. In February 1956 he shot a lioness that was thought to be attacking people and found it had three small cubs. He brought these home to Isiolo and the couple decided to adopt them. This was not unusual. Several Kenyan settlers provided homes for orphaned wildlife and developed strategies of rearing and feeding them. Joy had visited the farm of Raymond Hook, on the slopes of Mt Kenya near Nanyuki, who kept cheetah.[13] Carr Hartley was well known for the diverse collection of wild animals on his farm in Rumuruti – initially for export to zoos, but increasingly deployed for visiting film-makers.[14] The Adamsons themselves had kept a genet and a tree hyrax.

As soon as the cubs arrived, Joy and George started to photograph them. Entranced by these new arrivals, they also aimed to keep a record of their growth and behaviour. The early images depict domesticity: Joy hugging the cubs (published) or feeding them with bottled milk (unpublished); George playing with them as if they were domesticated kittens.[15] These captured the intimacy of bringing the animals into the household. At this stage, it seems they used only one camera. George played an important role in the photographic project; he took all the images that included Joy with the cubs, and perhaps others.

By July 1956, it was becoming increasingly difficult to manage three growing cubs; two were sent to Rotterdam Zoo but they kept the smallest, which they named Elsa. Joy still took outward-looking pictures of the sort that had interested her up to that point, for instance of African people and landscapes, as well as records of safaris. But the camera increasingly turns inwards. A striking unpublished image shows Elsa at about six months, her head almost the size of Joy's, still being fed by bottle, with intense concentration and affection on both faces.

From about eight months, Elsa became increasingly powerful and boisterous, but they also recorded attempts to train the lioness. For example, Elsa learned to retrieve dead birds and to retract her claws during physical interaction. A published image, captioned 'Elsa goes hunting with George', has them sitting together on a rock, George holding his rifle while Elsa has a paw on

Figure 7.1. Elsa and Joy. © Elsa Conservation Trust (Sunbury-on-Thames, England), reproduced with permission.

him; the lioness is clearly taking agency and the camera aims to capture this. There is a similar trust and intimacy in the picture of George on a camp bed with Elsa resting beside him. His eyes are closed with no defence against what was now an almost fully grown predator. The caption reads 'Elsa usually slept beside George and continually raised her paw to make sure he was still there'.

Figure 7.2. Elsa and Galgallo. © Elsa Conservation Trust (Sunbury-on-Thames, England), reproduced with permission.

Even in modest settler households such as the Adamsons', African staff were generally employed. They were involved in caring for the cubs from the start and are also recorded in the images – particularly Nuru, originally employed as a gardener, Ibrahim, employed as a driver, and Makedde Lobotio. A picture in *Born Free* has all three sitting around Elsa, who is on her back, with one of them tickling her stomach.[16] Like George, they sometimes carried rifles. An unpublished picture of Galgallo, another assistant, depicts him standing in casual clothing, long white slacks and a shirt, with a rifle slung over his shoulder, holding a dead bird, which Elsa reaches for on her hind legs. The photo is effective because of its symmetry as well as the contrast of light. The clothes, stance and rifle do not suggest a servant. Joy to some degree recognised this of Nuru. She wrote: 'we appointed him guardian and lion keeper in chief'.[17]

Images of Elsa as a Young Lioness, Hesitant Freedom

George was committed to his role as a colonial official.[18] Nevertheless, he had little means of enforcing authority in the large NFD and his work required negotiating with pastoralists to reduce hunting. He regularly left his base at Isiolo to traverse this area; his African assistants remember him as speaking Swahili and sitting with them at night.[19]

The Adamsons took Elsa with them on a trip to the coast, driving over 300 km on rough roads from Isiolo to Kiunga near the Somalia border. The reporting of their visit, both in text and images, emphasised the freedom characteristic of beach holidays, which were becoming more widespread in Britain. Five published photos in *Born Free*, chosen out of a longer sequence, show George in shorts and Joy in a bathing costume playing with Elsa on the beach.[20] They advertise that it was possible to manage a 1-year-old lion away from home as a member of the family.

In July 1957, they took Elsa, at sixteen months, on a long safari to Lake Turkana where George was monitoring wildlife and engaging pastoralist communities. Joy, who had visited there before, was clearly fascinated by the El Molo lakeside fishing communities of whom she took some informal photographs. George, as well as Makedde Lobotio, had started to walk with Elsa; images show the lioness both in front and behind them, the latter portraying an extraordinary degree of trust.

From late 1957, the sequence of photographs shows a subtle change in that there are fewer general photographs, and a greater number of repetitive images. The idea of publication was taking shape and Joy was in a sense operating more like a professional photographer. The sequence of photographs is thus a combination of animal portraiture, documentary record, snaps and a record of affection, with more strategic photography for publication.

There were no barriers around their Isiolo property and at 2 years old, from early 1958, Elsa's behaviour became problematic. Although put in a room at night, the lioness was free to roam by day and spent more time away from the Adamsons' house. Elsa had been fed on a diet of wildlife shot by George and the African staff, as well as purchased goats. Now the lioness was beginning to predate on livestock belonging to African communities. Elsa was also attracting the interest of other lions as she pushed the boundaries of her semi-domestication. Joy wrote: 'she had taken her future into her own paws'.[21]

Although they were losing control, this was not an entirely unwelcome development for the Adamsons. While they had tried to modify Elsa's behaviour, they were not aiming to suppress it. George kept a diary in which, as in the photos, they recorded and observed to a degree as scientists. Initially, they considered sending Elsa to Rotterdam, but they became convinced that the lioness could and should be released back into the 'wild' – 'an experiment well worth trying'.[22] They had to find a less populated place than Isiolo, and Maasai Mara, 500 km away in south-western Kenya, seemed the best option. They hoped that Elsa could connect with wild lions, learn to hunt more effectively and establish an existence free from the Adamsons.

Their aim of putting a semi-domesticated lion back into a wildlife reserve, which could now be seen as a forerunner of 'rewilding', echoed other experiments at that time. Young animals were being orphaned on some scale when wildlife was killed in hunting, land clearance and tsetse control. The Sheldricks at Tsavo, the largest national park in Kenya, had started to care for elephant and rhino orphans with the intention of returning them to the park. Joy Adamson had also advocated such orphanages.

The trip to Maasai Mara generated new photographs. Although they lived simply regarding personal possessions, the Adamsons' way of life was dependent on the more general technological underpinnings of colonialism such as vehicles and petrol, firearms, medical facilities and cameras. Land Rovers feature frequently in their photos. George took leave for a few months to settle Elsa in Maasai Mara – the lioness had little experience in hunting wild animals, nor the advantage of being part of a group. He continued to shoot in the reserve, with permission, for Elsa's food supply. After three months, Elsa fell ill, probably with a tick-borne disease. Maasai Mara was lower and wetter than Isiolo and it is likely that Elsa lacked the immunities acquired by local lions.

Had Elsa been able to integrate in this part of south-western Kenya, the Adamsons may well have lost their connection and produced only one book. But this failure gave a new impetus to their enterprise. They were so invested that they negotiated a new site on the shallow Ura River in Meru district, on the eastern side of Mt Kenya, where they set up camp in July 1958. This area became their centre of operations for much of the next decade. George was there sporadically, as he continued in post at Isiolo. Joy initially stayed for longer periods, in a small, tented encampment, with African staff, and she found time to write. She portrayed it as a particularly satisfying time: 'Sitting there with Elsa close to me, I felt as though I was on the doorstep of paradise: man and beast in trusting harmony; the slow-flowing river adding to the idyll.'[23] Few white women in Kenya lived in this way and it required a fearlessness about nature and wildlife. A number of images attempt to capture this peaceable kingdom.

Joy completed a typescript, and in February 1959 left for Europe to see friends and to find a publisher. After a number of rejections, Collins and Harvill, who specialised in natural history and animal books, saw potential; the images were particularly important. When she returned in July 1959, George took a sequence of ten photographs of Joy and Elsa embracing on the ground. It would have been impossible to construct them and proved to the Adamsons that a long-term bond had been created, manifest in a reciprocity of feeling after a four-month absence.

When George and Joy were both away, Nuru and Makedde took charge. Turkana by background, Makedde became the key figure who could cope with the difficult demands of Elsa as well as Joy's sometimes fiery temperament. He is regularly present in the photographs, often with a rifle, sometimes alone with Elsa. In a published image, he stands with George behind Elsa, both shirtless, both with rifles and suggesting an equilibrium – whatever the actual relationship.[24] The photographic record presents him as more than a servant. Makedde worked periodically with Joy for about fifteen years, through her experiments with cheetahs and leopards.

Figure 7.3. Human, nonhuman and affection. © Elsa Conservation Trust (Sunbury-on-Thames, England), reproduced with permission.

Though her health improved with the move north, Elsa was not yet an adequate hunter. A series of photographs depict a dead buffalo, shot by George, first being attacked by Elsa and then dismembered by African staff to make it possible for the lioness to eat fresh meat. George also hoped that the corpse would attract other lions so that Elsa could share in the consumption and 'make friends'. They were sufficiently open about such practices for some of this sequence to be published.[25]

Elsa again moved away for periods and partly learned to feed herself. In September 1959, she mated with a wild lion, and gave birth to three cubs in December (110 days average gestation period). This was cathartic for the Adamsons and central to their project because it proved that Elsa had renegotiated herself back into the society of wild lions and could reproduce successfully. The coverage of *Born Free* was extended to this period, with a postscript announcing the birth of the cubs. The final few photos in that book are in colour.

There is clearly an attempt to locate the book in Kenya and it is not, as is sometimes suggested in critiques of wildlife photography and film, simply a version of the myth of wild Africa without people.[26] Of the 110 photos in the *Born Free* hardback, twenty-one include African people, mostly the Adamsons' staff, but there are five of others. Twelve images are of Joy with the cubs, or Elsa alone, and nineteen of George. The photographs were not intended as examples of wildlife photography, but it is important to recognise that this well-established genre was expanding quickly during the 1950s in magazines and books. Popular new series were shown on British TV, such as David Attenborough's *Zoo Quest* and Armand and Michaela Denis's *On Safari* (with Des Bartlett). These aimed to introduce wild animals and their behaviour to mass audiences, so the images in *Born Free* fed into a growing visual literacy about African wildlife.[27]

Wildness and Separation

After *Born Free* was published, Joy Adamson wrote two further books detailing their relationship with the lioness and her cubs. They were also published by Collins and followed a similar format with substantial texts and over a hundred photographs in the initial hardbacks; although Joy had more colour images available, most photographs were black-and-white.

About six weeks after the birth of the cubs, in February 1960, Elsa started to bring them across the Ura River to the Adamsons' camp. This was an epiphany for them, showing that Elsa could transcend the boundaries between human and wild animal spheres. They took more photographs in 1960 than in any other year and many show the lions crossing the shallow river. Not only had the lioness maintained her affection for them, but they believed she was trying to transmit it to her cubs.[28] However, the cubs (although also named) never bonded with the Adamsons in the same way as Elsa. Food at the Adamsons' camp was one attraction, but the cubs were not dependent – as Elsa had been – on being fed and mothered by humans from within a week of birth. Arguably the relationship with Elsa was only possible because the lioness had imprinted on humans from the start.

The images from this time are still largely inward-looking, focused on their immediate environment, and reflect the emotional intensity of the Adamsons' experience. The publication of *Born Free* in March 1960 brought a substantial new flow of money, and this also partly explains the increasing number of photos. But even when they did get an additional camera and telephoto lens, Joy could not easily photograph Elsa and the cubs when they left camp; some shots are a blur of trees and grass.

One colour sequence shows publisher Billy Collins visiting in May 1960. He had an affair with Joy, and Elsa – perhaps out of jealousy – put her jaws around the top of his skull when he was asleep in his tent. He nearly lost his head. By this time the Adamsons' relationship was increasingly fragile, although they still worked closely together with the lions. One unpublished image shows Makedde, placed centrally with a rifle and equipment, between Collins and Joy. He is holding Elsa's tail and seems to be depicted as the figure of authority. Another from the sequence was published in the hardback of *Living Free*, but this does not include Joy, and Collins is made the central figure.[29]

Joy was writing *Living Free* as the events that it recorded were happening, knowing that it was eagerly anticipated by the publishers and that illustrations were invaluable. The Adamsons had also started to take their own film, and Collins and Harvill arranged for this footage to be edited into a ten-minute documentary to promote *Born Free*. It was collated by Harry Watt – an established feature-film maker who had directed *Where No Vultures Fly* (1951), set in Kenya.[30] This was an incentive for them to expand their own filming.

David Attenborough visited to make a documentary for the BBC. Up to this time, Attenborough had largely worked outside Africa for his series *Zoo Quest*.[31] His was the first extended film treatment of the Adamsons' experiment and brought further publicity in the UK. Although the film was largely sympathetic, Attenborough was sceptical about their experiment and formulated a critique. He later commented on the underlying sense of danger and violence in their relationship with Elsa and her cubs and he was also aware of the conflict between George and Joy.[32]

Attenborough's own filming had included interactions between people and wild animals, but he was increasingly concerned about semi-domestication. He also felt that the Adamsons put Elsa's needs above those of all other animals, and even other people: 'there are certainly contradictions in thinking one individual animal is so important that you will kill other individual animals in order to support it.'[33] The Adamsons recognised such tensions, and Joy did not hide the shooting that was needed to provide food for Elsa and her cubs. As noted, some of these photos were published.

At this time, the BBC was beginning to aim at wildlife image-making that was 'true to nature'.[34] The Natural History unit, launched in 1957, aimed to remove people from wildlife film and focus on the animals. They wanted to be guided by a scientific, educational approach, to avoid anthropomorphising or attributing human emotions and motives to animals, and to minimise a focus on human–animal relationships. While the BBC was not consistent in these aims, they nevertheless sat uneasily with the approach taken by the Adamsons. And while the latter also saw their work as in part scientific, and their longer-term goal as enabling wildlife to live in the wild, *Born Free* focused on a human–animal emotional relationship.

Adrian House, their biographer, suggested that the Adamsons recognised, at a time when some ethologists and scientists were becoming sceptical, that animals could have emotions of kinds. House himself wrote, in tribute to George Adamson: 'if we open our eyes we shall see . . . that we are sharing our world . . . [with] a galaxy of living creatures endowed with many of the same faculties as we are, and with others that we have lost or never possessed – all of which entitle them to freedoms and rights no different from our own'.[35]

The Adamsons believed that there could be new relationships between people and wild animals, based on empathy rather than fear and aggression. Joy's writings are suffused with this idea and she certainly interpreted Elsa's return with the cubs as more than an instrumental relationship in search of food and security. Such ideas have persisted in literature and feature film, and have been revived in the new animal-centric academic field of animal studies.[36]

Living Free covers 1960 – a shorter but intensively photographed phase. It focuses on Elsa raising her cubs, negotiating the space between the camp and the landscape around it, and the new relationship that the Adamsons tried to develop with the lions. The book lacks the same dramatic narrative, first of semi-domesticating a lion cub and then of trying to reduce its dependence – or, in their terms, setting it free. Nevertheless, the images of lions and people in the same frame remain powerful. Although *Living Free* started with the epiphany of the river crossings, it was more focused on gradual separation.

For a female lion not fully incorporated in a pride, a life in the wild was difficult, and Elsa, weakened by disease and by a fight with a wild lion in mid-1960, died in January 1961. The cubs were frequently photographed and they appear in the majority of pictures in *Living Free*. Most depict playfulness and interaction, but by then over 1 year old, they returned to the camp infrequently. There were Tharaka African communities to the south-west and *Forever Free* (1962) starts with a complaint by the local African district council about the danger posed by the cubs. One persistent critique of their experiment was that habituated lions were more dangerous to humans and livestock than wild lions, because they had lost their fear of people. Having arranged to move them to Serengeti National Park in Tanzania, the Adamsons captured the cubs. There is a dense photographic record of their translocation, in cages, which forms an important theme in *Forever Free*.

This was an irony, as the cubs had become increasingly independent of the Adamsons, but wildness for predators as dangerous as lions had its place. The idea of a 'wild' space without humans was, as the literature on East African environmental history shows, largely a figment of historical imaginations in that people and predators had long lived in some relation. By 1961, with the expansion of human population and settlement, all space in Kenya was to some degree circumscribed. The concept of wildness was in practice a relative one, indicating less populated spaces or those reserved as national parks.[37] Certainly there is an argument for a fluid concept of wildness and wilderness, as well as of boundaries between animals and people.

In Serengeti, the Adamsons retained a close relationship with Elsa's cubs for a short period, when George hunted for them (with reluctant permission from the park authorities) because they could not yet fend for themselves. Joy recorded this too in photographs. The final chapters are a rather forlorn chronicle of the Adamsons' attempts to keep in touch with the cubs as they learned to fend for themselves. The photographs in this section of the hardback first edition are increasingly general images of wildlife and landscape that differ significantly from the first two books. They lack the close focus or the emotional power of the earlier images.

In some senses, their experiment achieved their aim – the cubs' incorporation into a national park, without the need for human support. But they did not know for definite that the cubs survived, and the full separation caused them a great deal of personal sadness. Joy was shifting her role. She had worked intensively, writing three books in three years, and felt she should further publicise their enterprise. She left in 1962 on a long international tour to raise money, talk about their experiences and advocate wildlife conservation. George made some further unsuccessful forays in Serengeti to find the cubs. *Forever Free* was the least successful of the trilogy.

Promotion and Review of the Books

Born Free was successful because of its content. It was also heavily promoted by Collins and Harvill and widely reviewed. The *Sunday Times* included two extracts on consecutive weeks before the release date.[38] Serial rights were sold to the *Daily Mail*, and the *Reader's Digest* printed a condensed version. The publishers risked a large first print run of twenty-eight thousand copies in hardback, unusual for an unknown author. In the US, Pantheon Press published a separate edition, and the book was serialised. The films by Watt and Attenborough attracted further attention. Watt was keen to make a feature film in which Elsa would star, but the Adamsons refused because the lioness was so far down the road to independence.

From the first, the photographs were noticed and identified as essential to the books' impact.[39] Travel writer John Hillaby in *The Guardian*,[40] conservationist and writer Gerald Durrell in the *New York Times*[41] and biologist Bentley Glass acknowledged the images for bringing authenticity to a story that might otherwise have been less convincing. The latter suggested that the book 'might well have been regarded as a natural history hoax were it not for the extraordinary photographs'.[42] The images enhanced its appeal for readers across generations.

The first UK print run of *Born Free* sold out quickly; there were three further runs in April 1960 and nine by December. In the US the book spent thirteen weeks at the top of the *New York Times* Best Seller list, and nearly a year on the chart overall.[43] A newspaper article in New York recorded that by the release of the film in 1966, the books had been translated into twenty-one languages with an estimated readership of fifty million people.[44] Adrian House, who worked for Collins, noted in his biography (1993) that they rapidly sold five million in 'a dozen' different languages.[45] A later report claimed that within a year, *Born Free* had sold over six million copies and been translated into twenty-five languages.[46] The precise numbers may be exaggerated but it quickly reached a wide audience. Joy summarised the story in a children's book of photographs – *Elsa: The True Story of a Lioness*. The success of *Born Free* meant that the first run of *Living Free* (1961) was one hundred thousand.[47] When they lost touch with the cubs in Serengeti, Joy travelled on publicity and fundraising tours.

A striking feature of the *Born Free* saga was its repetition in different contexts. The visual record of the Adamsons' experience was dramatically expanded with the film, released in 1966. Director James Hill, producer Carl Foreman and the two stars, Virginia McKenna and Bill Travers, were determined to replicate the relationship between the Adamsons and Elsa as closely as possible. They employed George Adamson to work with a number of semi-domesticated lions, so that the actors could convincingly portray the Adamsons' earlier experiences. During the nine months of filming in Naro Moru, Kenya, the actors, wife and husband, developed a strong bond with George and some of the lions; this changed the course of their lives.[48]

On the release of *Living Free*, Gavin Maxwell tried to account for the popularity of *Born Free*. He had published *The Ring of Bright Water* in 1960, about an otter, which scored a similar success, though not on the same scale. He saw this as a result of the human need for a relationship with animals and nature: 'The enormous popular success of Mrs. Adamson's first book about the . . . lioness Elsa revealed some deep vital hunger in the reading public of at least three continents, a hunger that had remained unsatisfied, possibly unrecognised, for many years.'[49] This included both an emotional relationship and a widespread interest in animal behaviour. Reviewers noted that there seemed to be an honesty in the Adamsons' intentions.[50]

Julian Huxley gave scientific legitimacy to Joy's story in an introduction he wrote for *Living Free* and in an article in the *Sunday Times*. He praised Joy's contributions to ethology and to a study of lion development despite her lack of academic training.[51] Desmond Morris, zoologist and successful TV director, thought that *Born Free* had fundamentally changed a generation's attitudes to lions, and helped to query the morality of keeping animals in captivity in zoos and circuses.[52]

Others emphasised the way in which *Born Free* seemed to echo classical literary themes: 'Elsa, gay and loving, and the Adamsons, patient and perceptive, are model feline and human characters, ideal figures in a Peaceable Kingdom.'[53] House thought the book's success was a 'love story with a twist' that 'evoked a long history of archetypes and legends', and that 'many people suddenly felt it was important to bridge the widening gap between man and the natural world'.[54] It tapped deep into the stream of anglophone love of animals, which was also widely rehearsed in literature. Much of this related to pets and livestock, particularly horses. But there were important works depicting human relationships with wild animals – from Jack London on a hybrid dog-wolf to Henry Williamson on otters and Gerald Durrell's bestsellers published in the 1950s. Moreover, *Born Free* was, after careful editing, easy to read with strong central figures; readers identified with both Elsa and Joy – an adventure in which a woman participated as a central actor.[55]

With respect to more recent academic assessments, Gregg Mitman refers to Huxley's preface for *Living Free* and makes an analogy between *Born Free* and Kenyan independence.[56] Huxley, he thought, stressed the importance of patience and understanding in the relationship between the Adamsons and Elsa – and saw it as a model for decolonisation. I agree that *Born Free* carries an optimistic tone, in contrast to the highly negative portrayal of Kenyans during the Mau Mau rebellion, and there is also a relatively sympathetic, if paternalistic, approach to African staff – perhaps more so in the photographs than in the text. But I do not think that Joy aimed to develop this analogy. If there is a political resonance, it perhaps arose in readers' more general identification with the idea of freedom. This was a particularly popular concept at the time, among politicians and people, in the West and in Africa – though it had many different meanings, both personal and political.

The Images in an Era of Decolonisation

The visual record, together with Joy's writing, emphasised her emotional attachment to Elsa and her cubs, as well as Elsa's response. The literature on photography in Africa tends to highlight the power of the photographer and – following Sontag – the analogy between camera and gun.[57] This has been modified to some degree by work on African portrait photography, where the agency of the sitter is explored.[58] We can extend this idea to the Adamsons' lion photography, which conveys a more mutually constituted relationship. Different aspects of this approach can be found in analyses that discuss 'the power of images' and the idea that 'objects stare back'; metaphorically they say 'look at me'.[59]

As noted, Joy Adamson had long been conscious of this interaction in her painting and photography. The images she and George took were a reflection of their reciprocity with Elsa and had the capacity to surprise by placing people and a potentially dangerous predator in the same frame. They were different from most wildlife images because the latter tended to be generic shots of animals that were not known to the photographer. Such images could certainly capture the power and fascination of animals and, in film, their behaviour and movement. But much of the Adamson lion photography was close up and intense, even physical, and possible only because of their intimate relationship with Elsa, the other cubs and Elsa's cubs. This intimacy may also have given the Adamson images greater impact than the 'true to nature' version of wildlife, which attempted to exclude people from the frame.

Versions of the Adamsons' story were retold, retaining some of their interest and power. Both wrote autobiographies. Travers and McKenna remained in close contact with George and this produced further films and books – especially in connection with Christian, a cub born and initially raised in captivity in England and then sent to George in Kenya. A YouTube clip of Christian (2007) scored over sixty million viewers very quickly. In 2010, the fiftieth anniversary of the publication of *Born Free*, two British TV documentaries were made, and the Kenya Wildlife Service held an event in memory of George Adamson. The trilogy was reissued, with some new photographs.

For most of the colonial era, images and texts of hunting and capture of wildlife had predominated. Kenya was a well-established location for hunting safaris, fuelled not only by the British elite but by American celebrities such as Roosevelt and Hemingway. Hemingway's novella, *The Green Hills of Africa* (1935), was an unapologetic account of an obsession with safari hunting and killing.[60] Hunting and capture remained significant themes, especially in feature films, but by the 1950s conservationist approaches were rapidly gaining ground, with strong visual back-up.[61] This was coterminous with decolonisation and contributed to destabilising some of the cultural ideas and practices associated with high imperialism.

Attitudes to wildlife are not generally factored into an analysis of decolonisation. On the contrary, Africanists often dismiss wildlife conservation and an animal-centric view of Africa as part of the colonial ethos and postcolonial global agendas.[62] This is a complex and unresolved debate. The reservation of national parks did often result in some degree of exclusion of African people. Yet the destabilisation of the older imperial culture of hunting and aggression towards the natural world was in certain respects a profound reimagining of a postcolonial modernity. Nor was there simply a division between African and global interests.

Julius Nyerere's Arusha Manifesto in 1961 specifically saw wildlife as a global as well as a national asset, and its protection as requiring scientific and international support.[63] The expansion of protected areas in Tanzania was challenged, but it grew in part from a national policy.[64] The new Kenyan government supported the filming of *Born Free*, and Kenyatta was photographed with the film-makers and actors. Perez Olindo, the first African director of National Parks in the 1960s, met Joy when he was a student in Michigan and they kept in touch. He facilitated the transfer of orphaned leopards to her for rehabilitation and of lion cubs to George when he re-established his project in Kora; Olindo also encouraged some other filming enterprises.[65] Although the expansion of protected areas carried a variety of priorities, and was deeply contested, it arose in part from postcolonial, national approaches and interests.

Those for whom human relationships with animals, and the environment more generally, are fundamental may see *Born Free* as more than just a small footnote, coterminous with decolonisation in Africa. The strand of environmentalism associated with Rachel Carson in the United States was constructed around the intersection between science, pollution and conservation values; that around the Adamsons connected better with latent popular environmental ideas that drew on widespread affection for animals and attempts to value nonhuman species.

Born Free also linked to an expanding visual literacy in the mid-twentieth century when access to images was increasing in a huge range of newspapers, books and magazines, as well as in film and the new medium of television. Readers were attuned to images, and Collins and Harvill were generous in loading the original hardbacks with photographs. This helped to make the narrative accessible and give it authenticity. The images were very different from the visual record of hunting that had been ubiquitous in books about Africa for more than a century.

Joy Adamson's approach undermined this hunting ethos. Some readers expressed awe for the courage that it took to live with predators, rather than kill them at a distance with a firearm. Though not the first to develop such relationships, Joy succeeded in projecting her ideas to a global audience. The books told a story about a different kind of lion, the smallest and least aggressive of a litter, and – primarily – its relationship with a woman. It was a soft story of a lion, not entirely new, but fresh in its force and timing. It coincided with the beginnings of a shift in ideas about animal ethology and about wildlife conservation.

Born Free and the related books and images contributed to a significant phase in the development of a strand of popular, animal-centric conservationism, and to the use of media to expand the public reach of such ideas. The urgency of conservation for the Adamsons arose as part of a late colonial encounter between Western ideas and rapidly changing African political and environmental landscapes. This helped to destabilise and displace an old imperial culture of profligate hunting, which was already being challenged. It was in certain respects a profound reimagining of a postcolonial modernity. It has left a legacy in African countries and globally – a legacy that is both contested and available for appropriation.

William Beinart is the Former Director of the African Studies Centre, Professor Emeritus at St Antony's College (University of Oxford) and Research Associate at the University of Johannesburg. He has notably published *The Rise of Conservation in South Africa* (2003), *Environment and Empire* (2007, with Lotte Hughes), *Prickly Pear* (2011, with Luvuyo Wotshela), *African Local Knowledge* (2013, with Karen Brown) and *The Scientific Imagination in South Africa, 1700 to the Present* (2021, with Saul Dubow).

Notes

1. Gregg Mitman, *Reel Nature: America's Romance with Wildlife on Film* (Seattle: University of Washington Press, 2009 [1999]); William Beinart, 'The Renaturing of African Animals: Film and Literature in the 1950s and 1960s', *Kronos: Journal of Cape History* 27(1) (2001), 201–26.
2. William Beinart and Katie McKeown, 'Wildlife Media and Representations of Africa, 1950s–1970s', *Environmental History* 14(3) (2009), 429–53.
3. Joy Adamson, *Born Free* (1960), *Living Free* (1961), *Forever Free* (1962) (London: Collins and Harvill).
4. The black-and-white negatives are all in England, but until 2010, when I last visited, a number of albums with postcard enlargements were kept in cabinets at Elsamere, Joy's former home at Lake Navaisha, Kenya.
5. George Adamson, *Bwana Game* (London: Collins and Harvill, 1968). The conversion of hunters or poachers into game wardens was mentioned in 1951 in *Where No Vultures Fly*, an early conservation film made in Kenya by Harry Watt.

6. David Maugham Brown, *Land, Freedom and Fiction: History and Ideology in Kenya* (London: Zed, 2017 [1985]); Jopi Nyman, *Postcolonial Animal Tales from Kipling to Coetzee* (New Delhi: Atlantic, 2003).
7. When I refer to the Adamsons separately, I will use only their respective first names.
8. Joy Adamson, *Joy Adamson's Africa* (London: Collins and Harvill, 1972), 16.
9. Adrian House, *The Great Safari: The Lives of George and Joy Adamson* (London: Harper Collins, 1995), 55.
10. Joy Adamson, *Joy Adamson's Africa*, Chapter 5.
11. Ibid., 63.
12. Ibid.
13. House, *The Great Safari*, 55.
14. William Beinart and Dominique Schafer, 'Hollywood in Africa, 1947–1962: Imaginative Construction and Landscape Realism', in William Beinart, Karen Middleton and Simon Pooley (eds), *Wild Things: Nature and the Social Imagination* (Knapwell: White Horse Press, 2013), 44–66.
15. Joy Adamson, *Born Free*, 32.
16. Ibid., 96.
17. Ibid., 6.
18. George Adamson, *Bwana Game*.
19. Interviews, conducted in partnership with Neil Carrier, Isiolo and Meru, 2008.
20. Joy Adamson, *Born Free*, 48.
21. Ibid., 120–21.
22. Ibid.
23. Ibid., 110.
24. Ibid., 112.
25. Ibid.
26. Jonathan Adams and Thomas McShane, *The Myth of Wild Africa: Conservation without Illusion* (New York: W. W. Norton, 1992).
27. Beinart, 'The Renaturing'.
28. Joy Adamson, *Living Free*, 58.
29. Ibid., 103.
30. Beinart and Schafer, 'Hollywood in Africa'.
31. David Attenborough, *Life on Air: Memoirs of a Broadcaster* (London: BBC, 2002).
32. Ibid., 152.
33. Alex Parkinson (director), Harry and Laura Marshall (producers), *The Born Free Legacy* (London: BBC, 2011).
34. Chris Parsons, *True to Nature* (Cambridge: Patrick Stephens, 1982).
35. House, *The Great Safari*, 437.
36. For a brief summary: William Beinart, Karen Middleton and Simon Pooley (eds), *Wild Things: Nature and the Social Imagination* (Knapwell: White Horse Press, 2013), 1–23.
37. Harriet Ritvo, *Noble Cows and Hybrid Zebras: Essays on Animals and History* (Charlottesville: University of Virginia Press, 2010), 203–12.
38. Joy Adamson, 'Living with a Lioness', *Sunday Times*, 6 March 1960; Joy Adamson, 'The Huntress Awakens', *Sunday Times*, 13 March 1960.

39. House, *The Great Safari*, 232.
40. John Hillaby, 'Big Cat on the Hearthrug', *Guardian*, 18 March 1960.
41. Gerald Durrell, 'Living with a Lioness', *New York Times*, 22 May 1960.
42. Bentley Glass, 'Review of *Born Free*', *Quarterly Review of Biology* 37(4) (1962), 319–20.
43. John Bear, *The #1 New York Times Best Seller* (Berkeley, CA: Ten Speed Press, 1992), 78.
44. Anon., '"Born Free", Film about Lion Opens on June 22', *New York Amsterdam News*, 25 June 1966.
45. House, *The Great Safari*, 238.
46. Rachel Ray, 'Elsa's Legacy: The Born Free Story, Nature on PBS', *Telegraph*, 7 January 2011.
47. House, *The Great Safari*, 253.
48. Virginia McKenna, *Some of My Friends Have Tails* (London: Collins and Harvill, 1970); Virginia McKenna, *The Life in My Years* (London: Oberon, 2009).
49. Gavin Maxwell, 'The Lioness and the Lady', *Observer*, 8 October 1961.
50. Mercedes Mackay, 'Review of *Living Free*', *African Affairs* 61(243) (1962), 164–66.
51. Julian Huxley, 'The Importance of Elsa', *Sunday Times*, 10 September 1961.
52. House, *The Great Safari*, 266.
53. Anon., 'Display Ad 223', *New York Times*, 26 March 1961.
54. House, *The Great Safari*, 239.
55. C. L. B., 'Review of *Born Free* by Joy Adamson', *Oryx* 5(4–5) (1960), 307.
56. Mitman, *Reel Nature*, 200–1.
57. James Ryan, *Picturing Empire: Photography and the Visualization of the British Empire* (Chicago: University of Chicago Press, 1997).
58. Heike Behrend, '"I Am Like a Movie Star in My Street": Photographic Self-Creation in Postcolonial Kenya', in Richard Werbner (ed.), *Postcolonial Subjectivities in Africa* (London: Zed, 2002); Katie McKeown, 'Studio Photo Jacques' (Master's thesis, University of Oxford, 2007).
59. David Freedberg, *The Power of Images: Studies in the History and Theory of Response* (Chicago: University of Chicago Press, 1989); James Elkins, *The Object Stares Back* (San Diego: Harvest, 1996); Melissa Thackway, *Africa Shoots Back* (Oxford: James Currey, 2003).
60. Ernest Hemingway, *Green Hills of Africa* (London: Arrow, 2004 [1935]).
61. Beinart and Schafer, 'Hollywood in Africa'.
62. Dan Brockington, *Fortress Conservation* (Oxford: James Currey, 2002); Ruth Mayer, *Artificial Africas: Colonial Images in the Times of Globalization* (Hanover, NH: University Press of New England, 2002).
63. G. Wesley Burnett and Richard Conover, 'The Efficacy of Africa's National Parks: An Evaluation of Julius Nyerere's Arusha Manifesto of 1961', *Society & Natural Resources* 2(1) (1989), 251–60.
64. Roderick P. Neumann, *Imposing Wilderness: Struggles over Livelihood and Nature Preservation in Africa* (Berkeley: University of California Press, 1998).
65. House, *The Great Safari*, 265; Sue Hart, *Life with Daktari: Two Vets in East Africa* (London: Godfrey Bles, 1969); George Adamson, *My Pride and Joy* (London: Collins Harvill, 1986), 238.

Bibliography

Adams, Jonathan, and Thomas McShane. *The Myth of Wild Africa: Conservation without Illusion*. New York: W. W. Norton, 1992.

Behrend, Heike. '"I Am Like a Movie Star in My Street": Photographic Self-Creation in Postcolonial Kenya', in Richard Werbner (ed.), *Postcolonial Subjectivities in Africa* (London: Zed, 2002), 44–62.

Beinart, William. 'The Renaturing of African Animals: Film and Literature in the 1950s and 1960s'. *Kronos: Journal of Cape History* 27(1) (2001), 201–26.

Beinart, William, and Katie McKeown. 'Wildlife Media and Representations of Africa, 1950s–1970s'. *Environmental History* 14(3) (2009), 429–53.

Beinart, William, and Dominique Schafer. 'Hollywood in Africa, 1947–1962: Imaginative Construction and Landscape Realism', in William Beinart, Karen Middleton and Simon Pooley (eds), *Wild Things: Nature and the Social Imagination (*Knapwell: White Horse Press, 2013*),* 44–66.

Beinart, William, Karen Middleton and Simon Pooley (eds). *Wild Things: Nature and the Social Imagination*. Knapwell: White Horse Press, 2013.

Brockington, Dan. *Fortress Conservation*. Oxford: James Currey, 2002.

Brown, David Maugham. *Land, Freedom and Fiction: History and Ideology in Kenya*. London: Zed, 2017 [1985].

Burnett, G. Wesley, and Richard Conover. 'The Efficacy of Africa's National Parks: An Evaluation of Julius Nyerere's Arusha Manifesto of 1961'. *Society & Natural Resources* 2(1) (1989), 251–60.

Elkins, James. *The Object Stares Back*. San Diego: Harvest, 1996.

Freedberg, David. *The Power of Images: Studies in the History and Theory of Response*. Chicago: University of Chicago Press, 1989.

Glass, Bentley. 'Review of *Born Free*'. *Quarterly Review of Biology* 37(4) (1962), 319–20.

House, Adrian. *The Great Safari: The Lives of George and Joy Adamson*. London: Harper Collins, 1995.

Mayer, Ruth. *Artificial Africas: Colonial Images in the Times of Globalization*. Hanover, NH: University Press of New England, 2002.

McKeown, Katie. 'Studio Photo Jacques'. Master's thesis, University of Oxford, 2007.

Mitman, Gregg. *Reel Nature: America's Romance with Wildlife on Film*. Seattle: University of Washington Press, 2009 [1999].

Neumann, Roderick P. *Imposing Wilderness: Struggles over Livelihood and Nature Preservation in Africa*. Berkeley: University of California Press, 1998.

Nyman, Jopi. *Postcolonial Animal Tales from Kipling to Coetzee*. New Delhi: Atlantic, 2003.

Parsons, Chris. *True to Nature*. Cambridge: Patrick Stephens, 1982.

Ritvo, Harriet. *Noble Cows and Hybrid Zebras: Essays on Animals and History*. Charlottesville: University of Virginia Press, 2010.

Ryan, James. *Picturing Empire: Photography and the Visualization of the British Empire*. Chicago: University of Chicago Press, 1997.

Thackway, Melissa. *Africa Shoots Back*. Oxford: James Currey, 2003.

CHAPTER 8

Conservation in the Days of Independence

The Case of the Seychelles, 1968–1974

Grégory Quenet

The most frequent problem in conducting a small-scale global history of conservation is the limitations imposed by sources. It is hard to retrace discussions and circulations within a given colonial empire given that policies decided in colonial metropoles were then implemented locally. In the absence of any clearly established doctrine, it is no easy task to detect factors differentiating colonies, though we may guess at them given the varying degrees of negotiation with local authorities. NGOs and international organisations would shed a different light, being transnational by nature, but their archives are very defective, either because access is restricted due to the private nature of documents produced for internal use, or because these frequently voluminous papers seem to have disappeared or are currently impossible to track down.[1] Historical studies are thus often obliged to rely mainly on printed sources and grey literature, which has the effect of effacing all internal discussion and controversy, and of overstating the influence of colonial power, which for that matter was tested by the lack of personnel on the ground and by tensions between various conservation stakeholders.

By examining the context in which documents are produced, this chapter seeks to stave off the reality effect induced by narrative sources. Times of independence, especially when this seemed inevitable, constitute an exceptional moment in the archives, for disagreements caused the colonial routine – which tended to underestimate tensions in the colony – to fissure or crack open. Since it was a matter of preparing for an afterwards, and especially of conserving areas of European influence while accommodating new forces, it was also a time for redefining principles and policies: conservation issues were suddenly foregrounded in all their practical and effective dimensions, overlapping with other economic, military and political issues. When these failed to align, arbitration was required.

In the case of Britain, as of the mid-1950s advice on 'conservation problems abroad' was provided by the Nature Conservancy, a body set up in 1949 and replaced in 1973 by the Nature Conservancy Council.[2] For the Seychelles, this expertise work started in 1968, before being cut off in 1973 when the move towards autonomy and then independence was playing out. The two parties clashing over these issues had been set up in 1964: the Seychelles Democratic Party (SDP) and the Seychelles Peoples United Party (SPUP). After the SDP won the November 1970 elections, its leader, James Mancham, became prime minister, retaining the position after the 1974 elections. In 1975, the Seychelles were granted the status of an 'autonomous colony'. A rapprochement between the SDP and the SPUP then led to the islands' independence on 29 June 1976, with James Mancham as their first president, overthrown by France-Albert René, the leader of the SPUP, in July 1977.

The papers held by the British National Archives at Kew provide an exceptional observatory. The correspondences establish links with all the other protagonists and sources. We need to go over the events, constantly connecting the global and the situated, without prejudging an outcome that was in no way a foregone conclusion. To do justice to these dynamics and issues, which are nested like Russian dolls, the following analysis will proceed in reverse chronological order.

The Local Failure of International Conservation Networks (1973–74)

This sequence brought John Procter's time in the Seychelles to a close. He had been recruited in January 1972 as Conservation Adviser to the Seychelles government for a period of two to three years,[3] paid for as part of the Ministry for Overseas Development's (MOD) 'Special Commonwealth African Assistance Plan'. His job was to implement the report he had previously drawn up, in the form of three main missions: advising the new National Parks and Nature Conservancy Commission in charge of applying decisions made by the Seychelles government cabinet; advising the Ministry for Agriculture, Natural Resources and Trade on conservation measures for flora, wildlife and forest development for areas outside national parks and nature reserves; and selecting and training Seychelles staff.[4] Procter's action there proceeded at a brisk pace, as indicated by his many exchanges with the Nature Conservancy, which was advising the MOD on this matter. But two years later, his mission was abruptly cut short, and Procter and his family left the archipelago in January 1974.[5] What had happened?

John Procter's background and career are fairly typical of the circulation of conservation professionals in the colonial world.[6] He was born in 1927, and spent all his childhood in Kenya where his father was a schoolmaster. In 1946 he arrived at St Catherine's Society, Oxford to study forestry, which enabled him to fulfil his wish of returning to East Africa. He was recruited by the Tanganyika Forestry Department in 1950, and, apart from briefly returning to Oxford for postgraduate study in forestry, remained there until 1968, when the implementation of Julius Nyerere's new policy for 'socialism and autonomy' sparked the mass departure of Europeans.

While working as a conservation assistant then forestry assistant, Procter became highly proficient in Kiswahili and developed a pronounced taste for the field, heightened by his passion for photography and botanical collection, two pillars of safaris. Like many of his colleagues, he experienced difficulties on returning to the UK, where he was based in Edinburgh, and became a consultant on conservation matters abroad for the Nature Conservancy. The letters he exchanged with his future recruiters when they were looking for the ideal person for the Seychelles position show that he never felt completely at home in Scotland, being still irresistibly drawn to Africa. In discussions about the possibility of recruiting him, he openly admitted:

> My personal position is that while I am keen to do a good job here, and give this project any 'follow-through' that it needs, I am a continental ecologist and have no wish to become heavily involved in island ecology. I would like to get my teeth into an African project if one should offer itself.[7]

Besides, officials at the Nature Conservancy were opposed to the Overseas Development Administration's (ODA) plan to send Procter to Venezuela, deeming that 'Mr. Procter was not the sort of person Venezuela wanted', but would be well suited to the Seychelles.[8] It was his experience in Africa together with his expertise in forestry that qualified Procter for this job, for the archipelago was seen as part of the 'East Africa' area. The reasons for this were ecological (the origin of the fish in Seychelles waters, of which only 2 per cent were endemic), economic (studies of the Seychelles' trade potential were conducted by the British National Export Council's Africa Commission), demographic (it was a cosmopolitan society partly composed of people of African descent), logistical (boat and plane trips transited via Nairobi) and political (with the Organisation of African Unity (OAU) supporting the SPUP).[9]

This British reading of space did not have much historical depth, for in the days when sail had been the sole means of transport the Seychelles had not really been attached to the African coast, something that only became possible with the advent of fossil fuels.[10] In the early 1970s, most Seychellois still felt little affinity with the African continent, which many associated more readily with the Zanzibar revolution or with white power in Rhodesia.[11]

This initial reading leads us to posit that the failure was due to ideas – heavily tinged with colonialism – of an African Eden being projected onto this Indian Ocean territory by the British. This inevitably sparked distrust and hostility even among the burgeoning Seychellois elites. Indeed, Procter's departure resulted from a more general political crisis triggered in March 1973 by James Mancham's decision to hitch his party to calls for independence, turning his back on his prior objective of maintaining strong ties with the UK.[12] This had immediate repercussions for local conservation policies.

In the month of August, Procter commented on the 'government's slowness in making administrative decisions, especially on the structure and staffing of a conservation organisation'.[13] In September, the government decided to downgrade the planned position of Director of Conservation Services, a post that was meant to have been a new recruitment to strengthen the future director's staff, to that of Senior Conservation Officer.[14] This ruled out the solution Procter had advocated in his preparatory report, and which had been his main objective since arriving in the archipelago – namely, separating the forest and nature teams from those working on agriculture to form a new service with its own budget and enhanced capacities.[15] This failure triggered his decision not to prolong his contract as Conservation Adviser beyond the two-year minimum, and to return to the UK.[16]

Internal discussion between members of the Nature Conservancy sheds light on the terms of conflict with the local government, now heading down the path to independence. In July 1973, Dr Derek Frazer, in charge of external relations at the Nature Conservation Council headquarters, called in an independent observer to draw up a report on events in the Seychelles. To his mind:

> The latest reports are disturbing and conflicting and imply on the one hand that the government is more interested in short-term economic gains from tourism than in preserving its heritage for the future, and on the other that it is proposing to take rather drastic steps to preserve the endemic flora and fauna of Aride island.[17]

The warnings came from men of science, especially a letter sent a few weeks earlier by Russell George Dunnet to Sir Maurice Yonge, two well-known figures in British naturalist circles.[18] Passing through the Seychelles to see how Dr Chris Feare's research into sooty terns was progressing, Dunnett reported on the 'very considerable difficulty in having conservation ideas actually implemented', describing Procter as 'extremely disheartened by his inability to persuade Government through the Commission [for the conservation of national parks and nature] to take action, and everything is moving at a very slow pace indeed'.[19]

The accusations linked the government's wish to rapidly develop tourism to the hopes of certain speculators, losing Procter the arbitration he had requested to protect the mangrove zones on Mahé: 'there is tremendous pressure in the Seychelles brought about by a rapidly developing tourist industry and large sums of money are being spent building tourist hotels and the like.' Indeed, after winning the November 1972 elections, Mancham, in addition to carrying on as prime minister, awarded himself the ministerial portfolios for tourism, urban planning and territorial development, land, immigration, public relations, local government, and trade and industry, while appointing Robert Delorié – a member of his party who was head of the commission that Dunnett had criticised for failing to act on conservation matters – as minister without portfolio.[20] Dunnet ended up recommending applying 'any kind of pressure [that] can be brought to bear on the Seychelles Government through the Royal Society or any other British Agency'.[21]

If we adopt the framework of interpretation applied at the Nature Conservancy's headquarters in London and by academic networks based in the UK, which it regularly called on to produce reports on external territories that were or had been part of the British Empire, then the reason Procter was virtually fired originated in a conflict between two groups: on the one hand conservation experts based in Europe (academics, the Royal Society and more secondarily the WWF and IUCN), and on the other Seychelles businessmen and independentists seeking to undo conservation measures hampering their property developments. In short, on this interpretation, there was a confrontation between two nature policies, one from without based on the ideal of preserving nature, the other from within targeting the territory's economic development for the benefit of local populations.

However, close examination of all the discussions shows that the terms of the conflict are not reducible to a confrontation between a colonial reading and a local vision. What the archives describe is more elaborate, indirect and shifting.

The initial impetus came from a shift in British foreign policy. The UK no longer viewed the Seychelles as a strategic territory, first because it wished to get rid of the last traces of colonial empire, and also because other anchorages in the Indian Ocean seemed more essential (Diego Garcia, where the US was developing a base, and Mauritius, with which the UK was negotiating a mutual defence agreement).[22] The Nature Conservancy itself paid the price, for its demise in 1973 coincided with these new international priorities. Frazer referred to this shift in November 1973 in a letter mentioning 'all the changes with the guillotining of the old Conservancy, that Bob Boote is now Director of the NCC, and that Duncan Poore is going to [the] IUCN'.[23]

Duncan Poore, head of the Nature Conservancy from 1966 and 1973, had a career that would be of interest for a prosopography of global conservation protagonists, a work yet to be undertaken. He was born in 1925, and interrupted his Classics degree at Cambridge to take part in the Second World War in Sri Lanka. After his doctorate, he went to work in 1955 for Hunting Technical Services in Cyprus, Pakistan and the Middle East, before becoming Professor of Botany and Dean of the Science Faculty at the University of Malaya in Kuala Lumpur, where he specialised in tropical forests. The rest of his career was even more illustrious, for he became Science Director at the IUCN in 1973, then its Director-General (1976–77), before becoming Professor of Forest Science and Director of the Commonwealth Forestry Institute at Oxford University. He went on to play a central role both at the International Institute for Environment and Development (IIED) set up by economist Barbara Ward in 1972, and at the International Tropical Timber Organization (ITTO), an international organisation founded in 1983.[24]

Thus the 1973 shift did not correspond to a change in protagonists, so much as a redeployment of staff working for Britain's external services who, after an initial career in colonial affairs, joined the research programmes conducted by universities and international organisations. Duncan Poore's career illustrates this, as does that of John Procter who, after the Seychelles, became head of a research programme at the Dutch university of Wageningen in Suriname, where he died tragically in a car accident in 1977.

These two protagonists in positions of responsibility had another point in common: their experience during the Second World War in external theatres where awareness arose of the global nature of ecological problems threatening natural elements.[25] This was the case of Poore in Sri Lanka and of his successor at the head of the new Nature Conservancy Council, Bob Boote, who had fought in Greece and the Mediterranean. After the war, both joined the world of conservation, to which they had not been predestined.[26]

The failure of Procter's mission in the Seychelles thus needs to be seen against the global context of the end of colonisation, a phase when conservation objectives and institutional actors were being redefined, at the same time as the staff previously involved were re-employed in a new expertise set-up. Whereas the previous system had proceeded by internal negotiation between services involved at both the local and imperial level, this new one was characterised by pressure brought to bear from the outside, focusing on emblematic zones to be protected. Procter put this very clearly when explaining the causes of this failure to Derek Frazer, and envisaging new forms of action he planned on developing from the outside, rather than renewing his contract in the Seychelles:

> I have devoted a great deal of thought to possible reasons for this government pulling its punches in this way, but I am still at a loss. I am confident that it is not my ideas that are at fault, because they are so well endorsed by the various people in the world of conservation who visit the Seychelles. My hope is that I will be able to achieve more by trying to channel pressure and opinion in the UK than I have been able by direct action here. This place is extremely sensitive to certain kinds of opinion or pressure from overseas.[27]

The relationship with local authorities was changing and becoming more frontal. Consequently, the experts involved started to use internationally promoted conservation objectives to challenge the economic development interests of the places in question. The dichotomy between interior and exterior replaced the compromises between the metropolitan and colonial sides of empire. This new postcolonial era also saw non-governmental organisations acquiring a new and more important role, together with the mobilisation of international public opinion. The very nature of the documents produced by conservation protagonists illustrates this shift in stance. When the WWF and IUCN decided to send a marine biology specialist to the Seychelles to study its marine national park in March 1972, he was of course tasked with writing a report, but especially with penning an article for *World Wildlife*, the WWF magazine for the general public.[28] The topic of 'preserving the Seychelles' underwater treasures' was examined almost schizophrenically, with the connection between the global and the situated coming across as a radical divide between interior and exterior, rather than as interlocking scales. It was apparently up to the Seychelles population to become conscious of how valuable their nature was, to ensure that tourists – whose growing number posed a threat to it – would keep on wanting to come and enjoy it in situ:

> The people of the Seychelles themselves are now becoming conscious of the value of the natural beauty of their islands and the Government is losing no opportunity of making the people aware of its importance to maintain a successful tourist industry. This is one of the few occasions when a Government has the chance, and is taking it, to ensure the conservation of its natural resources before the tourist boom has fully developed, and the timely assistance of World Wildlife and [the] IUCN has once again been able to help in the preservation of these resources for present and, we hope, for future generations.[29]

This modern schizophrenia had such an impact on peoples and their environment that we need to go back in time to understand how the relationship between the global and situated had reached this point.

Reconciling Conservation and Development – Locally and Globally (1970–1972)

The tone of the news sent by Procter over the course of 1972 stands in sharp contrast to the failure of 1973, even though things were not progressing as rapidly as he might have wished. In October 1972, his progress report for the year referred to substantial progress on the national parks (Morne Seychelles,

Sainte Anne and La Digue), except for the Port Launay zone where there was a clash with development projects; additionally, discussions were under way for special reserves (Cousine, Desneuf, Aride, Bird, Fregate and La Digue) whose status was more complex as they were in private ownership.[30] This positive assessment was later confirmed by David White's unofficial report, which noted that among the documents sent in, 'the Progress report was particularly useful because it shows that most of Mr. Procter's recommendations have been accepted in principle and only relatively minor points have been rejected'.[31]

Over the course of 1972, Procter published three newsletters about conservation. The third, which came out in November, shows that his action was not limited to protected zones and threatened species, but also covered fires, sand extraction and mosquitoes. The population apparently supported or was at least interested in these efforts, for in the space of three days 7,600 people had visited the Nature Conservation and Forest stand, amounting to 14 per cent of the Seychelles population, and even 19 per cent of that on Mahé, thus attracting twice as many people as the Miss Seychelles pageant being held at the same event.[32]

The second newsletter, published in June, discusses ongoing negotiations with the owners of the private islands of Fregate, Silhouette and Desneufs to launch awareness actions (school education programmes, picking up empty cans and bottles) and to broker compromises between conservation measures and economic uses: conditioning the development of tourism on support for agriculture, and authorising the construction of jetties but equipped with anti-rat devices.[33] As for the first newsletter, in March 1972, it starts by signalling that 'events are moving rapidly'. This is followed by an account of Procter's arrival in the archipelago, then of the royal visit in which nature issues had played a major part. Procter re-transcribes in his own words the speech given by the Queen when opening the airport on 20 March 1971, in which she 'praised the government's policy of preserving the natural beauty and serenity of the islands for the Seychelles people so that they may benefit from the advantages of twentieth-century communication while not losing what isolation has given'.[34]

The Seychelles were in the international conservation spotlight, with the Queen's speech being just one example of this. Procter's newsletters and correspondence mention many researchers who visited the archipelago, together with the global scientific networks involved. For 1972 alone, these included: Dr Chris Feare, funded by the British Natural Environment Research Council, to study sooty terns and Bird Island; Dr Jack Frazier, working on turtles, who had requested funding from the IUCN, and was temporarily employed as a chief warden for the National Parks and Nature Conservancy; Carol Mather and Brian Harrison, conservation and development specialists; Prof. J. L. Jakubowski from the Warsaw Technical University, working on the conservation of terrestrial communities; the Swede Bengt Sjogren, who had written general-interest natural history books; Dr Harold Moore, who had come from Cornell University to produce a monograph on palm trees; Dr Julian Rzoska from the International Biological Program (IBP), based in London, who specialised in freshwater biology; Dr Raoul Serenc, an expert from the UNESCO Southwest Asia marine sciences unit; Dr Alec Forbes-Watson from the National Museum in Nairobi, who, with Don Turner, was studying birds in the western Indian Ocean; and Dr Schmidt from Zurich, in charge of analysing the ecology of an endemic species of Sarracenia plant.

The breadth of these scholarly networks was probably even greater than it seems. A quick search through other archives shows that it would be possible to add to this picture on various scales, by including the Sir Peter Scott papers held in Cambridge, those of the former MP Tufton Beamish, an avid ornithologist, and of course the archives of the Royal Society, the WWF, the IUCN and UNESCO (albeit with the difficulties mentioned above).[35] The archives allow us to work back up to the various private foundations contacted to fund research programmes, linking the global and local levels even more tightly. When the Nature Conservancy approached the Nuffield Foundation in London in 1971 to set up a marine biology research station in the Seychelles, discussions focused on comparisons between the biological diversity of several territories in the Indian Ocean and the various programmes already up and running in Aldabra and Dar-es-Salaam.[36]

This international attention dated back to October 1971, when the Seychelles government published a White Paper on conservation policies in the Seychelles. The printed booklet, which went on sale, and whose thirty-one detailed articles are exemplary of the genre, proclaimed in its introduction:

> The Government's policy statement on tourism development in Seychelles emphasised the overriding need to protect the natural beauty of the islands and their natural environment, which are among our greatest assets. It was envisaged that one of the most important ways of achieving this should be through the designation of National Parks and other reserves and through the protection of areas where characteristic wildlife could be conserved in its natural surroundings, for the enjoyment of the public.[37]

The articles that followed went over the existing legislation, then drafted the objectives of a conservation policy that involved establishing a National Parks and Nature Conservancy, whose action was to cover three types of zone to be put in place (national parks, special reserves and nature reserves proper). They also discussed forests, rainwater collection, underwater hunting and turtles, the administrative organisation to be put in place and the research policy to be conducted. The progress made in this White Paper was underlined by the US Secretary of the Interior, who sent a letter congratulating the Seychelles Minister of Agriculture, Natural Resources and Trade, backed up by the congratulations of the forty countries belonging to the Pacific Area Travel Association.[38] The White Paper also acknowledged in its introduction that it was based on the report submitted by John Procter in late 1970 – with in fact only minor differences that left the main principles of his approach unaltered.[39]

Yet all this in fact followed a more complex national and international political and geopolitical agenda, which is only visible in the archives, and often only by reading between the lines. First, with the opportunities stemming from the dismantling of the British Empire in the mid-1960s, the Indian Ocean had become a geopolitical stake for the US and the USSR, resulting in the Anglo-American wish to establish a mid-ocean military base.[40] The Seychelles were thus part of a two-stage strategy, starting with the plan for a British naval base on the island of Aldabra, followed by the secret decision in 1966 to enable the US to establish a major military site on Diego Garcia atoll in the Chagos Archipelago.

The point these two territories had in common was that they both belonged to a new colony created by the British in November 1965 especially for these geopolitical and military ends: the British Indian Ocean Territory (BIOT) including Aldabra, two Seychelles islands (Farquhar and Desroches) and the Chagos Archipelago (previously a dependency of Mauritius), all placed under the responsibility of the Seychelles Governor. That is why Britain wanted to prevent any independence for the Seychelles, which might benefit its adversaries. It therefore decided to finance the archipelago's economic development even though the guiding principle, ever since 1811, had always been to spend as little as possible on a territory of no interest.[41]

The reports written by development experts show how these global objectives connected with local conservation issues, which thus became central. In an archipelago with strong demographic growth and no notable natural resources, business activity or sizeable industry, the only solution was to develop international tourism by building an airport. But this created new threats to the fragile ecosystems that needed to be preserved, for they were the attraction pulling tourists in.[42]

This is precisely what was stated by the Queen and the White Paper alike: development had to be reconciled with conservation, openness with protection. That is why Procter was sent to the Seychelles. He had to put the necessary measures in place, but also broker compromises, such as for an extension to the port zone, which included remarkable mangroves that hindered access by large ships.[43] This agenda leads us to view in a new light the many passages in which Procter is not wholly declinist about the Seychelles ecology, while recognising that he really is fighting on multiple fronts. Further study is required of the available naturalist studies to gauge the ecological scope of these compromises. Nevertheless, we may note that they led Procter to an original stance: while deploring each arbitration that went against him, he recognised that the situation of certain species was improving at times – in contrast to the alerts issued by certain external experts about the risks of extinction and the Seychelles government's attempts to dismantle certain protection measures.[44]

Ultimately, the period 1970–72 leads us to reinterpret the failure of the 1973–74 conservation policy, focusing on three points. In 1970–72, the Mancham government had been in favour of conservation measures and compromises with economic objectives, leading it to maintain good relations with conservation experts. But when the British decided to abandon the Seychelles in 1973, its political elites in Mancham's SDP now lent their backing to development projects, and seemed henceforth to be opposed to the conservationists' point of view.

Second, threats to ecosystems did not initially come from a local wish for economic growth. They stemmed from secret UK–US negotiations, which led to tourism being developed to ward off independence: the SPUP's and SDP's demands to counter the cost of living called for British assistance, whether this entailed invoking the metropole's responsibility towards its colony, or going further and working towards the integration of the Seychelles into the UK.

Finally, certain indications show that the population and some of the local elites were receptive to nature conservation, though this did not preclude their observing the negative effects of changing consumption patterns among the population, who were not the first to enter modernity in contradictory manner. Nevertheless, this does not explain the pivotal role played by the Nature Conservancy, and we need to go back a bit further to see how the first expertise regime (1968–70) was connected to the two following regimes, that of 1970–72 (with the compromise between conservation and development at both the global and local level) and that of 1973–74 (with international conservation against local development).

Mobilising Conservation Experts Against the British Government (1968–70)

This period ushers into view new names and protagonists in the archives, such as the British Museum, the Smithsonian Institution, the Lindblad travel agency, Guy Lionnet, Albert René, Dr David Stoddart, Miss Barclay-Smith, Lord Percy and Viscount Ridley. Things came into focus in the wake of the British decision to create the BIOT and to build an airport that would have significant ecological consequences by allowing an influx of tourists.[45] The first draft of an ordinance on national parks in the Seychelles was written in 1968 under the governorship of Sir Hugh Norman-Walker. This generated a demand for conservation expertise, initiating dialogue with the Nature Conservancy, which was already more or less present in the Seychelles thanks to the scientific networks of Dr Poore, dating from his years at the University of Malaya.[46]

Poore's first stay in the Seychelles was in late 1968. On his return he was enthusiastic: 'I have rarely been in a place where I have felt that there was so much sympathy for a conservation policy and where I had more confidence that recommendations might be made effective.'[47] The tone of these early years was marked by a clear colonial ethos, transpiring in a certain sympathy for the Seychellois. This is true even of Poore, who explained that his optimism 'is, undoubtedly, due to the paternalistic form of Government but the peoples there are also very enthusiastic and sympathetic'.[48]

The IUCN played a central role in discussions of the draft national parks ordinance (intended to supplement legislation deemed insufficient given the pressure that would result from tourism), for it was a matter of projecting onto this territory a normative framework drawn up in Europe. The definition of a national park harked back to the 1933 London Convention for the Protection of African Fauna and Flora and to the 1942 Pan American Convention on Nature Protection and Wild Life Preservation in the Western Hemisphere.[49] Sir Hugh Elliott, the IUCN expert who was consulted, drew on his ongoing work on the UN List of National Parks and Equivalent Reserves to distinguish between nature reserves proper, national parks and special reserves.[50]

Nevertheless, within this sequence governed by international norms, a Seychelles dynamic was established after the arrival of a new governor, Sir Bruce Greatbatch, on 13 March 1969. Although the hundreds of items in the archives never refer or even allude to it, events on Diego Garcia clearly cast their shadow: Greatbatch set about managing the deportation of the Chagos population with great brutality, stripping them of their land and scattering them across Mauritius and the Seychelles without assistance or lodging, thus condemning them to many years of poverty.[51] He is remembered for one particular episode, embodying the paroxysm of this violence, which started with the inhabitants being forced off the island, and their houses destroyed along with all their personal effects. Greatbatch then had the hundreds of dogs they had abandoned trapped in copra furnaces, where they were gassed with carbon monoxide.[52]

In this tense context linked to the no longer secret agreement between the UK and the US to install the Diego Garcia military base, the governor wished to avoid any unrest that might reach British and international public opinion. But conservation matters had become a sensitive topic ever since the Aldabra affair, when the British Navy had wanted to transform this island of great ecological value – the last place in the Indian Ocean where giant turtles reproduced – into a military base. Protests by Dr David Stoddart (who was on close terms with Poore, at the Nature Conservancy) and Dr Christopher A.

Wright from the National History Museum, backed by the BBC and the Royal Society, managed to get this project definitively shelved in November 1967. Since then, Stoddart – now working at Cambridge University – had been consulted on all conservation issues concerning the Seychelles.[53] Stoddart, who continued to be much appreciated by the Seychellois, later expressed his regrets about Diego Garcia, together with the feeling that he had been betrayed by Greatbatch and the London authorities who, after backing down over the Aldabra turtles, were now making the Chagos islanders pay the full consequences of this decision.[54]

Shortly after arriving, Greatbatch was faced by two conservation controversies, directly targeting him, which are key to understanding what happened next. The first was launched by the International Council for Bird Preservation (ICBP), which was alarmed at the Seychellois' annual harvest of bird eggs and its consequences on bird populations on Desnoeufs. This concern dated back to 1955 and 1956, the years of the first missions by Lord Richard Percy and Lord Ridley, two keen zoologists.[55] It was an important topic, for these eggs were the main source of protein for the island's inhabitants, particularly and supposedly for the poorest among them, in a context of rising protests against the cost of living and social inequalities.

The controversy sparked much discussion until 1970, when both the naturalists and the governor wished to reach a compromise with local usage. Since nobody was envisaging a complete ban, discussions related to sustainable usage by introducing a limit on the quantities collected, rotating the collection zones and alternating them from one year to the next. This was the position of the Royal Society, which wrote to Percy and Stoddart:

> You may feel that a compromise is called for, and it has been suggested that the system you formerly recommended – that of halving the island and cropping each half alternately – might be adopted. We realise that in your 1966 report you recommended discontinuing this practice, but in view of present circumstances it may bear reconsideration.[56]

The second controversy was more damaging to Greatbatch, for it targeted development works he had commissioned in preparation for the arrival of tourists in the Vallée de Mai, the only site where sea coconut (*Lodoicea maldivica*) grew. The polemic was launched in 1970 by Dr Fosberg from the Smithsonian Institution, with the implicit support of the director of the Department of Agriculture, Guy Lionnet. Ecological protests were voiced against cleaning the site of dead leaves, introducing ornamental plants and building footpaths and viewpoints, together with a 'chalet' in the place of the existing kiosk.[57] When Forsberg spoke out on Seychelles radio, it was devastating for the governor, to whom he had already said in private:

> If people want to see a chalet, I guess they'll go to Switzerland . . . what you're going to get here is the better class of tourist who will have read about the scenic beauty and the biological interest, and won't want to find it all buried in new hotels when he gets here . . . [When asked what impressed him most about the Seychelles:] Mountain forest, nice people . . . and I was also highly impressed, and unfavourably, by the amount of destruction that has been going on.[58]

Percy also accused Greatbatch of claiming that these works had been requested by the Seychellois to contribute to the island's economic development, which was apparently denied by all his contacts there.

The two affairs were taken up by scholarly networks, creating a real stir in London where they fuelled each other, even though the first related to exploiting the environment and the second to awareness of natural beauty. On 19 January 1970, Sir Tufton Beamish, an MP since 1945, raised the Vallée de Mai affair in the House of Commons with the Secretary of State for Foreign and Commonwealth Affairs.[59] And it was in fact to neutralise these controversies that Greatbatch told the MOD that he needed a conservation adviser based in the Seychelles: this request was transmitted to the Royal Society, which asked the Nature Conservancy to fulfil this role, resulting in Procter's mission in 1970, followed by his recruitment.[60] As Stoddart noted in March 1970, 'the Conservancy will, I understand, soon be sending someone out to make a survey over a period of some months, and I think the Governor has been so impressed by the recent outcry that it is highly likely that any recommendations made will be accepted'.[61]

The concern with reconciling development and conservation here acquired a different meaning to that of the subsequent period: the practical questions of 1970–74 had been anticipated as an issue that was about to be thrown up by Britain's international decisions.

Here is how Poore described the situation in the Seychelles at the end of his short stay there in 1968: 'I found that the climate of opinion was most favourable to conservation and that all those with whom I discussed it knew that it was important to act before the Seychelles were opened up by the proposed airport.'[62] The conservation policies thus stemmed partly from the British promise of development to the Seychellois in order to ensure Britain's military and geopolitical conditions, and partly from knowledge and sensitivity to nature, although these were not the driving force in this instance.

The interplay of protagonists also became more complex and multifaceted, bringing into view a whole range of positions among British players. The Lindblad travel agency, which ran excursions to the Seychelles, sailing from Kenya, had served a first generation of tourists, scholars and elitists, marching hand in hand with the ornithologists and supporting their calls, even though their presence was not without consequences for the sites.[63] The IPCB represented another generation of conservationists, prior to that of the Nature Conservancy, all of whose studies of the Seychelles need to be factored into the analysis. The IPCB did not trust the Nature Conservancy, which it regarded as too accommodating towards development projects.

In several handwritten notes, Poore recommended saying nothing to Miss Barclay-Smith at the British Museum, the IPCB secretary, about ongoing negotiations between the Royal Society and the Nature Conservancy to entrust the latter with expertise about the Seychelles: Poore was 'unhappy at providing any inf. which could be used in the current dispute – esp to compromise the Government/MOD et al'.[64] Indeed, being under the oversight of both the Queen's Privy Council and the Ministry of Education until 1973, the Nature Conservancy was at the heart of government, leading it to operate compromises that differed in type from those of scientists, while being very close to them.

Poore thus adopted more conciliatory stances towards the development of the Seychelles than he had at first when writing to Stoddart in April 1969: 'I am horrified by the way this tourist threat is developing. They are so obviously ignoring protocol, that I feel they want to force our hand (and that of the Seychelles government) by presenting us with a fait accompli.'[65] Subsequent archives show, however, that a little later, in 1971, the Nature Conservancy and Miss Barclay-Smith at the ICBP looked favourably on Sir Patrick Crofton's suggestion of purchasing Fregate island to build a 150-room luxury hotel, backed up by bird conservation measures.[66]

In the face of these criticisms, Governor Greatbatch saw enemies everywhere busily throwing up ill-intentioned opposition to his development projects, and was unable to refrain from adding a colourful handwritten note to his letter to the ICBP: 'I would very much like you to let me know who have been sending you an untrue account of what is happening. (This is a fairly venal Seychelles trick by locals or expatriates who have been too long on the island).'[67] But there was still a piece missing to this paranoid interpretation, which though essential for understanding the history of conservation in the Seychelles only appears intermittently – namely, the role played by Albert René, the head of the other Seychelles party, who did not agree with Mancham's integration policy and was moving rapidly towards the path of independence.

Greatbatch was convinced that René was the one pulling the strings of the conservation protests, as Lord Ridley said in less diplomatic terms when faced with public pressure to not totally ban the collection of eggs on Desneufs: 'There are many disquieting facts behind this request (and not least the presence of Mr René on the scene).'[68] This colonial reading rejecting the idea that local resources were harmless to nature originated in diplomatic and government circles, as well as among ornithologists who followed a similar career path. But it was not shared by the Nature Conservancy and certain scientists because 'whatever its imperfections, the sea-bird egg industry in Seychelles is the indigenous conservation of a natural resource'.[69]

These divergences provide access to a history from a local perspective that is present in the sources even if not directly expressed, requiring even closer enquiry. During the London conference in March 1970 (known as the Constitutional Conference) about negotiations for greater political autonomy, Albert René chaired Commission no. II in charge, inter alia, of the issue of turtle and bird eggs.[70] Many documents other than those of the Nature Conservancy held in Kew indicate the importance René and his entourage attached to preserving Seychellois nature, even their opposition to overly rapid development, particularly of tourism, as called for by Mancham. The 1978 manifesto of the Seychelles People's Progressive Front (SPPF), which succeeded the SPUP, stated that 'tourism must not be our bread, but rather the butter for our bread. The bread will be what we will produce and exploit from our soil and ocean.'[71]

With the British position hardening, making the Seychellois responsible for threats against nature for which the British themselves had created the conditions, René set about building an environmentalism of Creole descent, reactivating the heritage of French late eighteenth-century conservation, rewritten around the figure of Malavois and his 1787 'Constitution de la nature'.[72] This is another topic, and beyond the scope of this chapter, but one that it is important to mention here, for defending one's own environment is inseparable from the capacity to write and rewrite one's own history by interweaving it with more global histories. And it is this capacity that was nullified by the narrative put in place in the wake of independence by international organisations cut off from the dialogue between colonial powers and their territories.

Here is what Rodney Salm, an independent consultant sent to the Seychelles at the request of its government, wrote in 1978:

> The conservation momentum, initiated by Procter's work and report, has slowed down. This is understandable. The movement started with good, but big ideas (there are those who say that these ideas were initiated by expatriates keen to make Seychelles the conservation show-piece of the world). The ideas flourished through the formative stages of planning and design, but when the time came for setting up the necessary infrastructure they proved too ambitious and broad for the existing funds and personnel. Many of the ideas are good, but Seychelles is a small nation with a proportionately small budget for conservation.[73]

The gap between this and the previously recounted history is obvious, together with the consequences it induced.

Taking the Seychelles as a case study, this chapter has reassessed the habitual interpretation of conservation issues and practices, bringing out how tensions ran through the Western camp and the local actors, rather than splitting them into two separate blocks. As these tensions were reconfigured over time, they engendered a dichotomy between global mobilisations and local obstacles, a dichotomy that in fact stemmed from the march towards independence and the way Britain managed it. This history is intended to be more symmetric, for it distances itself from moral readings posterior to independence, centred on the supposed environmental ignorance and incapacity of the 'natives', and instead makes them fully fledged protagonists in their history who entered postcolonial modernity with their own contradictions.

The place left to colonised peoples by colonial powers withdrawing from their former territories transpires thanks to two approaches. First, the contextualised study of the production and conservation of archives enables us to unravel the genealogies established subsequently, and thus to follow more closely the negotiations and power relations, justifying the plan adopted here.[74] Second, space no longer transpires as condensed time, as is often the case for historians, and instead partakes in a dual spatial and temporal matrix. In other words, it is not the action of protagonists in time that here links up a global and a local realm with stable contours, but the production of spatial scales that organises time frames that conjointly structure the differentiation of spaces.[75]

These considerations of method will perhaps only interest historians. But that would be a mistake. It was in 1972 that the famous Meadows report was published on the limits to growth, at a time when it was still possible to make a choice other than that which has led the planet to its current ecological situation.[76] It is thus worth understanding how the promise of development made to countries leaving colonialism took place in deep contradiction with the conservation of nature and resources, and that this was not preordained. Once we appreciate this, international tourism will certainly join the list of the great destructive monocultures, alongside wheat and palm oil.

Grégory Quenet is Professor of Environmental History at UVSQ-Paris Saclay University. He holds the Laudato Si' Chair *Pour une nouvelle exploration de la terre* at the Collège des Bernardins. Considered one of the pioneers of environmental history in France, he has worked on natural disasters in the modern era, written an intellectual history of environmental history and ecologised an emblematic place in French history, the Château de Versailles. His current work focuses on islands and globalisation.

Notes

1. Researchers working on the PANSER project had placed great hopes in the archives of the IUCN and the WWF, the two most powerful actors in global conservation, both headquartered in the same Swiss town (Morges then Gland) and today in the same building. Despite publications by former managers describing considerable volumes of archives, covering several decades, the only papers available for consultation are a very limited number of reports and grey literature.
2. 'Records Created or Inherited by the Nature Conservancy, the Nature Conservancy Council and English Nature', UK National Archives, Kew, https://discovery.nationalarchives.gov.uk/details/r/C134, retrieved 5 August 2024.
3. Letter from A. Blake to J. Linley, 'Establishment and Finance Branch Nature Conservancy', Natural Environment Research Council (NERC) headquarters, 6 December 1971, UK National Archives (UKNA), FT 3/615.
4. Ibid., Letter from the Foreign and Commonwealth Office, Overseas Development Administration to John Procter, 30 November 1971.
5. Ibid., Letter from John Procter to Derek Frazer, 6 December 1973.
6. P. J. Wood, 'John Procter', *Commonwealth Forestry Review* 59(1) (1980), 11–13. The *Commonwealth Forestry Review* replaced the *Empire Forestry Review*, revealing the organic, corporatist link between colonisation and conservation, traceable even in the obituaries.
7. Letter from John Procter to Morton Boyd, 9 June 1970, UKNA, FT 3/615.
8. Ibid., Letter from Duncan Poore to Bob Boote, 16 June 1971.
9. Ibid., WWF, 'Application for a grant from the World Wildlife Fund', 17 June 1971; ibid., The Africa Committee of the British National Export Council (BNEC) 'The Seychelles', London, 7 March 1971; W. D. Park, 'Note for the file', 9 March 1970, UKNA, FT 3/614.
10. Grégory Quenet, 'Une île des Lumières globales, les Seychelles: Retour sur le laboratoire environnemental', *Revue d'histoire moderne et contemporaine* (2024), forthcoming.
11. William McAteer, *To Be a Nation: The History of Seychelles 1920–1976* (Mahé: Pristine, 2008), 333.
12. Ibid., 363–73.
13. Letter from John Procter to Derek Frazer, 8 August 1973, UKNA, FT 3/615.
14. Ibid., Letter from John Procter to Derek Frazer, 3 September 1973.
15. Ibid., John Procter, 'Seychelles conservation newsletter', no. 3, November 1972.
16. Ibid., Letter from John Procter to Derek Frazer, 3 September 1973.
17. Ibid., Letter from Derek Frazer to D. A. White, 19 July 1973.
18. At the time, ornithologist George Dunnet was professor of zoology at the University of Aberdeen and worked on practical issues, such as dialogue with the oil industry and the environmental consequences of fisheries; P. A. Racey, 'Obituary: George Dunnet', *Independent*, 25 September 1995, https://www.independent.co.uk/news/people/obituary-george-dunnet-1602822.html, retrieved 5 August 2024. Sir Maurice Yonge was at the time professor of zoology at the University of Glasgow, and known for his work on the Great Barrier Reef; he was involved here as head of the Royal Society's Southern Zone Research Committee; B. Morton, 'Charles Maurice Yonge, 9 December 1899–17 March 1986', *Biographical Memoirs of the Fellows of the Royal Society* 38 (1992), 377–412.

19. Letter from George Dunnett to Sir Maurice Yonge, 4 July 1973, UKNA, FT 3/615.
20. McAteer, *To Be a Nation*, 332.
21. Letter from George Dunnett to Sir Maurice Yonge, 4 July 1973, UKNA, FT 3/615.
22. McAteer, *To Be a Nation*, 363. McAteer grounds his analysis in contemporary analysis of the *Times*, the memoirs of James Mancham, *Paradise Raped: Life, Love, and Power in the Seychelles* (London: Methuen, 1983), and Foreign Office papers held at the UK National Archives.
23. Letter from Derek Frazer to J. H. Hemsleu, 21 November 1973, UKNA, FT 3/615.
24. 'Vale Professor Duncan Poore, a Giant of Tropical Forestry', ITTO website, 6 April 2016, https://www.itto.int/news_releases/id=4777; 'Remembering Duncan Poore, a Leading Light in Sustainable Forestry', IIED website, 19 October 2016, https://www.iied.org/remembering-duncan-poore-leading-light-sustainable-forestry, all retrieved 5 August 2024.
25. This is what Sverker Sörlin shows by looking at published texts rather than individuals: post-1945, global awareness of ecological issues emerged, focusing on the problems of erosion and demography; Sverker Sörlin, 'Reconfiguring Environmental Expertise', *Environmental Science and Policy* 28 (2013), 14–24.
26. 'Robert Boote Obituary', *Times*, 19 December 2019, https://www.thetimes.co.uk/article/robert-boote-obituary-gfsch23pt, retrieved 5 August 2024.
27. Letter from John Procter to Derek Frazer, 10 September 1973, UKNA, FT 3/615.
28. Ibid., Letter from T. Scrope-Howe to C. Greenham, 'Public relations, Wildlife, London (The British National Appeal of the World Wildlife Fund), International Section, The Nature Conservancy', 10 March 1972.
29. Ian J. B. Robertson, 'Preserving the Seychelles' Underwater Treasures', *World Wildlife* (August 1972), 3–4.
30. John Procter, 'Conservation in the Seychelles. Progress report', October 1972, UKNA, FT 3/615.
31. Ibid., David White, 'An unofficial report on nature conservation in the Seychelles', 8 September 1973.
32. Ibid., John Procter, 'Seychelles conservation newsletter', no. 3, November 1972.
33. Ibid., 'Seychelles conservation newsletter', no. 2, June 1972.
34. Ibid., 'Seychelles conservation newsletter', no. 1, March 1972.
35. Cambridge University Library, Department of Archives and Modern Manuscripts, NCUAS (according to the online inventory): report by Sir Peter Scott on conservation in Mauritius, presented to the Mauritian Prime Minister, 87.8.99/D.390; study of conservation on Mauritius and in the Seychelles, 1973, D.386–D.390; general correspondence of Sir Peter Scott, M.1272–M.1277; correspondence between Sir Peter Scott and I. Frost on conservation in the Seychelles, 1967, D.775–D.780. Collection of articles by Tufton Beamish, Brighton (UK), archives of the East Sussex Record Office, CLW/1/5/22.
36. Discussions between the Nature Conservancy and the Nuffield Foundation, 1 December, 30 November, 17 November and 29 October 1971, UKNA, FT 3/615.
37. *Government of Seychelles: Conservation Policy in the Seychelles* (Mahé, October 1971).
38. 'Seychelles Leads the World', *Seychelles Bulletin*, 12 April 1973, UKNA, FT 3/615.
39. John Procter, *Conservation in the Seychelles* (Mahé: Government Printer, 1970).
40. On how these international stakes pertained to the Seychelles, see McAteer, *To Be a Nation*, 254–75.

41. William McAteer, *Rivals in Eden: The History of Seychelles 1742–1827* (Mahé: Pristine, 2008); William McAteer, *Hard Times in Paradise: The History of Seychelles 1827–1919* (Mahé: Pristine, 2008).
42. The Africa Committee of the British National Export Council (BNEC), 'The Seychelles', London, 7 March 1971, UKNA, FT 3/615.
43. Procter was in fact paid by the ODA, and not by the Nature Conservancy or a research programme (unlike the ornithologists depending on the NERC), nor even by the Seychelles government (thus unlike forest and park agents).
44. Re-examination is needed of all the files relating to Dr Feare and Dr Frazier, whose alerts were relayed to London by leading scientific figures in the UK.
45. Handwritten note by W. D. Park, the Nature Conservancy, in the margin of a letter by L. B. Chapman, Legal & Parliamentary Section, NERC, 'Aldabra is now party to the British Indian Ocean Territories', 25 November 1968, UKNA, FT 3/614.
46. Since October 1968 he had been in contact with Guy Lionnet, Director of Agriculture in the Seychelles, about ironwood seeds requested by his former colleague Kwiton Jong (Letter from Guy Lionnet to Duncan Poore, 30 October 1968, UKNA, FT 3/614).
47. Ibid., Letter from Duncan Poore to Sir Maurice Yonge, 7 January 1969.
48. Ibid.
49. Ibid., Letter from Sir Hugh Elliott (IUCN) to W. D. Park, 1 November 1968.
50. Ibid., Letter from Sir Hugh Elliott (IUCN) to W. D. Park, 30 October 1968.
51. McAteer, *To be a Nation*, 254–75.
52. Laura Jeffery, 'How a Plantation Became Paradise: Changing Representations of the Homeland among Chagos Islanders', *Journal of the Royal Anthropological Institute* 13(4) (2007), 951–68. The episode with dogs was taken up in Jean-Marie G. Le Clézio, 'Lavez l'injustice faite aux Chagossiens', *Le Monde*, 17 October 2009.
53. Stoddart's name crops up time and again in correspondence from the years 1968–80.
54. 'David Stoddart Obituary', *Guardian*, 9 December 2014, https://www.theguardian.com/environment/2014/dec/09/david-stoddart; Sharon Meriton-Jean and Hajira Amla, 'A Love Affair with Aldabra: Tribute to David Stoddart, the Champion of the World's Second-Largest Atoll', *Seychelles News Agency*, 20 January 2015, http://www.seychellesnewsagency.com/articles/2197/A+love+affair+w, both retrieved 5 August 2024.
55. Hon. M. W Ridley and Lord Richard Percy, 'The Exploitation of Sea Birds in the Seychelles', *Colonial Research Studies* 25 (1958).
56. Letter from R. W. J. Keay (deputy executive secretary, The Royal Society) to Lord Percy, copy to Dr Stoddart, 25 March 1970, UKNA, FT 3/614.
57. Ibid., Letter from Dr Fosberg (Smithsonian Institution, United States National Museum) to Miss Phyllis Barclay-Smith (Secretary, ICBP), 17 February 1970.
58. Ibid., Confidential excerpt from a letter from Malcolm Penny (Seychelles) to W. R. D. Bourne, 5 February 1970.
59. Ibid., Excerpt from Hansard, *Written answers to question*, 22, 19 January 1970. Beamish was an active member of the Royal Society for the Protection of Birds. His cousin, Tony Beamish, made programmes for the BBC that raised awareness about safeguarding Aldabra, which he used in a general-interest book he wrote; Tony Beamish, *Aldabra Alone* (London: Allen & Unwin, 1970).
60. Letter from Bruce Greatbatch to the Biological secretary (The Royal Society), 30 April 1969, UKNA, FT 3/614.

61. Letter from Dr David Stoddart to Bill Bourme, 5 March 1970, UKNA, FT 3/614.
62. Ibid., Duncan Poore, 16 July 1969.
63. Ibid., 'A unique series of expeditions to "the forgotten island of the Indian Ocean". Peter Scott and James Fisher invite you to explore with them the Seychelles and other Indian ocean bird islands', pamphlet, Lindblad Travel, Houlders World Holidays, BOAC.
64. Ibid., Duncan Poore, handwritten note, 23 February 1970.
65. Ibid., Letter from Duncan Poore to David Stoddart, 2 April 1969.
66. Letter from *Sir Patrick Crofton (Developments Ltd and hotel development) to Phyllis Barclay-Smith (Natural History Museum), 8 March 1971,* UKNA, FT 3/615.
67. Letter from Governor Greatbatch to W. H. Thorpe esq. (International Council for Bird Preservation, British Museum, Natural history), February 1970, AN FT 3/614.
68. Ibid., Letter from Lord Ridley to R. W. J. Keay (deputy executive secretary, The Royal Society), 1 April 1970.
69. Ibid., Nature Conservancy, 'Seychelles sea-bird industry', April 1970.
70. Ibid., Anon., 'Note for use by governing council delegation at the London Conference', March 1970.
71. Seychelles People's Progressive Front, *Onward to socialism. SPPF policy statement*, Mahé, 1978, Seychelles National Archives, no call number.
72. Grégory Quenet and Jan Synowiecki, 'Ce que conserver veut dire: Praxis et historicité de la nature. 1770–1810', *Annales historiques de la Révolution française* 399(1) (2020), 97–121. For these discussions on the 'Constitution de la nature', I wish to thank the director of the Seychelles National Archives, local scholar and historian Tony Mathiot, and the anthropologist Claude Mahoune.
73. Rodney V. Salm, *Conservation of Marine Resources in Seychelles: Report on Current Status and Future Management, Report of IUCN to the Government of Seychelles Prepared with the Financial Support of the United Nations Environment Programme and the World Wildlife Fund* (Morges, August 1978), 6.
74. This study originated in difficulties pertaining to the interpretation of the 'Constitution de la nature' (Seychelles National Archives, 'Règlement des îles Seychelles', 30 July 1787, A/25.5). Undoing the pattern of late eighteenth-century proto-environmentalism necessarily involved exploring the genealogy of contemporary archives.
75. Bernard Lepetit, 'Espace et histoire', in Bernard Lepetit (ed.), *Carnet de croquis: Sur la connaissance historique* (Paris: Albin Michel, 1999), 129–41.
76. For Dennis Meadows's response to this 1972 moment: 'Dennis Meadows: "Il faut mettre fin à la croissance incontrôlée, le cancer de la société"', *Le Monde*, 8 April 2022, https://www.lemonde.fr/idees/article/2022/04/08/dennis-meadows-il-faut-mettre-fin-a-la-croissance-incontrolee-le-cancer-de-la-societe_6121114_3232.html, retrieved 5 August 2024.

Bibliography

Jeffery, Laura. 'How a Plantation Became Paradise: Changing Representations of the Homeland among Chagos Islanders'. *Journal of the Royal Anthropological Institute* 13(4) (2007), 951–68.

Lepetit, Bernard. 'Espace et histoire', in Bernard Lepetit (ed.), *Carnet de croquis: Sur la connaissance historique* (Paris: Albin Michel, 1999), 129–41.

McAteer, William. *Rivals in Eden: The History of Seychelles 1742–1827*. Mahé: Pristine, 2008.
———. *Hard Times in Paradise: The History of Seychelles 1827–1919*. Mahé: Pristine, 2008.
———. *To Be a Nation: The History of Seychelles 1920–1976*. Mahé: Pristine, 2008.
Morton, Brian. 'Charles Maurice Yonge, 9 December 1899–17 March 1986'. *Biographical Memoirs of the Fellows of the Royal Society* 38 (1992), 377–412.
Quenet, Grégory. 'Une île des Lumières globales, les Seychelles: Retour sur le laboratoire environnemental'. *Revue d'histoire moderne et contemporaine* (2024), forthcoming.
Quenet, Grégory, and Jan Synowiecki. 'Ce que conserver veut dire: Praxis et historicité de la nature. 1770–1810'. *Annales historiques de la Révolution française* 399(1) (2020), 97–121.
Racey, P. A. 'Obituary: George Dunnet'. *Independent*, 25 September 1995. Retrieved 5 August 2024 from https://www.independent.co.uk/news/people/obituary-george-dunnet-1602822.html.
Sörlin, Sverker. 'Reconfiguring Environmental Expertise'. *Environmental Science and Policy* 28 (2013), 14–24.
Wood, P. J. 'John Procter'. *Commonwealth Forestry Review* 59(1) (1980), 11–13.

CHAPTER 9

Tracking Wildebeests

The Technological Mediation of Spaces for Humans and Wildlife in the Serengeti since 1950

Simone Schleper

> So great are the advantages of a light aeroplane for all aspects of game management, including anti-poacher operations and research, that the effective management of the 5,000-square-mile Serengeti National Park, in Tanzania, now depends on the plane.[1]

In 1965 Myles Turner, Chief Game Warden of the Tanzanian National Parks (TANAPA), and Murray Watson, a doctoral student at the University of Cambridge with a pilot's licence, published a paper on the usefulness of aircraft for both game management and wildlife research. This paper in the British conservation journal *Oryx* gives a telling account of how field research in the early 1960s was conducted in East African national parks. Research and park management demanded that researchers and game wardens move freely over vast distances. In addition, aerial mobility allowed game wardens such as Turner, who had at first been critical of the use of aircraft, to fulfil core conservation tasks including monitoring the placement of traps and snare-line fences, which potentially threatened herds of animals on the move, as well as recovering lost tourists.[2]

For many decades, the East African Serengeti has been a focus for local wildlife authorities, researchers and international conservation organisations interested in the negotiation of space for the movements and territorial claims of animals and humans. Today, the 40,000 km^2 of the larger Serengeti-Mara ecosystem include a number of areas with different protection statuses, most of them in Tanzania, such as the Serengeti National Park, the Ngorongoro Conservation Area, the Maswa, Grumeti, Ikorongo and Kijereshi Game Reserves, the Loliondo Game Controlled Area and the Ikona and Makao Wildlife Management Areas. In Kenya, the Masai Mara National Reserve is counted as part of the ecological system.

Early on in the history of the Serengeti National Park, ungulate migration – the 'synchronous large-scale return movement' of the Serengeti wildebeests, zebras and Thomson's gazelles across wide distances – played a crucial role in strategies to protect Tanzanian and Kenyan environments.[3] Indeed, the Serengeti has been under one form or another of protection since late nineteenth-century German colonial rule,[4] and for Western visitors evokes images of how other vast landscapes, such as the North American plains, must have looked before the introduction of commercialised hunting in the mid-nineteenth century. Discourse about East African nature as world heritage in need of international protection and aid has been embraced by the Tanzanian government and wildlife authorities, who have likewise used the spectacle of the Great Migration to buttress the country's extensive wildlife tourism industry.[5]

Over the seventy years of ecological research in the Serengeti, the methods and techniques to monitor wildlife and collect, store and circulate the resultant data have undergone significant changes, especially after the intensification of research in the 1960s and 1970s. From the 1950s onwards, ecologists like Watson relied on a range of instruments such as aeroplanes, cameras, radio trackers and computer simulations to understand the movements and health of animal populations.[6]

In this chapter, I examine the intellectual and technological history of the ways in which migratory wildebeests have been researched since the 1950s as part of a landscape inhabited by both animals and humans. With this I aim to add hitherto missing insights into the role that individual technologies have played in mediating researchers' spatial and temporal distance to, and their experience of, their research objects, including animals, landscapes and people.

These technologies to assist ecological observation, just like laboratory instruments, were not simply extensions of the researchers' senses. By bridging the distance to the observed animals, monitoring technology supported wildlife ecologists' 'authority to speak on their behalf'.[7] Technological tools, however, do more than buttress professional authority. By shaping the ways in which we experience, understand and communicate the world around us, their use has ontological consequences too.[8] In this chapter, I zoom in on different technologies to bring out important differences in the ways that both animal migration and human land use were experienced. These had direct consequences for the type of conservation advice formulated.

In this analysis, I draw on the concept of technological mediation from the field of science and technology studies (STS).[9] The post-phenomenological concept of technological mediation suggests that our relationship with the world, including our moral judgement, is fundamentally shaped by our use of particular technologies, and that no such relationship exists independently of the technologies we use. Phenomena, such as animal migration or human population pressure, only emerge in the interplay between observer and observation technology.[10] As such, the concept holds valuable insights for understanding the role that particular observational technologies may play in ecological fieldwork and in making recommendations about the management of protected areas for wild animals and people. Analysing the reports, scientific publications and personal correspondences by Serengeti researchers and members of international zoological societies from the 1950s to the 2000s, I show that in the 1960s, direct tracing of migratory wildebeest by aeroplane allowed for more flexibility in local land use and conservation recommendations. This changed in the 1970s, when aerial photography supported a return to colonial ideas about spatial conflict between animals and humans. When in the 1980s and 1990s preservationist views made room for early forms of community-based conservation, the introduction of computer modelling partly inhibited approaches that truly benefitted local pastoralist communities.

Tracking: Direct Aircraft Observations and Migration Patterns, the 1950s and 1960s

The first published accounts on animal migrations originated in the context of colonial game management. When in the 1910s and 1920s game wardens of the Kenyan Game Department noticed a fluctuation of wildebeest counts on their territory, they speculated about the animals' seasonal movement to German territory in today's Tanzania.[11]

In 1955, what had been a problem for local game management attracted the interest of the international scientific community when the administration of the British Territory of Tanganyika announced a decision to partition the Serengeti National Park, which had been established in 1951. The political dimension of this plan and its colonial heritage have recently been discussed in great detail by Thomas Lekan.[12] The separation of the park into two conservation areas would have lifted the strict conservation status of the central plains, an important grazing area for ungulates. International preservation societies such as the International Union for the Preservation of Nature (IUPN, from 1956 onwards IUCN) and the Fauna Preservation Society (FPS) expressed their dismay, fearing that land use by pastoralists would lead to overgrazing, erosion and desertification, a fear that dominated colonial conservation discourse in East and South Africa.[13]

In the post-war period, colonial landscape and wildlife protection had become an international issue, and the members of the IUCN and the FPS feared that the proposed partition might encourage other East African governments to rethink and potentially downsize the areas set aside for conservation. After international disapproval, the Tanganyikan administration allowed William Harold Pearsall, a botanist at the University of London and fellow of the Royal

Society, to enquire into the possible ecological consequences of the suggested changes to the park's boundaries.[14] Despite the preservationist fears fuelling the international debate, some of the subsequent research into wildebeest migration in fact supported mixed forms of land use, encouraged by experiences mediated by direct tracking of migratory ungulates from small aircraft.

When Pearsall visited the Serengeti in 1956, the national park was shaped like a triangle, located in the north of today's Tanzania, east of Lake Victoria, with hilly land in the west, central plains, and volcanic crater highlands towards the east. Pearsall linked the movement of wildebeest to different patterns of precipitation during the wet and the dry season. Ungulate movement, as he described it, took place mainly between the plains, the western hill lands close to Lake Victoria, and the north-east at the border with Kenya, with some movement in the Ngorongoro crater area. Pearsall understood the protection of the Serengeti plains as requiring extensive management and planning, because the high mobility of many hoofed species meant that population counts could be affected by changes at all stages along their journey, while a changed population size would affect migration routes in return.

Based on Pearsall's findings, the partition was halted, but the borders of the park were moved to include the northern region between Banagi and the Kenyan border in 1959, two years before Tanzanian independence. The Ngorongoro crater region in the south-east was turned into a conservation area where the protection of flora and wildlife was less strict. Yet with the new northern extension the protected territory had been increased overall, to the satisfaction of the international conservation community.[15]

Pearsall's study stimulated new research interest in the ecological linkages between animal movement and the spatial design of protected areas. Between 1959 and 1963, another study of the Serengeti ecosystem and its mammal populations was conducted by two American biologists, Martha and Lee Talbot. Lee Talbot had briefly visited Tanganyika in 1956 for a survey of the Serengeti plains for the Wildlife Management Institute and the American Committee for International Wild Life Protection. In 1959, Lee and his wife Martha returned to continue this initial research. The Talbots' project on the Serengeti's ecology was sponsored by the Foreign Field Research Program of the National Academy of Sciences, the National Research Council of the United States, with additional funding by the New York Zoological Society (NYZS) and the Government of Kenya.

The objective of the project was to study the interrelations of various wild animal species with soil and climate, as well as with the Maasai and their livestock, which was of increasing interest to the conservation community against the backdrop of colonial fears of overgrazing. In order to conduct a manageable study of the ecosystem, the Talbots – like Pearsall – decided to put the seasonal migration of ungulates at the centre of their inquiry, focusing on the Western white-bearded wildebeest, which was particularly dominant when it came to interests in 'biomass, migrations, predation, contact with man, and impact on the landscape'.[16] After several months, the Talbots found that there were indeed seasonally recurring patterns, but that the wildebeests' migratory behaviour was irregular, with 'no two seasons' movements . . . the same'.[17]

In the early years of Tanzanian independence, between 1962 and 1966, these irregularities observed by the Talbots informed another international research project, the Serengeti Research Program, funded this time by the United Nations Food and Agricultural Organization (FAO), and coordinated by John Owen, a Ugandan-born British conservationist, at the time director of TANAPA. The study focused on migration as an important parameter for park management and for the potential 'utilisation of living natural resources'.[18] Watson was hired as a consultant for the FAO study to research the ecology of the park's wildebeest population more closely.

Watson was part of an emerging strand in British conservation with utilitarian ambitions that understood conservation as resource management, firmly rooted in science. Between 1962 and 1965, Watson observed large parts of the wildebeest population in the Tanzanian part of the ecosystem, flying twice a week for several hours. Watson was not the first to use aerial observation. Bernhard Grzimek, the director of the Frankfurt Zoological Society (FZS), and his son Michael had used a zebra-striped Cessna to conduct animal censuses in the Serengeti plains in the 1950s for their campaign to keep the park as large as possible. Pearsall had likewise drawn on aerial observation methods, as had the Talbots a few years later.[19]

Watson's emphasis on live tracking and direct recording of observations was particular, however. Whereas the Grzimeks had used photography and film in their endeavour to capture the aesthetics of the plains, Watson recorded ecological data in context. Watson's observations thus represented a particular interpretation of animal migration and landscapes, mediated by a particular set of methods and technologies, namely real-time observation by aircraft.

In the 1960s, direct aerial observation had become important in zoology. Thanks to aerial observation it became possible to study the migration 'of large and fast-moving animals in remote and difficult terrain'.[20] Real-time observation of wildlife migration was closely related to game management, for which the use of aircraft was essential to reach difficult terrain, track poachers and locate injured animals. Naturally, large national parks where game management and scientific research were closely intertwined, such as those in the East African savannah, were the testing grounds for aerial observation in zoology.[21] By the mid-1960s, the 'warden's flying movement' – pioneered according to Watson by Kenyan game ranger Denis Zaphiro and by Turner in the Serengeti – had become widespread.[22]

For researchers like Watson, using small aeroplanes enabled the detailed and constant plotting of the route of migratory mammals in their spatial contexts. Watson's method included reconnaissance flights that he conducted with Turner to understand the geography of the area, followed by a stratification of the surveyed terrain into different sub-regions, taking into consideration animal density, geography and landscape.[23] The emphasis was on using the same qualitative empirical methods as in the field, and on questions relating to wildlife management. After all, the 'best results [were] obtained when reconnaissance [was] planned in relation to the question awaiting solutions'.[24] Out of the four seasonally migrating species in the Serengeti, direct aerial observation focused predominantly on wildebeests, which were the easiest to spot from an aeroplane. While the striped pattern of the zebras in the African sun made it hard to spot them from the air, wildebeests had a light patch on their back, facilitating observations of individual animals from above.[25]

Thus wildebeest migration, experienced as a constant but irregular flow, determined the spatial management strategies used in the Serengeti National Park and its surrounding areas, which tied in with an understanding of conservation as resource management. Live observation bridged the spatial and temporal distance to the animals, allowing direct data to be gathered on the ecological circumstances of the animals and the land. Watson's publications about his experiences in the 1960s stressed that real-time aerial observations were essential to wildlife ecology for conservation purposes. A variety of indicators linked to wildebeest herds – their routes of movement, the landscape through which they moved, as well as any human activities or interaction with wildlife in the area – could be directly integrated into the interpretation of routes and population counts. Photographs, in contrast, taken through the polycarbonate windows of the aircraft, were often blurry and lacked this type of immediate and everchanging context.[26]

Live migration research by aeroplane, then, mediated a particular form of ecological expertise as management of change. The Talbots, through their membership in international organisations such as the IUCN, were connected to influential figures in the early postcolonial network who acted as a link between African park authorities and international research communities. An important contact was the British ecologist Hugh Lamprey, at the time head of the Tanganyika Game Department. Lamprey gave the Talbots access to the resources of the national parks in Tanganyika and Kenya, including the use of aeroplanes and pilots.[27]

The aerial studies from the 1950s and early 1960s were thus marked by close cooperation between individual researchers and local park employees, often of European descent – such as that between Watson and Turner – who exchanged ideas and functioned as research teams. In fact, game wardens who resided locally were considered excellent observers, as they usually stayed in the park for long periods.[28] The close-knit airborne cooperation that Watson depicted of game warden–researcher teams might have been an idealisation. Watson later described how observation and counting were often done by the pilot alone, as Turner frequently suffered from motion sickness.[29] Yet for the type of research questions being asked, local game wardens possessed the necessary skills and knowledge of the observed terrain.

Game wardens held a crucial position for applying the joint observation findings to conservation, land development and other modernisation measures in accordance with British conservation practices. Since the May 1947 Conference on the Fauna of British Eastern and Central Africa held in Nairobi, long-standing local hunting traditions had been increasingly suppressed. The expansion of the national parks and protected areas had led to the loss of local communities' traditional hunting and grazing rights in more and more areas.[30] Individual game wardens began to promote a more people-centred approach to conservation.[31]

Researchers like Watson – though reasoning partly from false ideas about pastoralism – believed that local hunting was the 'most effective form of direct human exploitation' of the park's resources, and should be legalised, albeit with certain restrictions, as long as it did not obstruct the annual migration of the park's ungulates and with it the health of the overall ecosystem.[32]

By the mid-1960s, a number of European and American expats and African-born game wardens of European descent were experimenting with forms of land use that combined conservation and cultivation. By observing wildebeest routes, game wardens could steer the grazing of cattle to areas outside the annual migratory corridors, and could ensure that hunting in controlled game areas did not exceed the 10 per cent of migratory animals considered unproblematic. In contrast to international agendas to preserve the Serengeti as a supposedly pristine place without human interference, projects like Watson's tolerated human activities such as pastoralism and to some extent agricultural cultivation and hunting provided these did not disturb the annual migration patterns of wildebeests in the park and its direct surroundings.

Mapping: Photographs and the Ecological Recording of Habitats, the 1970s

Between 1969 and 1971, scientific interest in migratory wildebeests culminated in the institutionalisation of ecological research at the Serengeti Research Institute (SRI). The institute – founded after a long struggle between African and British scientists over the need for fundamental versus applied research – sought to maintain international access to the Serengeti area after the slow but steady Africanisation of government institutions, including the wildlife sector, under Julius Nyerere's Tanganyika African National Union (TANU).[33] This struggle remained evident throughout the institute's early years.

In 1967, Owen described the institute's objective as the study of the Serengeti region to provide the evidence base for the rational management and conservation not only of the Serengeti ecosystem, but also of other national parks in Tanzania.[34] This agenda was supported by influential figures in the Tanzanian wildlife regime, such as Lamprey, who was appointed the first scientific director of the institute. At the same time, a recognised research institute such as the SRI could improve the new independent government's international standing. TANAPA likewise hoped to benefit from the institute's international prestige. Already in 1969, Mr Maeda, the Tanzanian middleman between the SRI and TANAPA, stressed that the institute needed to expand its international research programme.[35] Concerns about international prestige soon led to a focus on fundamental rather than applied research. Studying the Serengeti as a representative ecosystem for the East African savannah resulted in a different type of access to and experience of the land and its animal and human inhabitants, which was mediated by aerial photography.

Thanks to improvements to camera equipment, cheaper film and higher-resolution images, by the late 1960s aerial photography was being used in various disciplines concerned with mapping and recording large topographies, such as geology, geography, soil science, archaeology, plant ecology and increasingly zoology. The use of photography by SRI researchers soon led to a return to ideas of spatial competition between wildlife and humans, for it captured land occupation in a seemingly permanent way.

One of the institute's core programme points was to expand on Watson's work on ungulate migration. This ecological monitoring programme was however more ambitious than the individual projects by the Talbots and Watson, which had only formulated initial hypotheses on the deeper ecological causes of ungulate migration. Now, the quest to understand the underlying biological laws of animal migration and its ecological consequences formed the core of SRI-hosted research projects on precipitation, vegetation, wildfires and predators.

In 1969, a new ecological monitoring programme began. The institute hired Michael Norton-Griffiths and Henry Croze, two recent graduates from Oxford, to set up a regular aerial survey of the ecosystem, covering about 26,000 km^2 with systematic flights.[36] Monthly flights were used to record animal distribution and movement, as well as rainfall, grazing, cultivation by human inhabitants in the park's surrounding areas and land erosion.[37] In addition, a radio tracking project was launched in 1971 to experiment with tracking large mammals by aeroplane.[38] Norton-Griffiths and Croze's main aim was however not the real-time observation of wildebeest movements for ad hoc land use management decisions, but the creation of maps for the long-term study of land inhabitation by animals and people. To create a more complete spatial overview of the ecosystem being mapped, they drew heavily on photographic methods.[39]

In 1968, Alexander MacFarlane, a geographer working at the SRI, had used aerial photographs to create a set of maps based on 5 km^2 squares, to be used with a 'mechanical data recording system' to be introduced the following year in the form of a card punch computer.[40] In spring 1969, Norton-Griffiths mounted motorised cameras on the SRI aircraft to take aerial photographs of landscapes and vegetation within which migratory species were present or absent at the moment of data collection.

The idea was that over time – twenty to thirty years, it was estimated – data recorded on the photographic grid would not only provide information about ungulate movements, but also make it possible to draw conclusions about the environmental causes and 'stimuli' behind these animal movements. These stimuli could relate both to natural causes and human activities.[41] Vegetation, which could be recorded using repeated series of photographs, was regarded as the 'most sensitive indicator of any . . . changes in the environment such as climate, soils, or animal influence'.[42] Setting up the new programme with a focus on vegetation cover and occupancy maps, including signs of agriculture or overgrazing by domesticated cattle, meant leaving behind the focus on individual species and their changing movements.[43]

The focus on photography and vegetation mapping in the long-term monitoring project had consequences for the way human mobility and land use were apprehended by individual researchers. Aerial photography helped coordinate the work in large and diverse research teams on long-term projects, as photographs could be stored and shared across projects.[44] As such, this form of data recording was useful for the SRI, because from the outset externally funded short-term projects for fundamental ecological research outweighed the efforts spent on the ecological monitoring programme.[45]

In the late 1970s, it was decided to split up the permanent team's working areas, based on MacFarlane's maps. The ecosystem was thus divided into the Plains, the Western Corridor and the Northern Extensions, to be studied by three teams of scientists.[46] Several scientific advisors were concerned that fundamental research with a long-term focus would result in a lack of applicability to park management. Nevertheless, both Lamprey and Norton-Griffiths agreed with the plan to leave behind the 'one-man-one-animal' approach, to accelerate the collection of ecological data without 'following your problems around the ecosystem'.[47]

After the division of the Serengeti into smaller research areas, aerial photography became even more important in reassembling spatial overviews of the three separate areas. Population counts were continued using aerial photography of sample areas.[48] In order to infer any information on the causes of movement, however, one needed to wait for ecological monitoring data from the rain gauges and the vegetation studies, as well as the results from individual experiments with captured animals on physiology and digestion.[49]

By the mid-1970s, then, the earlier focus on ungulate migration to understand basic ecological processes and to devise successful land use strategies had disappeared. In the 1970s, photographic mapping encouraged a return to a discourse on human settlements, cultivation and agriculture as being in spatial competition with wild animals and as being the main causes for the degradation of soils and natural environments, a view that had dominated colonial conservation policies.

Although there had previously been room for pastoralists to follow migratory animals through the park and for experimental agriculturalism on the fringes of the protected areas, aerial photography, by suggesting permanence rather than flux, did not allow for this flexible form of land use. For instance, Feroz Jafferali Kurji, one of the few Tanzanian scientists at the institute, studied human settlements in and around the park. Using aerial photography and district population counts, Kurji found that in more than half of the game reserves around the Serengeti, human settlements had emerged and grown since 1959. The Ngorongoro Conservation Area in particular had seen an increase from 8,500 human inhabitants in 1966 to over 18,000 in 1976.[50]

Other smaller research projects at the institute likewise supported the underlying fear of human 'encroachment', not only by agricultural practices but also by pastoralists grazing their cattle close to the park. Ethological research into ungulate behaviour led to physiological comparisons between wild wildebeests and Maasai livestock. After the eradication of rinderpest in the 1970s, this research suggested direct competition for grazing spaces and water between cattle and migrating wildebeests and buffalos.[51]

In many ways, the fragmentation of data collection during the 1970s had consequences for recognised forms of expertise. Photography allowed the synthesis of data at a later stage. This type of data collection and the need for a posteriori synthesis resulted in expert knowledge shifting to visiting international experts as opposed to locally based wardens. Although local administrators such as Owen and Lamprey had originally intended the monitoring programme to feed into park management, the observations of the early 1970s had little applicability. In fact, daily management was left to TANAPA, which lacked the SRI's levels of staffing and funding.

This lack of cooperation between SRI researchers and local park authorities was compounded by the Africanisation of the Tanzanian National Park authorities and Nyerere's policy of national self-reliance.[52] In turn, visiting researchers conducting fundamental research for international prestige and the advancement of their careers were often sceptical about working with African park wardens and researchers.[53] At the same time, few of the visiting scientists fulfilled their teaching duties at the University of Dar-es-Salaam.[54]

Scientific specialisation was however at least as big a driver of their distanced relationship. The research institute depended on a steady inflow of individual scientists with their own funding. According to Lamprey, it was easier 'to find people to carry out individual short-term projects' than to enter into long-term research agreements with the Tanzanian authorities.[55] There was also concern among the scientific council that investing too much in long-term migration studies would mean that the institute would lose its pioneering role, 'no longer obtain[ing] the very best people in its field'.[56]

Overall, the interest in experimental forms of shared land use between agriculturalists and migratory mammals diminished.[57] By the time the monitoring programme was closed when war broke out between Tanzania and Uganda in 1977, research into the migration of the Serengeti mammals differed markedly from the applied conservation research conducted by Watson.

Projecting: Computer Models and Population Densities, the 1980 to the 2000s

In the 1960s and 1970s, Nyerere's government continued to follow preservationist ideas about wildlife management dating from the colonial era, based on the state controlling farmers and herders. When between 1973 and 1976 the socialist government relocated members of all ethnic groups to comply with the Ujamaa policy of villagisation, this led to new settlements and new forms of agropastoralism by the Maasai, both in the national park and in the surrounding game reserves. The simultaneous prohibition of cultivation in some of the park's surrounding areas, such as the Ngorongoro Conservation Area in the wake of the 1974 Wildlife Conservation Act, stimulated international concern that poaching would increase.

In fact, during the war with Uganda, at a time of economic instability that weakened the strict Tanzanian protection regime, poaching became a more widespread issue, observed critically by international conservation NGOs.[58] At the same time, international organisations had less access to Tanzanian wildlife due to stricter regulations by Nyerere's government. Having established nine national parks, Tanzania's commitment to conservation was seen by the 1980s as exemplary for other independent African states. The international conservation community was worried to see the experiment fail, as the economically stricken state struggled to finance its wildlife sector.[59]

Following the IUCN's World National Parks Congress in Bali in 1982, initiatives by international organisations and NGOs questioned the viability of strict preservation.[60] In the early 1980s, for instance, IUCN members were involved in developing regional conservation strategies to demonstrate the local benefits of conservation as sustainable resource use. The Serengeti was the first region on the African continent for which such a strategy was published. On first sight, the programme of the Serengeti Regional Conservation Strategy (SRCS) resembled calls in the 1960s for more experimental land use. During the 1980s and 1990s, however, a new concern for animal population densities inside the park, mediated by the use of computer modelling, resulted in only half-hearted attempts at what came to be called community-based conservation.

After the SRI research was discontinued, a new international attempt at creating a monitoring programme was launched in 1985, involving the FZS and the World Wildlife Fund (WWF), the IUCN's sister organisation that partly funded its work on regional conservation strategies. Again, the idea was to design a long-term ecological monitoring programme for the Serengeti (called SEMP).[61] At the same time, the New York Zoological Society (NYZS) together with former SRI researcher Tony Sinclair, now working at the University of British Columbia, set up a monitoring project in the Ngorongoro Conservation Area.[62] Research into the Serengeti ecosystem thus resumed in 1986.

In both programmes, computers were to replace additional field workers by synthesising old data and making new projections. The motivation for the new programmes was, first, monitoring for management (the failed ambition of the SRI's research during the 1970s) and second, the centralisation and synthesis of SRI data that had thus far been of little practical use for TANAPA. Kenneth Campbell, an ecologist and programmer, was hired for this purpose and to train Tanzanian monitoring specialists, who were considered essential for the long-term implementation of the SRCS.[63]

However, as most scientists working at the SRI in the 1970s had taken their data with them, little came of the synthesis of fundamental research for more applied purposes originally envisioned for SEMP. Starting afresh, the aerial reconnaissance flights and occupation mapping of the early 1970s were resumed, using the old 5 km grid system.

These flights were conducted with an aircraft financed by the FZS to support TANAPA's anti-poaching efforts.[64] Again, aerial photographic samples rather than observers' records were used for wildebeest censuses and for vegetation and land-use data.[65] New data suggested that wildebeest numbers had remained stable at 1.3 million, which was interpreted as a sign that the park's ungulate population had reached a natural density.

Systematic research into migration routes picked up again in the 1990s. In 1999, the Swiss zoologist Markus Borner, who coordinated the FZS's research efforts in East Africa, working with Sinclair, used GPS collars to track eight wildebeests from a Cessna 182 aircraft.[66] Comparing their findings to data from the 1970s, they concluded that migration routes had shifted westwards and partly outside the national park.[67] An aerial survey of human populations in the Serengeti and surrounding areas was also conducted. In 1988, the survey concluded that human land use had increased in all these areas.[68]

This data fed into schemes for community-based conservation, which during the 1980s and early 1990s investigated if hunting and game cropping (the culling of animals for meat) could be allowed in the areas surrounding the park, to decrease the predicted human 'pressure' on the park itself.[69] Game cropping had already been part of the Talbots' and later of Watson's ideas for resource utilisation in the park, in the spirit of conservation as science-based resource management that had emerged in the 1960s.

However, while researchers returned to older methods, ad hoc management of land use was no longer a key concern. New data collected in the 1990s and the use of computer models to understand the relationship between different ecological variables shifted research interests towards understanding long-term trends.[70] In 1991, Sinclair and the marine biologist Ray Hilborn organised a workshop at the Serengeti Wildlife Centre in Arusha to synthesise existing knowledge and to identify remaining data gaps. At the workshop they discussed a QuickBASIC computer model, using data from the 1960s and 1970s, and newly collected data from SEMP, to understand the interrelation between vegetation growth, ungulate populations and predator numbers, as well as human activities inside and outside the national park, such as tourism and hunting. The computer model calculated animal population trends through to 2020, based on variables such as rainfall patterns, disease outbreaks and poaching policies.[71]

Computer simulations also focused on wildebeests. The wildebeest was not only the ecosystem's most dominant species in terms of biomass; by now it was also the best studied species, providing an abundance of data for model making. This understanding of the wildebeest as a variable in future predictions of population densities was in conflict with wildebeest cropping schemes. In the mid-1990s, Simon Mduma, a Tanzanian doctoral student at the University of British Columbia, used Hilborn's predictive model in a project supervised by Sinclair to calculate life tables for wildebeest populations, based on mortality and growth factors, rainfall, food, adult and calf survival, the possible effects of hunting, and potentially sustainable hunting rates. Mduma concluded that despite theoretically viable hunting rates, there was no guarantee of any significant reduction in poaching. Thus while it was calculated that (legal) hunting of forty thousand wildebeests per year would potentially be sustainable, researchers like Mduma, Hilborn and Sinclair feared that a failure to prevent illegal hunting would decrease the population density to an extent that affected all other species in the ecosystem, and so experimentation was not encouraged.

In work with computer models on future population trends in the 1990s, wildebeests were viewed as key factors in the overall dynamics of the ecosystem. As a keystone species they dominated all the Serengeti ecosystem's natural processes.[72] For researchers like Sinclair, the Serengeti ecosystem presented a 'baseline' area to study and model ecological processes in animal populations without human interference.[73] The intact migration routes of migratory wildebeests, buffalos, gazelles and zebras within the Serengeti National Park made the area especially valuable for this type of research. It was therefore necessary that the Serengeti remained under strict protection status, excluding all human activities other than research from the park and surrounding buffer zones.[74]

Sinclair's reasoning, while based on a scientific line of argument, tied in with the preservationist agendas of the Tanzanian government and its ambitions to expand national park tourism.[75] At the same time, the focus on human population pressure was in direct contrast to calls in the SRCS to consider the needs of local human communities.

The use of computers also influenced the type of recognised expertise. Despite SEMP's original ambitions to include and train African staff, the focus on modelling populations, keystone species and the Serengeti as a baseline for modelling elsewhere contributed to the continued exclusion of African researchers. During the 1980s and 1990s, the need for computer skills and additional aerial surveys had consequences for the organisation of the monitoring programme and for recognised forms of expertise.

Again, research was not done with the help of local park authorities, but rather with that of international organisations, especially the Germany-based FZS, which provided researchers and technology. In 1980, the Tanzanian government had established the Serengeti Wildlife Research Institute (SWRI) led by Karim N. Hirji of the University of Dar-es-Salaam, an overarching institute with five research centres of which the former SRI was one. Yet, as Hirji had to spend much time travelling between his family in Dar-es-Salaam and the institute in Seronera, research progressed only slowly. It was not until 1988 that two Tanzanian colleagues joined the team, Asukile Kajuni, who had studied range management at Texas Technical University, and G. Mngongo, with a bachelor's degree from Dar-es-Salaam University and administrative experience in the Regional Wildlife Office.

While Kajuni improved the communication with TANAPA, according to Borner and Campbell both Tanzanians lacked the experience to take over more technical or leadership roles within the programme. Campbell noted that their training progressed only slowly, especially in the use of cameras and computers. Technological ability was here used both to exclude and to legitimise exclusion. The cooperation between the FZS and TANAPA staff improved only after a negative report by Sinclair and B. N. N. Mbano from the Wildlife Division of the Ministry of Tourism, Natural Resources and the Environment. In 1988, SEMP became part of the Tanzanian Wildlife Conservation Monitoring (TWCM) project, which cooperated more closely with the SWRI and TANAPA.[76] Yet through to the present day, the TWCM depends on the financial assistance and computing expertise provided by such organisations as the FZS and the NYZS (renamed the Wildlife Conservation Society in 1993).[77]

Thus in the 1980s and 1990s, the resumed Serengeti monitoring programme, aided by better cameras and stronger computing power, functioned as a 'watchdog' over people and wildlife. Regular and systematic long-term 'surveillance' of projected animal and human population densities was considered necessary to understand any potential long-term trends and threats 'in time'.[78] Research into the underlying socio-economic problems of poverty in the park and surrounding areas was not part of the TWCM.[79]

By the mid-1990s, community-based forms of wildlife management were still mostly limited to the reluctant cropping of migratory mammals in the controlled game areas surrounding the park. Due to the short length of time migratory wildebeest spent in the cropping areas and the costs of purchasing game meat, the benefits of these community-based initiatives went to the more affluent households, while illegal hunting in and around the park continued.[80] Since the late 1990s, the FZS has started to invest in qualitative surveys and questionnaires in order to understand local motivations behind hunting or participation in conservation. However, as mediated by computer projections, many of these surveys also start from the assumption that human populations outside the park are a potential threat to wildlife populations and to stable migration routes within the park, if not now then in the future.[81]

This chapter has used the concept of technological mediation to investigate how differing observation and monitoring techniques, the type of data they produced and the type of expertise they required led to different experiences of the Serengeti as a space shared by animals and humans. By examining scientific and technological approaches to monitoring the movement of migratory wildebeests in the Serengeti from the 1950s until the 2000s, I have shown for three historical phases that key technologies and associated practices sustained particular forms of research cooperation and mediated different normative ideas about human–wildlife interaction and the sharing of space in and around protected areas. Over time, these technologies have contributed to the increased perception of the incompatibility of human and wildlife interests in protected areas. Yet zooming in on three forms of technological mediation has enabled me to point out noteworthy differences.

Overall, developments in observation technology correspond to larger narratives of postcolonial fortress conservation, yet this chapter has shown that different types of observation technology can bridge the distance between researchers, animals, land and local people, as well as create gaps that are both spatial and temporal in nature. This remains true for present-day observation technologies. A recent FZS project, for instance, seeks to develop smart algorithms for automated population counts, based on real-time images taken in a number of sample areas in the Serengeti National Park.[82] In this case, the real-time imaging has the potential to increase our understanding of the dynamic character of animal movement and distribution across the ecosystem. Yet while the algorithm is trained by ecological experts with experience in the field, the focus is on animal counts as a single indicator of a healthy ecosystem, and thus does not embed animal activity and interactions into their ecological environments.

Other projects that rely less on computational modelling pay comparably little attention to the ecological and social embedding of their collected data. The 'Snapshot Serengeti' citizen science project by researchers from Minnesota, Oxford, Chicago and Harvard uses volunteers to identify snapshots of animals taken by photo traps installed throughout the park.[83] While the project is an attempt to democratise the ecological research process and allows for various types of camera encounters, here too snapshots are analysed remotely and with little knowledge of either the ecological or the social context in which they were taken.

Observation technologies and the 'views from above' that these generate are here to stay, and so it remains important to pay attention to the potential roles of technologies in creating additional distances between researchers and research subjects. Historical insights, such as those presented here, can help us reflect on how various forms of remote sensing may mediate normative ideas on human–wildlife interaction and conservation expertise, and may consequentially affect local livelihoods.

Acknowledgements

This chapter is a revised version of a more comprehensive article by the same author: Simone Schleper, 'Airplanes, Cameras, Computers, Wildebeests: The Technological Mediation of Spaces for Humans and Wildlife in the Serengeti since 1950', *Environment and Planning E: Nature and Space* 5(2) (2022), 740–61.

Simone Schleper is Assistant Professor for the global history of science, technology and environment at Maastricht University. She is a historian of twentieth-century science and the environment. As part of the NWO (Dutch Research Council) Moving Animals project, she is currently researching the management of migratory species and the ways in which migration and animal movement evolve through negotiations between animals and the spatial, economic and industrial expansion of human societies. Her most recent work

(*Planning for the Planet*, Berghahn, 2019) focuses on the history of environmental and conservationist expertise in international institutions.

Notes

1. Myles Turner and Murray Watson, 'Game Management and Research by Aeroplane', *Oryx* 8(1) (1965), 13–22, here 13.
2. Myles Turner, *My Serengeti Years: The Memoirs of an African Game Warden* (New York: W. W. Norton, 1988).
3. John Hopcraft, 'Ecological Implications of Food and Predation Risk for Herbivores in the Serengeti' (PhD diss., University of Groningen, 2010).
4. Bernhard Gissibl, *The Nature of German Imperialism: Conservation and the Politics of Wildlife in Colonial East Africa* (New York: Berghahn, 2016).
5. Thomas Lekan, *Our Gigantic Zoo: A German Quest to Save the Serengeti* (Oxford: Oxford University Press, 2020); Benjamin Gardner, *Selling the Serengeti: The Cultural Politics of Safari Tourism* (Athens: Georgia University Press, 2016).
6. Donna Haraway, *When Species Meet* (Minneapolis: University of Minnesota Press, 2007); Astrid Schwarz, 'Die Ökologie des Sees: Diagramme als Theoriebilder', in Horst Bredekamp et al. (eds), *Bildwelten des Wissens: Kunsthistorisches Jahrbuch für Bildkritik* (Berlin: De Gruyter, 2003), 64–74; Jeremy Vetter, 'Field Stations and the Problem of Scale: Local, Regional, and Global at the Desert Lab', in Raf de Bont and Jens Lachmund (eds), *Spatializing the History of Ecology: Sites, Journeys, Mappings* (London: Routledge, 2017), 70–98.
7. Etienne Benson, *Wired Wilderness: Technologies of Tracking and the Making of Modern Wildlife* (Baltimore: Johns Hopkins University Press, 2010), 190.
8. Lorraine Daston, 'On Scientific Observation', *ISIS* 99(1) (2008), 97–110; Jennifer Gabrys, *Program Earth: Environmental Sensing Technology and the Making of a Computational Planet* (Minneapolis: University of Minnesota, 2016); Finn Arne Jørgensen, 'The Backbone of Everyday Environmentalism: Cultural Scripting and Technological Systems', in Dolly Jørgensen, Finn Arne Jørgensen and Sarah Pritchard (eds), *New Natures: Joining Environmental History with Science and Technology Studies* (Pittsburgh, PA: University of Pittsburgh Press, 2013), 69–86.
9. Robert Rosenberger and Peter-Paul Verbeek, 'A Field Guide to Postphenomenology', in Robert Rosenberger and Peter-Paul Verbeek (eds), *Postphenomological Investigations: Essays on Human-Technology Relations* (London: Lexington, 2015), 9–42.
10. Karen Barad, 'Posthumanist Performativity: Toward an Understanding of How Matter Comes to Matter', *Signs* 28(3) (2003), 801–31; Haraway, *When Species Meet*; Don Idhe, *Technology and the Lifeworld: From Garden to Earth* (Bloomington: Indiana University Press, 1990).
11. Lee Talbot and Martha Talbot, *The Wildebeest in Western Masailand, East Africa* (Bethesta: Wildlife Society, 1963), 53.
12. Lekan, *Our Gigantic Zoo*.
13. See for example Jean Paul Harroy, *Afrique, terre qui meurt: La dégradation des sols africain sous l'influence de la colonisation* (Brussels: Marcel Hayez, 1944).

14. William Pearsall, 'Report on an Ecological Survey of the Serengeti National Park Tanganyika', *Oryx* 4(2) (1957), 71–136, here 71.
15. Ibid., 74, 78–79 and 127.
16. Talbot and Talbot, *The Wildebeest*, 10.
17. Ibid., 53–56; Lee Talbot and D. R. M. Stewart, 'First Wildlife Census of the Entire Serengeti-Mara Region, East Africa', *Journal of Wildlife Management* 28(4) (1964), 815–27, here 819.
18. Murray Watson, 'Game Utilization in the Serengeti: Preliminary Investigations. II. Wildebeest', *British Veterinary Journal* 122(1) (1966), 18–20, here 18; Anthony Sinclair, *Serengeti Story: Life and Science in the World's Greatest Wildlife Region* (Oxford: Oxford University Press, 2012), 231.
19. Talbot and Stewart, 'First Wildlife Census', 823.
20. John Milton and Frank Fraser Darling, 'Air Photography and Zoology', in J. St Joseph (ed.), *The Uses of Air Photography* (London, John Baker, 1977 [1966]), 114–23.
21. Murray Watson, 'Air Photography in East African Game Management', in J. St Joseph (ed.), *The Uses of Air Photography* (London, John Baker, 1977 [1966]), 124–34.
22. John St Joseph, 'The Scope of Air Photography', in J. St Joseph (ed.), *The Uses of Air Photography* (London, John Baker, 1977 [1966]), 12–34, here 14.
23. Murray Watson, G. H. Freeman and George Jolly, 'Some Indoor Experiments to Simulate Problems in Aerial Censusing', *East African Agricultural and Forestry Journal* 34(1) (1969), 50–56.
24. St Joseph, 'The Scope'.
25. Milton and Darling, 'Air Photography'.
26. Murray Watson and C. I. Tippett, 'Examples of Low-Level Aerial Surveys Conducted in Africa from 1968 to 1979: One Firm's Experience', in International Livestock Centre for Africa, Kenya Ministry of Environment and Natural Resources, and United Nations Environmental Programme (eds), *Low-Level Aerial Survey Techniques: A Workshop within the Global Environmental Monitoring System* (Kenya: International Livestock Centre for Africa, 1979), 33–57, and here 46–49 and 57.
27. Talbot and Talbot, 'The Wildebeest', 14–15.
28. Kay Turner, *Serengeti Home* (Nairobi, 1977); Myles Turner, *My Serengeti Years*.
29. Watson and Tippett, 'Examples of Low-Level Aerial Survey'.
30. Jafari Kideghesho, 'Co-Existence between the Traditional Societies and Wildlife in Western Serengeti, Tanzania: Its Relevancy in Contemporary Wildlife Conservation Efforts', *Biodiversity and Conservation* 17(8) (2008), 1861–81, and here 1866.
31. Alistair Graham, *The Gardeners of Eden* (Lanham, MD: Rowman & Littlefield, 1973); Ian Parker, *What I Tell You Three Times is True: Conservation, Ivory, History and Politics* (Kilbryde: Librario, 2004).
32. Murry Watson, 'Game Utilization in the Serengeti: Preliminary Investigations, Part I', *British Veterinary Journal* 121(12) (1965), 540–6, and here 545.
33. Jeff Schauer, *Wildlife between Empire and Nation in Twentieth-Century Africa* (Cham: Palgrave Macmillan, 2019), 186–90.
34. Hugh Lamprey, 'Director's Report to the Scientific Council II. January – November 1967', Archives of the Frankfurt Zoological Society (FZS), box 'Stadt Frankfurt a.M. Serengeti Research Institute', 1.

35. Ibid., 'SRI Minutes of the Meeting of the Scientific Council. January 1969'.
36. Ibid., 'SRI Progress Report. April – June 1969', 5.
37. Ibid., 'SRI Minutes of the Meeting of the Scientific Council. January 1970'.
38. Ibid., pp. 15 and 23.
39. Linda Maddock, 'The "Migration" and Grazing Succession', in Anthony R. E. Sinclair and M. Norton-Griffiths (eds), *Serengeti: Dynamics of an Ecosystem* (Chicago: University of Chicago Press, 1979), 104–29, and here 108.
40. 'SRI Minutes of the Meeting of the Scientific Council. January 1969', FZS, box 'Stadt Frankfurt a.M. Serengeti Research Institute', 6.
41. Ibid., 'SRI Progress Report. April – June 1969'.
42. Ibid.
43. Ibid., 'SRI Minutes of the Meeting of the Scientific Council. January 1970'.
44. Ibid., Watson, 'Air Photography in East African Game Management', 124–26.
45. Ibid., 'SRI Minutes of the Meeting of the Scientific Council. January 1969'.
46. Ibid., 'SRI Quarterly Report. December 1970', 33.
47. Hugh Lamprey, Hans Kruuk and M. Norton-Griffiths, 'Research in the Serengeti', *Nature* 230(5295) (1971), 497–99; 'SRI Minutes of the Meeting of the Scientific Council. December 1970', FZS, box 'Stadt Frankfurt a.M. Serengeti Research Institute'.
48. Ibid., 'SRI Progress Report. April – June 1969'.
49. Ibid., 'SRI Quarterly Report. December 1970'.
50. Feroz Jafferali Kurji, *Conservation and Demographic Settings in Tanzania* (Dar es Salaam: Bureau of Resource Assessment and Land Use Planning (BRALUP), 1976).
51. Anthony Sinclair, 'Food Selection and Competition in the East African Buffalo (Syncerus Caffer Sparrman)', *East African Wildlife Journal* 10 (1972), 77–89.
52. Jeff Schauer, '"We Hold It in Trust": Global Wildlife Conservation "Africanization" and the End of Empire', *Journal of British Studies* 57(3) (2018), 516–42.
53. 'SRI Minutes of the Meeting of the Scientific Council, January 1970', FZS, box 'Stadt Frankfurt a.M. Serengeti Research Institute'.
54. Ibid., T. Mcharo, 'Quarterly Report. January – March 1973'.
55. Ibid., Hugh Lamprey, 'Director's Report to the Scientific Council II. January – November 1967'.
56. Ibid., 'SRI Minutes of the Meeting of the Scientific Council. December 1970'.
57. Ibid., 'SRI Minutes of the Meeting of the Scientific Council. January 1970'.
58. Stephen Makacha, Michael J. Msingwa and George W. Frame, 'Threats to the Serengeti Herds', *Oryx* 16(5) (1982), 437–44, and here 439.
59. IUCN, *The World National Parks Congress, Bali, Indonesia 11–22 October, 1982* (Bali: IUCN and Indonesian Directorate-General of Forestry, 1982), 30–31.
60. Robert Malpas and Scott Perkin, *Toward a Regional Conservation Strategy for the Serengeti: A Report of a Workshop Held at Serengeti Wildlife Research Centre, Seronera, Tanzania, 2–4 December 1985* (Dar-es-Salaam: Ministry of Natural Resources & Tourism, IUCN Regional Office for Eastern Africa, 1986), 5.
61. Markus Borner, 'Project Outline Serengeti Ecosystem Monitoring & Training Programme,' FZS, box 'Tanzania 1071_86 Serengeti Monitoring SEMP', 1985, 4.
62. Ibid., Letter from Richard Faust to William Conway (New York Zoological Society – NYZS), 25 March 1987.

63. Ibid., Kenneth Campbell, 'Serengeti Ecosystem, Monitoring and Training Programme. Project Proposal to World Wildlife Fund and Frankfurt Zoological Society', 1986, 2–3.
64. Ibid., Letter from Richard Faust to William Conway, 25 March 1987; ibid., Letter from Bernhard Grzimek to P. J. P. Kitomati (financial manager, TANAPA), 6 March 1979.
65. Ibid., Frankfurt Zoological Society, 'Serengeti Ecological Monitoring Programme. Conservation Monitoring in the Serengeti Ecosytem. Continuation of the Joint WWF/FZS Project. New Division of Focal Areas. A Proposal Based on Discussions Between WWF, FZS, SWRI, TANAPA and WD', July 1989, 6.
66. Simon Thirgood et al., 'Can Parks Protect Migratory Ungulates? The Case of the Serengeti Wildebeest', *Animal Conservation* 7(2) (2004), 113–20.
67. Jared A. Stabach, *Movement, Resource Selection, and the Physiological Stress Response of White-Bearded Wildebeest* (Fort Collins: Colorado State University, 2015), 114–19.
68. Anthony Sinclair and B. N. N. Mbano, 'Review of the Serengeti Ecological Monitoring Programme, 1986–1988, February 1988', FZS, box 'Tanzania 1071_86 Serengeti Monitoring SEMP'.
69. Martin Loibooki et al., 'Bushmeat Hunting by Communities Adjacent to the Serengeti National Park, Tanzania: The Importance of Livestock Ownership and Alternative Sources of Protein and Income', *Environmental Conservation* 29(3) (2002), 391–98.
70. Anthony Sinclair, 'Natural Regulation of Ecosystems in Protected Areas as Ecological Baselines', *Wildlife Society Bulletin* 2(3) (1998), 339–409.
71. Ray Hilborn, 'A Model to Evaluate Alternative Management Policies for the Serengeti-Mara Ecosystem', in A. R. E. Sinclair and Peter Arcese (eds), *Serengeti II: Dynamics, Management, and Conservation of an Ecosystem* (Chicago: University of Chicago Press, 1995), 617–34.
72. Simon Mduma, Anthony Sinclair and Ray Hilborn, 'Food Regulates the Serengeti Wildebeest: A 40-Year Record', *Journal of Animal Ecology* 68(6) (1999), 1101–22.
73. Sinclair, 'Natural Regulation'.
74. Anthony Sinclair et al., 'Long-Term Ecosystem Dynamics in the Serengeti: Lessons for Conservation', *Conservation Biology* 21(3) (2007), 580–90.
75. Gardner, *Selling the Serengeti*, 25–26.
76. Sinclair and Mbano, 'Review of the Serengeti'.
77. Mara Goldman, 'Partitioned Nature, Privileged Knowledge: Community-Based Conservation in Tanzania', *Development and Change* 34(5) (2003), 833–62.
78. Markus Borner, 'ZGF-Vorhaben 1093/89: Naturschutz-"Monitoring" in Tanzania', *ZGF Jahresberichte* 2 (1990), 9–10.
79. John Boshe, 'The Commercial Utilization of African Wildlife', *Swara* 9 (1986), 20–22.
80. Tomas Holmern et al., 'Uneconomical Game Cropping in a Community-Based Conservation Project Outside the Serengeti National Park, Tanzania', *Oryx* 36(4) (2002), 364–72.
81. Mduma, Sinclair and Hilborn, 'Food Regulates the Serengeti'.
82. Colin Torney et al., 'Assessing Rotation-Invariant Feature Classification for Automated Wildebeest Population Counts', *PLoS ONE* 11(5) (2016), 1–10.
83. Alexandra Swanson et al., 'Snapshot Serengeti, High-Frequency Annotated Camera Trap Images of 40 Mammalian Species in an African Savanna', *Scientific Data* 2(150026) (2015), 1–14.

Bibliography

Barad, Karen. 'Posthumanist Performativity: Toward an Understanding of How Matter Comes to Matter'. *Signs* 28(3) (2003), 801–31.

Benson, Etienne. *Wired Wilderness: Technologies of Tracking and the Making of Modern Wildlife*. Baltimore: Johns Hopkins University Press, 2010.

Borner, Markus. 'ZGF-Vorhaben 1093/89: Naturschutz-"Monitoring" in Tanzania'. *ZGF Jahresberichte* 2 (1990), 9–10.

Boshe, John. 'The Commercial Utilization of African Wildlife'. *Swara* 9 (1986), 20–22.

Daston, Lorraine. 'On Scientific Observation'. *ISIS* 99(1) (2008), 97–110.

Gabrys, Jennifer. *Program Earth: Environmental Sensing Technology and the Making of a Computational Planet*. Minneapolis: University of Minnesota, 2016.

Gardner, Benjamin. *Selling the Serengeti: The Cultural Politics of Safari Tourism*. Athens: Georgia University Press, 2016.

Gissibl, Bernhard. *The Nature of German Imperialism: Conservation and the Politics of Wildlife in Colonial East Africa*. New York: Berghahn, 2016.

Goldman, Mara. 'Partitioned Nature, Privileged Knowledge: Community-Based Conservation in Tanzania'. *Development and Change* 34(5) (2003), 833–62.

Haraway, Donna. *When Species Meet*. Minneapolis: University of Minnesota Press, 2007.

Hopcraft, John. 'Ecological Implications of Food and Predation Risk for Herbivores in the Serengeti'. PhD diss., University of Groningen, 2010.

Idhe, Don. *Technology and the Lifeworld: From Garden to Earth*. Bloomington: Indiana University Press, 1990.

Jørgensen, Finn Arne. 'The Backbone of Everyday Environmentalism: Cultural Scripting and Technological Systems', in Dolly Jørgensen, Finn Arne Jørgensen and Sarah Pritchard (eds), *New Natures: Joining Environmental History with Science and Technology Studies* (Pittsburgh, PA: University of Pittsburgh Press, 2013), 69–86.

Kideghesho, Jafari. 'Co-Existence between the Traditional Societies and Wildlife in Western Serengeti, Tanzania: Its Relevancy in Contemporary Wildlife Conservation Efforts'. *Biodiversity and Conservation* 17(8) (2008), 1861–81.

Lekan, Thomas. *Our Gigantic Zoo: A German Quest to Save the Serengeti*. Oxford: Oxford University Press, 2020.

Loibooki, Martin, Heribert Hofer, Ken Campbell and Marion East. 'Bushmeat Hunting by Communities Adjacent to the Serengeti National Park, Tanzania: The Importance of Livestock Ownership and Alternative Sources of Protein and Income'. *Environmental Conservation* 29(3) (2002), 391–98.

Maddock, Linda. 'The "Migration" and Grazing Succession', in Anthony R. E. Sinclair and M. Norton-Griffiths (eds), *Serengeti: Dynamics of an Ecosystem* (Chicago: University of Chicago Press, 1979), 104–29.

Mduma, Simon, Anthony Sinclair and Ray Hilborn. 'Food Regulates the Serengeti Wildebeest: A 40-Year Record'. *Journal of Animal Ecology* 68(6) (1999), 1101–22.

Rosenberger, Robert, and Peter-Paul Verbeek. 'A Field Guide to Postphenomenology', in Robert Rosenberger and Peter-Paul Verbeek (eds), *Postphenomological Investigations: Essays on Human-Technology Relations* (London: Lexington, 2015), 9–42.

Schauer, Jeff. '"We Hold It in Trust": Global Wildlife Conservation "Africanization" and the End of Empire'. *Journal of British Studies* 57(3) (2018), 516–42.

———. *Wildlife between Empire and Nation in Twentieth-Century Africa*. Cham: Palgrave Macmillan, 2019.

Schleper, Simone. 'Airplanes, Cameras, Computers, Wildebeests: The Technological Mediation of Spaces for Humans and Wildlife in the Serengeti since 1950'. *Environment and Planning E: Nature and Space* 5(2) (2022), 740–61.

Schwarz, Astrid. 'Die Ökologie des Sees: Diagramme als Theoriebilder', in Horst Bredekamp, Claudia Blümle, Katja Müller-Helle and Matthias Bruhn (eds), *Bildwelten des Wissens: Kunsthistorisches Jahrbuch für Bildkritik* (Berlin: De Gruyter, 2003), 64–74.

Stabach, Jared A. *Movement, Resource Selection, and the Physiological Stress Response of White-Bearded Wildebeest*. Fort Collins: Colorado State University, 2015.

Thirgood, Simon, et al. 'Can Parks Protect Migratory Ungulates? The Case of the Serengeti Wildebeest'. *Animal Conservation* 7(2) (2004), 113–20.

Vetter, Jeremy. 'Field Stations and the Problem of Scale: Local, Regional, and Global at the Desert Lab', in Raf de Bont and Jens Lachmund (eds), *Spatializing the History of Ecology: Sites, Journeys, Mappings* (London: Routledge, 2017), 70–98.

CHAPTER 10

Conserving Nature in Mozambique

Relaying Conservation Practices and Imaginaries since Colonial Days

Rozenn Nakanabo Diallo

K. spent the first part of his career as a hunting inspector (*fiscal de caça*) in game reserves in the provinces of Gaza, Sofala and Zambezia, before being posted to the central Ministry of Agriculture in 1984. He is one of a handful of Mozambican specialists in conservation issues since the colonial period. Although now retired, until 2013 he was the Ministry of Agriculture official in charge of the Convention on International Trade in Endangered Species of Wild Fauna and Flora (CITES).

> There was a convention in London in 1933. It placed wildlife obligations on colonial powers, via effective protection. It revolutionised wildlife around the world. And that is what led to national parks being set up. . . . There was a game warden corps. I come from that. It was in the 1970s. Before there were Europeans who did the work issuing fines. Afterwards, they accepted Blacks. . . . The way game wardens functioned was that we were like policemen, not technicians. . . . There were a lot of trials: the Portuguese didn't protect, they hunted! Most of the people who stood trial were Portuguese. . . .
>
> Unfortunately things changed after that. We Mozambicans didn't have a culture of conserving wildlife. For us, it was a matter of killing, selling and having money. Especially the soldiers. What they said was: we won the war, so it all belongs to us. We lost authority. By 'we' I mean the Portuguese and the wardens who depended on them. There was a period when the law didn't function, from 1974 to 1980. The soldiers hunted a lot. Times were hard. I was part of the colonial army, and was more used to obeying the Portuguese. That was a very sad time for us. For example, the soldiers would go hunting rhinoceros from helicopters, that happened right across the country. Often we couldn't go out into the field, otherwise the army would assault us.
>
> In 1980 new legislation was drawn up. Samora Machel was the first person we won over. There had been an exhibition of wildlife items in a European

> city, Bucharest. Mozambique exhibited ivory teeth and was awarded a gold medal. Samora Machel was very proud of that, and got a directive issued: people living in national parks had to leave; the army had to stop hunting. After that, things started reverting to normal a bit. But we had difficulty enforcing the directive, for we didn't have cars or fuel. At the time, wildlife belonged to the National Veterinary Directorate. There was a service in charge of wildlife. It was only a minor part of what it did. In the 1980s, there was talk of setting up a National Wildlife Directorate, where there would be areas, hence national parks, national reserves and game reserves. . . . But what complicated things was the war between us. . . .
>
> We had the laws, but we couldn't enforce them. We couldn't go out into the field. Nor could tourists and hunters. The service was weak, we were there doing nothing. The Ministry of Agriculture decided to merge the Forestry Directorate and the Wildlife Directorate, giving rise to the Forestry and Wildlife Directorate in 1986. Since then we've been together, even if there have been changes. After October 1992 we went back out into the field.[1]

As this testimony shows, the history of nature conservation is both global and situated. In the South, the conservationist model gradually spread through and between colonial empires.[2] Representations of the world and ways of proceeding thus exhibit varying degrees of family resemblance: the quest for a garden of Eden, the idea of nature as untouchable or the stigmatisation of local populations' practices perceived as deeply destructive of the environment.

Within this framework, the case of Mozambique is original. Due to Portuguese colonisation, which until the 1920s was based on granting concessions to large companies, conservation policies arose later than in neighbouring countries such as South Africa. However, the same general dynamics may be observed, including significant violence towards local populations, particularly in the form of evictions from protected areas.

Mozambique acquired independence from Portugal in 1975, that is, after the UNESCO Intergovernmental Conference for Rational Use and Conservation of the Biosphere held in Paris in 1968 and the UN Conference on the Environment in Stockholm in 1972. In other words, when Mozambique became independent, other independent African countries had already been thinking about and implementing international conservation policies for over a decade. Mozambique rapidly took up the major guidelines instigated by the United Nations Environment Programme (UNEP) and the International Union for Conservation of Nature (IUCN). And amid the many changes brought about by independence and the postcolonial era more generally, the legitimisation of protected areas acquired (partly) new garb: it was no longer a question of protecting nature for the benefit of the few (a white colonial elite), but in the name of biodiversity and especially of (sustainable) development, by developing tourism to help combat poverty among populations living in the vicinity of protected zones.

Furthermore, like in most Southern states, these areas, and especially national parks, were seen as a way of enhancing territorial coverage and building the national story.[3] It was in this context that Samora Machel, the first president of independent Mozambique, gave a speech in 1981:

> The conscious destruction of our flora and fauna damages our most precious assets. Forests are one of our greatest treasures. They provide precious and essential wood, purify the air we breathe, balance the climate and river flow, fertilise the land and protect from the destructive action of winds. . . . Our wildlife is of incalculable cultural, scientific, ecological and economic value. In colonial times, wildlife was reserved for the pleasure of rulers and a few privileged individuals. Nowadays wildlife serves the people of Mozambique. . . . If we are to give material form to all the programmes for the conservation and correct use of our wildlife resources, we need to educate our people about the environment and nature conservation.[4]

This speech shows that while the new regime pursued a similar line to that of the Portuguese colonial state, it presented its nature conservation as innovative (it was a matter of working 'for the people of Mozambique'), and endorsed international discourse about battling the destruction of natural resources, particularly via protected areas. While maintaining links with the Eastern Bloc, where its wildlife veterinarians continued to train through to the late 1980s, Mozambique ratified CITES in 1981, when it also became a member of the IUCN, then in 1994 it signed the new Convention on Biological Diversity.

Thus after having led the country to independence in 1975, FRELIMO (Frente de Libertação de Moçambique) complied with the international rhetoric being relayed in the country by the World Bank, the FAO (the UN Food and Agriculture Organization), the WWF (World Wildlife Fund), the IUCN, USAID (the United States Agency for International Development), the AFD (Agence française de développement) and KfW (Kreditanstalt für Wiederaufbau, the German development bank).

Drawing on conservation archives and interviews conducted in Mozambique between 2009 and 2014, this chapter argues that even though governments insisted they were promoting conservation to serve Mozambicans, we must not be misled by the break with the colonial past. The regime, the configuration of stakeholders and the ideas proclaimed may well have changed, yet this was also a matter of (re)actualising prevailing trends in worldwide nature conservation.

What we may observe is an interplay between colonial legacy and postcolonial reappropriation. Protected areas continued to act as an instrument of social control, while beliefs and practices instilled by the colonial period persisted at the heart of the Mozambican administration. These dynamics were rooted in circulations linking up different scales: ideas and policies devised in international spheres reverberated in the capital, whose decisions produced effects in protected areas at the local level. In other words, there was a perpetual toing and froing between scales and periods, in which ideas were constantly transposed and tweaked. This is what the account by game warden K. tells us, and what I wish to illustrate by focusing on two aspects: changes to the state's nature regulations since colonial days, and management practices at Gorongosa National Park, the flagship for Mozambican conservation policy since the early twentieth century. These two cases show how conservation practices and imaginaries have been relayed and perpetuated throughout the lengthy contemporary period.

In Maputo: A Short History of Conservation Regulations, and the Influence of Internal and External Factors

In Mozambique, nature conservation is a small world, in which an administrative elite of twenty or so people based in Maputo oversee protected areas out in the field.[5] The first generation of veterinarians have known each other since childhood or the beginning of their studies in the 1970s and 1980s. They have pursued careers in the same ministerial departments, calling on each other to fill certain management positions. In an informal interview, the assistant manager of the TFCA unit (Transfrontier Conservation Areas)[6] exclaimed: 'Those three (L., M. and E.) have been together since secondary school, they're inseparable!'[7] All five veterinarians from this first generation now hold senior government positions and/or act at the daily interface between the state and international development and conservation agencies such as the FAO, the WWF, the AFD and the World Bank.

As for the second generation, they all studied in the 1990s and 2000s, and have a background in biology or the anthropology of development, rather than veterinary science. Having said that, these conservation professionals cultivate an esprit de corps, characterised by beliefs and practices inherited from colonial days: colonial nomenclature for nature presents certain affinities with present-day ways of devising conservation regulations.

In Mozambique, like elsewhere in Africa,[8] the earliest conservation policies grew out of measures to supervise hunting, taking the form of hunting reserves, some of which were turned into national parks in the 1950s and 1960s when the colony started to be properly governed with a single administration and an organised nationwide economy. Thus the first 'Regulations for hunting in the province of Mozambique' were published in 1909, but it was not until

1955 that they truly took shape, shortly after Portugal ratified the 1933 London Convention Relative to the Preservation of the Fauna and Flora in their Natural State, which emphasised the importance of establishing protected areas.

Decree no. 40040, published by the Portuguese Overseas ('Ultramar') Ministry in 1955, was foundational in this respect. It covered all Portuguese overseas territory and regulated most hunting activities, viewed as 'the most generalised form of wildlife destruction', laying down a categorisation of 'protection zones' where hunting was forbidden, including national parks. Irrespective of age, the Mozambican administrative elite still regularly cite this decree. 'There was decree 40040 for the protection of natural resources, hence wildlife, forests, water and soil, via regulations for each domain', a member of the Ministry of Agriculture's Wildlife Department explained. 'This decree was a doctrine, our religion.'[9]

This decree established a single body in each province of the Portuguese empire to manage natural resources. Veterinary Services were responsible for wildlife, and Agricultural Services for the land and flora, with protected zones being categorised into national parks, integral nature reserves, partial reserves and special reserves.

Two important events occurred in 1972, shortly before independence. First, a 'Gathering to study problems relating to wildlife and protecting nature in Portugal's overseas provinces' was held in Angola, a few months after the UN Conference on the Environment in Stockholm (the first 'Earth Summit'). This was attended by representatives of each province, including veterinarians from the provincial directorates.

The emphasis was on big game tourism and on the 'rational use' of wildlife to help develop Portugal's overseas provinces. Private sector involvement was mentioned on several occasions, including in point XIII in the final conclusion, which states for example that 'tourism and protecting nature are in no way incompatible'. Equally, the question of the link between conservation areas and resident populations is raised in point XVIII, in which the emphasis is placed on boosting local economies, especially in terms of employment, populations being 'entitled to benefit from the revenue produced by tourism'.[10] Point XXIII addresses the issue of the institution in charge of nature policies, which was to be granted legal personality together with administrative and financial autonomy.

The second key event of 1972 was the publication by the Province of Mozambique Veterinary Services of what has since been widely referred to as the *Livro Azul*, literally, the blue book. This document of three hundred or so pages is a compilation of the conservation legislation – still based on regulating hunting – applicable in the remaining reserves. Most of the regulations date from the 1960s. The *Livro Azul* thus gives us a snapshot of the regulations applicable in colonial Mozambique. For example, 'Resolution' (*Diploma Legis-*

lativo) no. 2496 of 4 July 1964, published by the Governor General of the Province of Mozambique, makes provisions for five categories of conservation area: national parks, integral nature reserves, partial reserves, special reserves and zones subject to special vigilance, hence one more – zones subject to special vigilance – than decree no. 40040 of 1955. The definition of national parks is the same as in decree no. 40040,[11] with fishing, camping and transiting outside officially established places and paths now being forbidden. The categories and practices instigated in the late 1950s thus affected subsequent periods, despite the legislation undergoing minor – though at times more radical – modifications. Things continued alike after independence.

On reaching power in 1975, FRELIMO wished to retain the existing nature conservation policies, and conservation areas thus continued to be managed as prime national assets. Between the late 1970s and early 1980s, the authorities stipulated the conditions under which hunting was allowed, adopting the broad outlines established at the third FRELIMO party conference in 1977: 'the action of the state and party in the field of wildlife should seek to secure ecological equilibrium via its rational use, by defining the task of regulating hunting.'[12]

The first National Wildlife Gathering was held at Gorongosa National Park, in the centre of the country, in 1981, when Mozambique also became a member of the IUCN and ratified the CITES convention. The following years were marked by the escalating civil conflict between FRELIMO and RENAMO (Resistência Nacional Moçambicana), the rebel group that fought FRELIMO from the late 1970s until the peace agreements of 1992, and which has since become a political party (though without disarming fully). Conservation was neglected, and ten years passed before regulations started being produced once again. Thus in 1990 the 'Strategy for Developing Forestry and Wildlife' was published, a declarative text laying down principles for action. Emphasis was placed on re-establishing the network of parks and nature reserves, with the objective of renovating infrastructure and training managers. The main missions of parks and reserves were to conserve flora and wildlife assets, to provide the population with a rational quantity of animal protein and to generate foreign currency from hunting, tourism and research.

The Convention on Biological Diversity was ratified in 1994, then the National Forestry and Wildlife Programme (1995–2000) was published in 1995, drawn up with the financial and technical assistance of UNEP, the FAO and the World Bank. This programme accords a place to 'rural communities' who are meant to benefit – alongside 'consumers' and the 'private sector' – from the 'sustainable use of resources'. This was the first time the issue of involving rural communities and the private sector appeared in writing, even though there was nothing new in terms of narratives, as we have seen with the 1972 summit and Machel's speech in 1981.

The emphasis placed on rural communities was due to the re-emergence of 'traditional power' with the end of the civil conflict: increasingly, the authorities insisted on the role of 'local traditional knowledge' in resource management. This shift also echoed mainstream institutionalised conservation in the wake of the Rio Earth Summit in 1992. International conservationist discourse (re)attributed environmental virtues to local traditional communities, and after having been described as potentially harmful for the preservation of natural environments, they were progressively viewed as key sustainable development partners.[13] This emphasis on communities was further linked to FRELIMO's rehabilitation of 'tradition' after having roundly condemned it in the 1970s and 1980s, a move that had sparked resentment, especially among rural populations, and caused some to rally to RENAMO.

Having said that, the central administration also referred to 'rural populations invading' certain conservation areas described as in the process of being 'salvaged'. Hence the call for reflection about 'rural populations' and communities' right to use the land', followed by the drawing up of an objective to set up community pilot projects, overseen by the Ministry of Agriculture, which was in charge of protected areas at the time. The year 1995 also saw the publication of the National Environmental Policy, immediately after the Ministry for the Coordination of Environmental Action was founded. This policy specified that the government recognised that development and environmental protection were interdependent.

Then in 1996, at the Seventh National Forest and Wildlife Gathering, the authorities pointed to two major factors behind the 'threats' to natural ecosystems: absolute poverty (causing the uncontrolled use of resources) and institutional weakness – picking up once again on the watchwords of international environmentalism and developmentalism in the wake of the UN Conference on Environment and Development in Rio in 1992. Centralisation was explicitly criticised by the national head of forestry and wildlife; what was now needed was to 'seek and implement solutions for administrative and financial autonomy at the institutional level'.

This last point was taken up in the Wildlife and Forestry Development Strategy and Policy published in 1997, which chimed with the 'third phase of conservation',[14] that of 'integrated conservation' in which it was internationally recognised that 'natives' needed to 'participate' in devising and implementing conservation policies.[15] The year 1997 also saw Mozambique adopt its first (formal) decentralisation policy: the state was to share decisions with local communities and the private sector. From the same perspective, the idea gathered strength that the administration in charge of forestry and wildlife issues should be self-financing, which may be regarded as a neoliberal meta-referent: irrespective of the sector, all donor policies were affected by the principle that administrations were to be self-financing and run along business principles to reduce costs. These objectives were an implicit criticism of centralised political decision-making: they were formalised in this declarative text, which once again explicitly echoed major international watchwords.

Then in 1999 the law on forestry and wildlife genuinely prioritised conservation areas. Article 10 defines three types of protection zone: national parks, national reserves and usage zones of cultural historic value.[16] There were henceforth only two categories of natural conservation areas – national parks and national reserves – as against four in the colonial period. Yet the real shift occurred the following year, in 2000, when the Ministry of Tourism was set up. It was tasked with developing the country's unexploited potential, its '3,000 kilometres of beaches', now called 'conservation areas for tourism purposes', which included game reserves. The Ministry of Agriculture remained in charge of wildlife outside these conservation areas.

This was, in short, a paradigm shift, one that senior officials at the Ministries of Agriculture and of Tourism in Maputo often raised during interviews: previously, tourism had been marginal to legislation and to summits held to discuss nature protection, but henceforth conservation was very officially rooted in the country's tourism policy. This proved controversial, for nature policies were now clearly associated with a mercantile approach combining conservation with economic development. This was emphasised by E. and M., Mozambican veterinarians formerly employed at the Ministry of Agriculture and now working at the Ministry of Tourism:

> The transition was difficult, hurtful . . . . Parks are preoccupied with spaces to protect biodiversity and manage the environment. But the ministry obliged us to view parks as business centres.[17]

> Conservation areas 'for tourism purposes'. This expression caused a lot of confusion and disputes. The priority isn't tourism, or use: it is conservation itself. That is the primary objective. None of the conservation areas were created for tourism purposes. Sometimes it was to conserve species. Nowadays, it is an expression that causes a tremendous amount of confusion in the country.[18]

Creating the Ministry of Tourism thus endorsed an approach connecting conservation to tourism. This was not novel per se, for the policy lines adopted in colonial days and in national programmes following independence had already insisted on this association. What was novel, however, was the institutionalisation of the alliance between conserving biodiversity and economic development *through* tourism. Nature was not conserved in itself and for the benefit of a few, but *in order to* attract tourists and compete with the neighbouring countries dominating this market segment, with South Africa in pole position.

Thus after being based on regulations from the colonial period, conservation policies gradually shifted towards protecting 'natural ecosystems', primarily within conservation areas. Official discourse referred less to hunting than to conservation areas. But wildlife remained central, for it was what drove tourism: safaris and recreational hunting (in dedicated hunting reserves) were the two sources identified for tourism revenues. This change was propelled by the head of state in person, who took up and emphasised the idea that tourism was an international issue, for it generated foreign currency, a point made by many donors. This shift was also linked to a change in narrative model: the usage-based conservation model, which had emerged in southern Africa (especially South Africa) in the 1940s and 1950s, was now superseded by a more scientific and preservationist one in which reserves and parks could be more lucrative as tourist destinations than via direct exploitation of the wildlife.[19] The 2003 'Tourism Policy' sanctioned this change, referring to the need to 'maximise earnings [in] conservation areas for tourism purposes'.

The final innovative text was the 2009 Conservation Policy, the fruit of several years' 'dialogue' between ministries (Tourism, Agriculture, Environmental Coordination and Fisheries), conservation bodies (the WWF and the IUCN) and donors (the AFD, World Bank and the GIZ (Deutsche Gesellschaft für Internationale Zusammenarbeit, the German international development agency)). This document suggested introducing three new policy instruments: local management councils, a national administration for conservation areas, and sustainable financing mechanisms, particularly a National Conservation Foundation.

This policy was the first in the history of Mozambican conservation to focus almost exclusively on conservation areas. The emphasis placed on the alliance between tourism and conservation was confirmed, as was the link between conservation, development and eradicating poverty. The state, private sector and local communities were urged to work hand in hand to promote and properly manage conservation areas, with mention of 'PPC' (Public, Private and Community) partnerships. A set of 'strategic actions' was to be carried out, including drafting a resettlement procedure should populations' continued residence in an area be deemed incompatible with conservation, and overhauling the categories of conservation area in accordance with IUCN classification. The Mozambican authorities partly confirmed these policy directions over the following years, first in 2011, when a decree established an administration to manage conservation areas (ANAC), then in June 2014, when a law to protect biodiversity was passed that emphasised public–private partnerships.

The great watchwords have been repeatedly wheeled out, in what one might call a triptych of development: promoting development in which conserving biodiversity is associated with tourism and combating poverty – an approach that came to dominate public action. First, for example, the tourism–conservation pair had been mentioned as early as the 1972 summit. Second, the participation of local populations was mentioned – though not perhaps valued – as a way of managing natural resources. And third, the involvement of the private sector was invoked – and continues increasingly to be so – once the viability of conservation areas ceased to be envisaged independently of the private sector's necessary involvement via contracts and concessions. This then gave rise to the commercialisation of conservation areas.

This change results from the subtle interplay between external influence (embodied by the international aid regime) and long time frames (illustrated by the colonial legacy and its postcolonial reappropriation). These two processes are perpetually interconnected, mutually constraining one another in a process that underpins the regulations and overarching narratives that enact change – the new texts adopted and new institutions created – though without ever being wholly unprecedented, thus revealing the persistence of singular practices and world views.

For example, the emphasis placed on the commercial aspect of nature conservation is still hotly contested by veterinarians, who have overseen the sector since colonial times. References to old norms, such as the *Livro Azul*, show that adopting new regulations and instruments does not necessarily transform stakeholders' practices. One of the members of the TFCA explained:

> In 1997, the law was revised, what we call the *Livro Azul*. This book was very well made, and was devised over the years during the colonial period. But for some reason, it was decided that it was important to revise this text. And this 1997 law introduced a lot of confusion. For example, A. W. [the head of Limpopo National Park] still uses the *Livro Azul*. It provides for various categories of conservation areas, unlike the new 1997 law [on forestry and wildlife], under which there are only two, namely parks and reserves. So things are very confused.[20]

As observed by Thomas Bierschenk and Jean-Pierre Olivier de Sardan, when concepts, procedures or institutions are imported, this is rarely to the detriment of what is already in place.[21] Pre-existing arrangements tend to be perpetuated, for they are not replaced but overlaid by the imported ones. Thus the *Livro Azul* continues to influence the practices and mindsets of certain national stakeholders: this code of procedures is regularly invoked and reactivated in contemporary public action, with stakeholders still referring to it via their actions, abstentions, omissions or misunderstandings. And the ethos of Mozambican officials is implicitly constituted by the interplay of mutual constraints between inherited beliefs (the influence of the lengthy time frame) and the international agenda (the influence of the outside world).

In Gorongosa: Perpetuating the Narrative of Environmental Destruction from Colonial Days to Those of Philanthropic Capitalism

Since colonial times, there has been an increasing number of legal texts about managing the natural environment. However, it is not always possible to detect any clear intention on the part of the legislative, for the laws, policies and programmes have often contradicted one another. Nevertheless, broad trends may be identified, as revealed by the history of Gorongosa National Park. Gorongosa lies in the centre of the country, and since the early twentieth century has been the showpiece of Mozambique's conservation policy. The park is home to varied wildlife, and its ecosystem combines savannah, woodland and watercourses (with Lake Urema in the middle), flanked to the west by a mountain range called the Serra.

The importance of this protected area may be measured not solely in terms of its environmental treasures. Indeed, it has held a special place in the country's political imaginary since the end of the colonial period, the Serra having played a strategic role during the armed conflicts waged in the run-up and subsequent to independence. The mountain range was one of FRELIMO's bases during the twenty-year war of national liberation against the Portuguese, and RENAMO established its main base there during the civil conflict against FRELIMO.[22] Despite changing configurations, this political and military framework displays real continuity reaching back to colonial times, a continuity that also characterises the alarmist discourse about environmental destruction focusing on the Serra: the American philanthropic foundation in charge of the park since the 2000s perpetuates this discourse, despite simultaneously echoing the communitarian and participative narratives.

The Gorongosa wildlife reserve was set up in 1920 by the Mozambique Company, which had held a concession in the north and centre of the colony since 1892. The local population was constrained from hunting, fishing and farming in this territory dedicated to exploiting wildlife for meat and ivory. These restrictions became even keener when, as a result of the 1933 London Convention, the reserve increased in size from 1,000 km^2 to 3,200 km^2. The colonial government took direct control of administering it in the early 1940s, and the Wildlife Commission in Beira (the capital of Sofala province) recommended to the governor of the province, José Ferreira Martins, that hunting be more effectively regulated to stamp out poaching.

This emphasis on Gorongosa, which went on to become the crown jewel of Mozambique's conservation policies, was largely due to recognition of its 'potential'. Thus in 1948, the head of South Africa's Kruger National Park, Colonel J. A. B. Sandenberg, visited the site accompanied by the Governor General of Mozambique, Gabriel Texeira. On this occasion, Sandenberg declared that 'Gorongosa could be the finest wildlife reserve in Africa'.[23] Banking on the reserve's attractiveness meant opting for the model of a park emptied of its inhabitants. Indeed, the first resettlement measures (without compensation) were put in place: according to Rosemary Galli's estimates,[24] in the 1950s twenty-five thousand inhabitants lived in the territory before a quarter of them were obliged to leave the reserve. They resettled on the eastern and southern slopes of the Serra da Gorongosa mountain range, on the edge of the park.

Since then, the Serra ecosystem has given rise to alarmist discourse. In July 1960, the Gorongosa reserve became a national park and its size was increased to 5,300 km^2, before being reduced to around 4,000 km^2. The new park administrator, a veterinarian from Lisbon, wished to develop safaris and the tourist industry. The number of visitors rose steadily, from 1,361 in 1952, to 6,096 in 1960, then 8,203 in 1965.[25] It was in this context that the idea of the population's harmful effect on the Serra first arose. At the time, the park administrator referred to the mountain being 'attack[ed]' by peasants who were farming at ever higher altitudes, inevitably causing the deforestation of the mountain range.[26]

In fact, this 'attack' came in the wake of the abolition of forced cotton cultivation in 1961. The Mozambican Cotton Institute was founded, and the administration appropriated the most fertile land, especially to the south and west of the mountain range, where it settled colonist cotton farmers. Many peasants left to go and settle on the flanks of the Serra, looking for farmland no longer available elsewhere. The administrator's fear was that the reduction in forest cover might affect the region's climate, by reducing rainfall in the mountains and, in turn, the park. The link between the hydrological equilibrium of the park and mountain range was thus established by the colonial administration, resulting in the first bans on logging on the slopes of the Serra.

However, in the mid-1960s, the administration was referring to a 'major risk of environmental destruction'. According to anthropologist Todd French, it was a Portuguese agronomist, Macedo, who came up with this narrative of environmental destruction back in 1964,[27] institutionalising a rhetoric of threat and danger:

> In his first study of Gorongosa in 1964, which lasted only eighteen days, Macedo constructed 'a narrative of environmental destruction' that blamed the deficient moral characteristics of Africans for the 'ecological crisis'. The main problem was the 'attack on the mountains' by African agriculturalists practicing agriculture at progressively higher altitudes. While he recognised the evictions from the park as a contributing factor, the ultimate cause was the 'greed' of African peasants. . . . Besides their greediness, Africans were accused of primitive and environmentally wasteful livelihood practices, including . . . the use of fire to clear vegetation and to hunt, wanton destruction of the forests, and lack of conservation practices.[28]

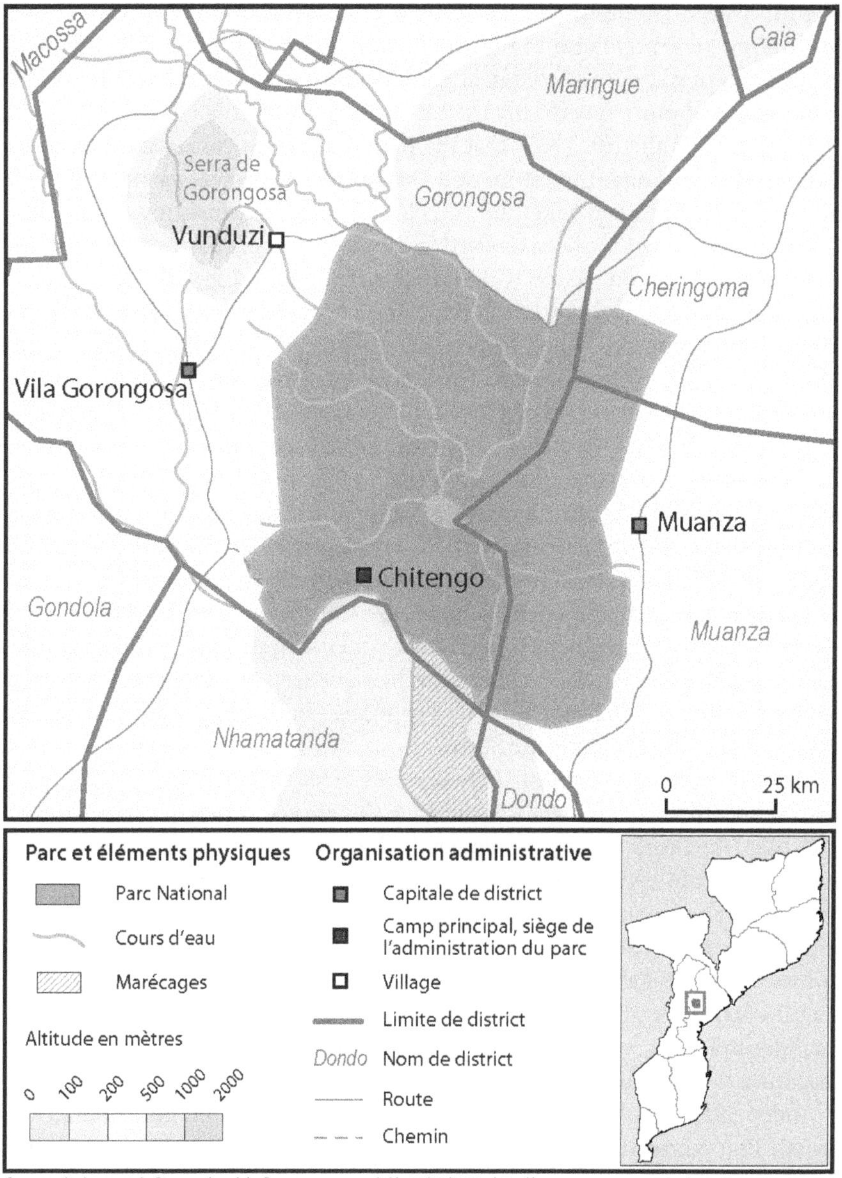

Source : site internet du Parc national de Gorongosa, consulté le 01/12/2012 - http://www.gorongosa.net/en/page/maps/maps.
Réalisation : Violaine Jurie - Adobe Illustrator CS3.

Map 10.1. Gorongosa Park. © Violaine Jurie, inspired by data created from Gorongosa National Park, 2012. Published with permission.

The idea was of an imminent catastrophe, and the solution was clear: closely controlling local populations' practices using rational and scientific methods, in order to preserve the ecological equilibrium of the central region of Mozambique, and more specifically that of Gorongosa National Park.

It was also in the late 1960s that the park came to be conceived in terms of its scientific value and interest. In other words, the ecological narrative initially devised by Macedo was taken up in colonial discourse, insisting on a defect: Gorongosa Park had not been devised for its ecological equilibrium, particularly regarding its close connection with the Serra. This idea arose with Macedo, then took root thanks to work by South African ecologist Kenneth Tinley. In the late 1960s, he was commissioned to produce a study of the park's wildlife by the Mozambique Veterinary Services. This study was of major importance, for in the late 2000s the managers and employees I interviewed were still referring to it.[29]

In this study, Tinley insisted on the idea of an ecosystem, with the objective that the park be 'a viable ecological unit'.[30] Following Macedo, Tinley stated that the greatest threat to Gorongosa National Park's ecosystem was the uncontrolled cultivation of (mainly food) crops on the flanks of the Serra. He therefore proposed extending the park. Although these propositions were favourably received by the scientific staff, the colonial authorities did not back the project, insisting rather on protecting water sources and the need to 'educate' populations. The war of liberation being waged by FRELIMO was escalating, no doubt explaining why the Portuguese authorities were reluctant to expand the park to include the Serra.

After the peace agreements between FRELIMO and RENAMO in 1992, a first eighteen-month restoration plan for the park was put in place in 1994. The plan, financed by the European Union and the African Development Bank, sought to expand the park to include the wooded zones above 700 m on the Serra, deemed more ecologically sensitive and thus strategic for the mountains' role as a water reservoir. This idea has become a touchstone in discourse by new stakeholders involved in operating the national park, indicating that the narrative of a viable ecological unit is still strong. And it is primarily farming practices, viewed as driving deforestation, that are targeted by such discourse, which has persisted since colonial days.

Independently of whether such accusations are true or not, French observed that conservation policies are objectively responsible for engendering these farming practices on the mountain slopes. For it was precisely policies to 'safeguard' the protected area that largely inspired the decision to evict and resettle the population, triggering the changes in farming and land ownership subsequently decried by conservationist discourse. This discourse stemming from 'scientific' studies in the 1960s has been reactivated by international development and conservation agencies, and to a certain extent has been given the central state's blessing.

In 2008 this vision was endorsed by the public–private partnership (PPP) – named the Gorongosa Restoration Project (GRP) – that took over management of the national park. The president of the Carr Foundation, Greg Carr,[31] decided to devote about $40 million to 'restoring' the park. While the Portuguese considered Gorongosa to be the 'crown jewel' of their overseas conservation policies, the PPP describes it as 'Africa's Lost Eden', or more recently as 'Africa's greatest wildlife restoration story'.[32]

Restoring the national park to its former glory involves several levers. To help 'rewild the world' – one of the conservation world's objectives at the time[33] – hippopotamuses, elephants and buffalo were reintroduced to Gorongosa. A tourist amenities investment plan was drawn up for the park, which in the 1960s and 1970s had attracted tens of thousands of European tourists. Above all, it was deemed necessary to carry out development projects to 'educate' local populations, that is, to 'raise awareness' of the environmental cause and 'incite' them to reside outside the park, hence to no longer live there (about five thousand people lived within its boundaries when I was conducting my enquiry between 2009 and 2014), and to not fish and hunt, or 'poach', inside the park.

The social aspect of the PPP thus fitted neatly with the model of a conservation fortress, which though not explicit was well and truly omnipresent. As Greg Carr stated in a documentary, conserving the protected area meant people were no longer to live within it:

> We need to help these people lift themselves out of poverty, because we care about them, and that's a reason enough. But the second reason is that poverty is a threat to the ecosystem, and when these people don't have enough to eat, of course they are going to come to the park and hunt animals: who wouldn't? And so we need to help them with their agriculture, with their economic development . . . . That's the dream we can achieve: full development and biodiversity protection at the same time.[34]

The emphasis here is placed on the prospect of economic benefits for the population. Like in many very contemporary conservation projects, it is a matter of converting some inhabitants into employed workers as part of a 'modern' cash economy,[35] producing dry fruits and organic coffee, working in tourist concessions – and reforesting the Serra, as Carr has indicated in several press articles:

> In an ideal world, 20 years from now, the park would be surrounded by a nice middle-class – by African standards – community that loves the park and understands that the park is helping them. We have to create that culture through conservation education.[36]

> There are people who believe that conservation and human development are two different things. That's wrong. Our project targets both and intends to make them complementary .... [Tourists] want to be awed by a magnificent place, but at the same time we're creating new jobs. If you look at our project, you'll see we are reforesting and helping the 200,000 people living in what we call the buffer zone.... A large tourism industry creates jobs and improves living conditions.... We think that if living conditions around the park improve thanks to schools, electricity, and health centres, and if farming techniques improve too, then people will end up leaving the park.[37]

The PPP thus echoes major development agendas, insisting for example on the importance of reducing poverty among those living in the vicinity of the park. But the vision promoted is clearly that of a conservation fortress, based on the US model of Yellowstone, the first contemporary national park, created in 1872. The objective is to expel habitants from the park. Admittedly, they are encouraged rather than compelled to leave. It is a matter of exhorting people to remain outside the park, and to attract those still living in the 'heart' of the park to settle in the 'buffer' zone by making it economically attractive to do so.

The rhetoric of the project is strikingly American in its ethos. Greg Carr often refers to Yellowstone, having grown up nearby: 'The point of Yellowstone Park was to recover it, and a hundred years later it's back. I look at Gorongosa that way.'[38] He also often quotes fellow Americans Roderick Nash, the first historian of the US wilderness,[39] and Edward O. Wilson, a world-renowned biologist who has visited the park, and whose backing endows the PPP's scientific research with legitimacy.

This US ethic is part of a tradition of national parks devoid of inhabitants, glorified as intact, untouchable nature. The underlying discourse is that the local population is responsible for the fragile ecological equilibrium. This population corresponds to the few thousand inhabitants living inside the park, plus the two hundred thousand people in the buffer zone, especially on the slopes of the Serra. And it is on the Serra da Gorongosa that this discourse comes into focus, presented once again as scientifically justified, just like in colonial times: 'scientists for Carr's project estimate that unless something is done, within five years the mountain's ecosystem will degrade to a point from which it cannot recover.'[40] Although wildlife surveys have been conducted since the early days of the PPP, and although renowned researchers (mainly biologists, from Harvard University for example) are regularly involved in studies, throughout my enquiries I never had access to any precise figures about the deforestation hypothesis grounded in documented studies. Such discourse seems more a matter of opinion than of science.

More broadly, this representation of the population as being 'naturally' destructive has come in for sharp criticism from academics, particularly American anthropologists who spent several years with villagers living on the slopes of the Serra and the descendants of those resettled during the colonial period. After a lively debate between Greg Carr and some of these academics on H-Africa, an informal but well-regarded scientific group, a panel was held at the African Studies Association's annual conference in November 2010. This panel, called 'Protecting an African Eden? Conservationists, Communities, and Collaboration in Mozambique', provided the Gorongosa Restoration Project head of community relations with an opportunity to defend the PPP's activities from allegations about its discourse.

But the narrative remains alarming and alarmist. The declinist rhetoric resembles that of the 1960s and 1970s. In other words, the GRP adopts and actualises the discourse of environmental destruction that first arose in colonial times. Thus the central authorities are intensely lobbied by its teams, first and foremost by Greg Carr. One objective has been to include the Serra – or at least its summit, that is, zones above 700 m in altitude – within the park. In parallel to this, in the mid-2000s, information campaigns and 'consultations' about including the summit within the park were launched with the traditional authorities (*régulos*) on the slopes of the Serra. Two communities in particular were in open and complete disagreement with this idea. Finally, the Mozambique Cabinet issued a decree in summer 2010 endorsing this decision to include the peak within the park.[41] Opposition from the population has continued to run strong, and since then certain observers have mentioned deforestation as a novel form of protest.[42] While such instances appear to be sporadic, it is notable that they have been mentioned in the literature. They no doubt partly explain why over recent years the GRP has significantly expanded its work on community participation programmes (hence reforestation) targeting much of the Serra population.

What may we conclude from the foregoing? There has been a clear break with the colonial period, for nature is now thought of as being for the benefit of the Mozambican people. But there is also continuity: the authorities consider conservation areas, particularly national parks, to be symbolic instruments of domination as well as economic assets. Nature becomes an instrument for economic growth by tourism, and as donors become ever more involved, nature is increasingly privileged as a development domain. This is not an unprecedented policy. On the contrary, adopting a long time frame shows how rhetorical constructs associating nature with the private sector, or nature with development, have been regularly reactivated, inscribed in regulations and implemented at the local level. At times coinciding with struggles over how to define and manage nature, the joint effects of the lengthy time frame and external influence continuously fashion discourse, imaginaries and practices, often independently of the shifting political context and realignments among stakeholders.

Rozenn Nakanabo Diallo is Associate Professor of political science at Sciences Po Bordeaux and researcher at LAM (Les Afriques dans le monde). She works on issues of state-building and transnational public action, mainly in a context of international aid. She is particularly interested in developing a biographical approach to these issues by considering the key role administrative elites have played in the environmental sector in Africa.

Notes

1. Interview, Ministry of Agriculture, Maputo, 12 June 2009.
2. Richard Grove, 'Science coloniale et naissance de l'écologie', *Écologie et politique* 56(1) (2018), 83–102.
3. Guillaume Blanc, *Une histoire environnementale de la nation: Regards croisés sur les parcs nationaux du Canada, d'Éthiopie et de France* (Paris: Publications de la Sorbonne, 2015).
4. Samora Machel, 'Diretivas sobre a conservação da natureza' Republica Popular de Moçambique, Ministério da Agricultura, Departamento de Florestas e Fauna Bravia, 7 August 1981 (Machel pronounced this speech after winning the gold medal at the World Hunting Exhibition in Bulgaria).
5. This figure does not include administrative staff employed in ministries and field staff working in protected areas.
6. Project financed by the World Bank.
7. Interview, Ministry of Tourism, Maputo, 3 March 2011.
8. John M. MacKenzie, *The Empire of Nature: Hunting, Conservation and British Imperialism* (Manchester: Manchester University Press, 1988).
9. Interview, Ministry of Agriculture, Maputo, 12 June 2009.
10. 'Da Contemplação e da Caça de Animais Selvagens', in *Reunião para o Estudo dos Problemas da Fauna Selvagem e Protecção da Natureza no Ultramar Português* (Sá da Bandeira, 1972).
11. 'National parks are areas subject to public management and surveillance, reserved for the propagation, protection and conservation of wildlife and spontaneous vegetation, and also the conservation of objects of aesthetic, geological, prehistoric, archaeological or other scientific interest, for the benefit and recreation of the public, and in which it is forbidden to hunt, kill or capture animals and to destroy or pick plants, except at the initiative and under the surveillance of the authorities'; Ministério do Ultramar, 'Decreto nº 40040 (Estabelece preceitos destinados a proteger nas províncias ultramarinas o solo, a flora e a fauna)', *Boletim Oficial*, I(8), 24 February 1955.
12. Manuel Carlos Serra and Manuel Cunha, *Manual de Direito do Ambiente* (Maputo: Ministry of Justice, 2008), 402.
13. Florence Pinto, 'Savoirs traditionnels et territoires de la biodiversité en Amazonie brésilienne', *Revue internationale des sciences sociales* 178(4) (2003), 667–78.
14. Estienne Rodary and Christian Castellanet, 'Les trois temps de la conservation', in Estienne Rodary, Christian Castellanet and Georges Rossi (eds), *Conservation de la nature et développement: L'intégration impossible?* (Paris: GRET and Karthala, 2003), 5–44.

15. Bill Cooke and Uma Kothari, *Participation: The New Tyranny?* (London: Zed, 2001).
16. República de Moçambique, 'Lei nº 10/99 de 07 de Julho (Lei das Florestas e da Fauna Bravia)', *Boletim da República* I(27), 12 July 1999.
17. Interview, Ministry of Tourism, Maputo, 19 May 2009.
18. Interview, Ministry of Tourism, Maputo, 18 June 2009.
19. Todd French, 'Like Leaves Fallen by Wind: Resilience, Remembrance, and the Restoration of Landscape in Central Mozambique' (PhD diss., University of Boston, 2009).
20. Interview, Ministry of Tourism, Maputo, 13 May 2009.
21. Thomas Bierschenk and Jean-Pierre Olivier de Sardan (eds), *Les pouvoirs au village: Le Bénin rural entre démocratisation et decentralization* (Paris: Karthala, 1998).
22. The conflict was between FRELIMO, led by urban elites from the south, and RENAMO, a 'warrior social corps' based mainly in rural zones: Christian Geffray, *La cause des armes au Mozambique: Anthropologie d'une guerre civile* (Paris: Karthala and Credu, 1990). Nowadays, there are still non-demobilised RENAMO troops present on the flanks of the Serra, and the new RENAMO leader, Ossufo Momade, was still in hiding there in the late 2010s.
23. French, 'Like Leaves Fallen', 196.
24. Rosemary Elizabeth Galli, *People's Space and State Spaces: Land and Governance in Mozambique* (Oxford: Lexington, 2003).
25. Armando José Rosinho, 'O parque Nacional da Gorongosa', in *Anais dos Serviços de Veterinária de Moçambique* 16 (Maputo: Ministry of Agriculture, 1968), 161–74.
26. French, 'Like Leaves Fallen'.
27. J. M. de Aguiar Macedo, 'Serra da Gorongosa: Necessidade e Bases da Sua Proteção', *Comunicação no. 44, Instituto de Investigação Agronómica de Moçambique* (Lourenço Marques, 1970).
28. French, 'Like Leaves Fallen', 221.
29. And more specifically what subsequently became his thesis on 'wildlife management': Kenneth Lochner Tinley, 'Framework of the Gorongosa Ecosystem' (PhD diss., University of Pretoria, 1977).
30. Ibid.
31. Greg Carr made his fortune in voicemail technologies in the 1980s. He sold his company, Boston Technologies, in the late 1990s, and created his foundation for the arts, human rights and the environment.
32. This rhetoric used in the PPP's marketing documents is extensively adopted in the press, as well as in documentaries such as the National Geographic film released in 2010: James Byrne (dir.), *Africa's Lost Eden*, https://www.imdb.com/title/tt1612227/, retrieved 04 September 2024.
33. Caroline Fraser, *Rewilding the World: Dispatches from the Conservation Revolution* (New York: Metropolitan Books, 2009).
34. Byrne, *Africa's Lost Eden*, 2010.
35. Roderick P. Neumann, 'Primitive Ideas: Protected Area Buffer Zones and the Politics of Land in Africa', *Development and Change* 28(3) (1997), 559–82.
36. Keith Bellows, 'Saving the Wild Kingdom', *National Geographic Traveler*, 23 October 2010.
37. *Focus* 540 (2010), 13–17.
38. Bob Shacochis, 'The Gardener of Eden', *Outside*, July 2009, 104.

39. Roderick Nash, *Wilderness and the American Mind* (New Haven, CT: Yale University Press, 1967).
40. Stephanie Hanes, 'Greg Carr's Big Gamble', *Smithsonian Magazine*, May 2007.
41. For an in-depth study of the political and military issues the site poses for the central authorities, which partly explain their reluctance to include the summit of the Serra within the park, see: Rozenn Nakanabo Diallo, 'Sortie(s) de guerre et conservation de la nature: Trajectoire d'un parc national au Mozambique', *Gouvernement et action publique* 8(4) (2019), 97–118.
42. Christy Schuetze, 'Narrative Fortresses: Crisis Narratives and Conflict in the Conservation of Mount Gorongosa, Mozambique', *Conservation and Society* 13(2) (2015), 141–53.

Bibliography

Bierschenk, Thomas, and Jean-Pierre Olivier de Sardan (eds). *Les pouvoirs au village: Le Bénin rural entre démocratisation et décentralisation*. Paris: Karthala, 1998.

Blanc, Guillaume. *Une histoire environnementale de la nation: Regards croisés sur les parcs nationaux du Canada, d'Éthiopie et de France*. Paris: Publications de la Sorbonne, 2015.

Cooke, Bill, and Uma Kothari. *Participation: The New Tyranny?* London: Zed, 2001.

French, Todd. 'Like Leaves Fallen by Wind: Resilience, Remembrance, and the Restoration of Landscape in Central Mozambique'. PhD diss., University of Boston, 2009.

Diallo, Rozenn Nakanabo. 'Sortie(s) de guerre et conservation de la nature: Trajectoire d'un parc national au Mozambique'. *Gouvernement et action publique* 8(4) (2019), 97–118.

Fraser, Caroline. *Rewilding the World: Dispatches from the Conservation Revolution*. New York: Metropolitan Books, 2009.

Galli, Rosemary Elizabeth. *People's Space and State Spaces: Land and Governance in Mozambique*. Oxford: Lexington, 2003.

Geffray, Christian. *La cause des armes au Mozambique: Anthropologie d'une guerre civile*. Paris: Karthala and Credu, 1990.

Grove, Richard. 'Science coloniale et naissance de l'écologie'. *Écologie et politique* 56(1) (2018), 83–102.

MacKenzie, John M. *The Empire of Nature: Hunting, Conservation and British Imperialism*. Manchester: Manchester University Press, 1988.

Nash, Roderick. *Wilderness and the American Mind*. New Haven, CT: Yale University Press, 1967.

Neumann, Roderick P. 'Primitive Ideas: Protected Area Buffer Zones and the Politics of Land in Africa'. *Development and Change* 28(3) (1997), 559–82.

Pinto, Florence. 'Savoirs traditionnels et territoires de la biodiversité en Amazonie brésilienne'. *Revue internationale des sciences sociales* 178(4) (2003), 667–78.

Rodary, Estienne, and Christian Castellanet. 'Les trois temps de la conservation', in Estienne Rodary, Christian Castellanet and Georges Rossi (eds), *Conservation de la nature et développement: L'intégration impossible?* (Paris: GRET and Karthala, 2003), 5–44.

Schuetze, Christy. 'Narrative Fortresses: Crisis Narratives and Conflict in the Conservation of Mount Gorongosa, Mozambique'. *Conservation and Society* 13(2) (2015), 141–53.

Tinley, Kenneth Lochner. 'Framework of the Gorongosa Ecosystem'. PhD diss., University of Pretoria, 1977.

CHAPTER 11

Catfights and Crocodile Tears

Conflict, Charismatic Species and Nature Professionals in India's Conservation History

Meera Anna Oommen

Nature conservation in India is dominated by discourse about iconic large vertebrates.[1] However, as reflected in the large numbers of human (and animal) casualties, many high-profile species viewed as conservation flagships are embroiled in widespread conflict with people. It is estimated that several hundred people lose their lives to elephants and large carnivores in India each year, and that over 1.2 million snake bites result in thirty to forty thousand human fatalities.[2] Despite these grim statistics, local-scale human engagements with many of these species have been emphatically multifaceted and multidimensional. Historically, despite conflict, such species brought a range of economic and cultural benefits for traditional and rural societies, and were often an integral part of kinship and reciprocity networks. Species preservation and veneration were common elements of local relationships, as was the removal or elimination of problem animals.[3] While the benefits and disadvantages were distributed within different local constituencies in some cases, in other instances state agencies or outsiders with disparate goals secured benefits or disrupted local dynamics. Consequently, even during the colonial and pre-colonial phases, local and extra-local views remained in a state of constant tension. This situation continues.

Over the last century and a half,[4] colonial and postcolonial nature governance have increasingly exerted top-down control via a range of overarching legal and political instruments, including laws governing extractive practices, sport hunting and the preservation of species and wild spaces. As India itself was not a unified entity during the colonial period, injunctions from the rulers of princely states and regional laws pertaining to presidencies were the first to address conservation (some of these were in fact in place even before the colonial conservation era). Excessive hunting and harvesting prompted colonial laws such as the Nilgiri Game and Fish Preservation Act, 1879, to redress the emerging shortage of game. In the case of iconic species such as the elephant, the initial laws were region-focused (e.g. the Madras Wild Elephant Preservation Act of 1873) and reflected regional requirements; later, these were adopted across broader areas or else for the country as a whole (e.g. the Elephant Preservation Act of 1879). Large-scale exclusionary conservation and forest reservation was strengthened by the Indian Forest Act (1927), which consolidated similar legislation passed in 1865 and 1878.

The overarching post-independence milestone is the Indian Wildlife (Protection) Act (WLPA), 1972,[5] which had immense consequences and provided extensive levels of protection to a large number of species. As the country's most protectionist conservation legislation, the WLPA and its amendments progressively prohibited the hunting and harvesting of most Indian species. The last three decades of the twentieth century also saw large-scale initiatives for flagship species, starting with Project Tiger in 1973 and followed by Project Elephant in 1992. These pieces of legislation are significant in that they resulted in the formation of zones of anomaly in which homogenising demands by the state and conservationists disrupted local or region-specific usage, rules and traditional practices.

The most recent phase involves the conservation politics of the new millennium, which pitches legislation to favour Indigenous people such as the Scheduled Tribes and Other Traditional Forest Dwellers (Recognition of Forest Rights) Act, 2006, against interlinked preservationist and animal rights agendas driven by urban proponents (influenced by emerging discourse from the Global North about compassion, animal sentience and personhood). Although the Forest Rights Act facilitated opportunities for many communities to regain lost historical land rights, the long-term impacts of this controversial legislation have been mired in complex on-the-ground realities and continuing contestation from both sides.

To explore these issues further, I present several historical and contemporary contexts relating to wildlife species that have played dual roles as conservation flagships and battleships of conflict,[6] together with the nature professionals – naturalists, foresters, scientists, conservationists and animal advocates – influencing these power games. I explore trajectories of engagement with three species – the tiger, the Asian elephant and the saltwater crocodile – that partake in conflict and coexistence, as well as being the subject of deliberations about species conservation or preservation at different points of time. All three species receive protection under international conservation arrangements such as CITES. At a subcontinental scale, interactions with all three species can be characterised as multifaceted. Yet their dynamics vary spatially and temporally, and are also largely determined by the practices and world views of different groups of local stakeholders and outsiders. For each species, I highlight a primary regional-level scenario that is illustrative of a particular aspect of historical or contemporary conservation politics, while at the same time attempting to give a brief summary of the conservation history of the species as a whole. Following this, I provide a discussion of a more generic, birds-eye view of India's conservation trajectory, which displays similarities to and differences from that of individual species.

Tigers and Tribals: Local Lives vs Animal Rights in Contemporary Central India

Yavatmal is located in the Vidharbha region of Maharashtra, the hotspot of agrarian distress and farmer suicides in India. Over an eighteen-month period beginning in 2016, a tigress, probably from adjoining protected areas (Tipeshwar Wildlife Sanctuary or Tadoba National Park), carried out a spate of attacks that resulted in an attributed thirteen human fatalities in Yavatmal villages. Media reports also mention that the tigress was preying on cattle. The human victims were mainly marginal farmers and grazers from the tribal hamlets; it has been suggested that a few were victims of accidental encounter, whereas others succumbed to deliberate ambush attacks.[7] In addition to the loss of lives and their consequences for the affected families (several of the victims were sole breadwinners), the attacks had significant local repercussions including increased crop-raiding by wild boar, together with disruption to crop harvests due to the atmosphere of fear and anxiety constraining these and other everyday activities for the region's local citizens over a prolonged period of time.

After several lives were lost, T1 (as the tigress was designated) was declared a man-eater, and efforts were made by the local Forest Department to capture the animal (many field biologists and tiger conservationists likewise hold that tigers that kill people should be eliminated or removed). As the human casualties mounted, the Forest Department deployed scores of workers and redoubled its efforts, deploying camera traps, ground crews and elephants for tracking, thermal imaging drones and baited traps. Even unconventional experiments such as spraying tiger urine and liberal quantities of Calvin Klein's Obsession for Men were reportedly attempted.[8] A pair of Italian Cane Corso mastiffs were brought in by a public personality in the hope of tracking the tigress, as was a well-known figure who had been involved in the successful removal of man-eaters elsewhere in the country. Most of these efforts came to nought due to the hilly, undulating terrain, inclement weather and the presence of lantana weed impeding the tracking teams' mobility and lines of sight. The situation was complicated by T1's own wariness and elusive habits, and by the fact that she had cubs during the later phase of operations. While the most desired option was to tranquilise the tigress, it soon became evident that this was an increasingly futile exercise. After several months the tigress was eventually shot dead in November 2018 by the private hunter during a night drive. The exact circumstances of the story remain unclear, and it is mired in claims and counterclaims.

Coincident to the developments in Yavatmal, a parallel drama unfolded in some of India's urban centres that was to have wide-reaching consequences. Animal rights activists, incensed by the possibility of T1 being exterminated, conducted a series of formal litigations, vigils and protests targeting the Forest Department as well as the individuals involved in the operation. T1 was rechristened 'Avni' (trans. 'Earth/of the Earth'), her status as a mother was highlighted to gain sympathy and the use of the term 'man-eater'[9] to describe an innocent mother looking out for her cubs was condemned. Impoverished tribal communities living in Yavatmal were portrayed as encroachers into the forest. The Forest Department was depicted as corrupt, and, without any attempt to understand its ground-level limitations, branded inefficient. Maneka Gandhi, a well-known animal activist and Indian politician from the central ruling party, berated the government authorities and condemned the role of the private hunter. The hunter, Shafath Ali Khan, and his son (who is believed to have eventually shot T1) were targets of a vilification campaign that highlighted their hunting heritage and portrayed them as bloodthirsty killers. Local celebrations marking the death of the tigress were denounced. Enquiries relating to the 'unlawful' shooting of the tigress were initiated by animal activist groups on a range of procedural and ethical grounds. A feature film including a high-profile celebrity cast was produced. Many aspects of this case are currently ongoing. While much of this is reminiscent of colonial and postcolonial continuities in the form of conflict between local communities and wildlife advocates, animal advocacy driven by urban animal rights proponents, celebrities and social media personalities including those from abroad is an increasingly common feature of conservation stories involving large charismatic vertebrates such as the tiger.

The conservation history of India's national animal, the tiger, is a long and complicated one. As pointed out by Rangarajan,[10] human relationships with the species during the colonial period were contingent on context, and it would be erroneous to uniformly characterise these as either conflict or harmonious cohabitation. Tigers were variously regarded as vermin, as objects of the hunt, as personifications of local deities and malevolent spirits and as embodiments of 'gentlemanly virtue'.[11] Indian royalty and colonial officials routinely engaged in opulent tiger hunts well into the twentieth century. Despite their propensity for attacking humans and livestock, local communities incorporated tigers into religion and mythology, occasionally even justifying their depredations on livestock and attacks on people. In other instances, traditional hunting tribes and colonial officials were paid bounties to remove tigers; at the same time, certain local communities (Baghel Rajputs and Gond communities in Central India, the Khonds in Ganjam, the Mikirs in Assam) refused to co-operate or else actively resisted as they revered tigers or claimed kinship with them.[12] By and large, however, the norm during the colonial period was to destroy man-eating tigers. In the Terai and the lower Himalaya, hunters such as Jim Corbett (1875–1955) contributed immensely to the removal of man-eating tigers and leopards. Corbett himself was an avid naturalist who in his later years refrained from hunting tigers unless they were dangerous to people or livestock. India's first national park established in 1936 (initially named Hailey National Park) was in 1955 rechristened the Jim Corbett National Park in his memory. A suggestion to rename this park to move away from Jim Corbett's imperial legacy has also been in the offing.

The early phase of postcolonial conservation leading to the launch of Project Tiger, India's flagship conservation programme, in 1973 was characterised by political one-upmanship and subterfuge, though it eventually resulted in the tiger being granted formal protection. This exercise involved a range of players including government officers (from the prime minister down to members of the newly formed Indian Forest Service); international conservation agencies (the WWF and the IUCN) conducted extensive lobbying and, in the initial phases, mobilised funding, while Indian non-governmental organisations (Bombay Natural History Society (BNHS)[13] and WWF-India) campaigned alongside US partners (e.g. the Smithsonian Institution). Prominent individuals in these representations included Guy Mountfort, John Seidensticker, S. Dillon Ripley, George Schaller, Salim Ali and Zafar Futehally. Key individuals in the Indian government included the father-daughter duo Jawaharlal Nehru and Indira Gandhi, prime ministers known for their love of wildlife and nature,[14] and representatives from the erstwhile princely elite holding positions in the Indian administration such as Karan Singh and the architect of the WLPA, M. K. Ranjitsinh. Indian conservationists such as Salim Ali represented the BNHS in alliance with American organisations such as the Smithsonian Institution.[15]

In the catfights to conserve the tiger, multiple battle lines were drawn between these groups, the most significant perhaps being that of the government lobby – led by scions of erstwhile royal families such as Karan Singh and M. K. Ranjitsinh, and Kailash Sankhala, the first director of Project Tiger – actively scuttling attempts by a coalition of US[16] and Indian NGOs to partake in scientific research or decision-making. Smarting from their experiences of colonialism, this highly nationalistic government ensemble hoped to develop a home-grown 'Indian science for Indian tigers' devoid of parachute science and emerging American cultural (scientific) imperialism.[17] Needless to say, due to the existing power differentials, the odds were heavily stacked against any real involvement or decision-making by local communities, who were regarded as the primary constituency that tigers needed protection from. In remote Indian landscapes such as the Sundarbans, a long period of killing the tiger as vermin was overturned in favour of protection, with local inhabitants being subject to restrictions and resettlements. As pointed out by Chakrabarti,[18] in such landscapes coexistence with tigers is thrust upon local communities whose traditional knowledge and strategies are disregarded. A postcolonial conservation hangover of the politics of the 1970s and 1980s continues to this day in the ideas behind the notification of protected areas, in setting aside inviolate areas for tigers, in disputes between government scientists and NGO researchers concerning tiger census and research methods, and in tussles between the Forest Department and researchers in general.

Travancore's Treasure: Elephants in Colonial India's Deep South

Unlike the tiger, which induced extensive catfights among several generations of conservationists, elephant conservation in post-independence India has been a comparatively less volatile affair, but nevertheless one that has seen occasional flare-ups between 'musth bulls'.[19] Its salience, however, is not so much in the politics between individual conservationists as in the diversity of complex regional trajectories from the colonial era, which were determined by how elephants were valued or utilised. To understand one such regional context, we shall turn to the hills of the Western Ghats, which are mentioned as an ancient hinterland for elephants and their ivory in some of the earliest sources of information on Indo-Roman trade reaching back to the fourth century BC.[20]

Almost two millennia later, we find that elephants played an equally significant role in the affairs of erstwhile Travancore, which remained under indirect British rule until the formation of the modern state of Kerala. Since the kingdom remained indebted to the British for protection from other feudal entities, several generations of Travancore's rulers entered into various one-sided treaties of 'perpetual friendship and alliance' with the East India Company, which assumed the 'tiresome responsibility' of protecting Travancore, simultaneously undermining its sovereignty and draining its resources through the imposition of unfair monopolies in forest products.[21] To improve revenues, taxation was encouraged, as was the extraction of products from the forests, including ivory and elephants. Elephants were declared the property of the state and were accorded protection long before the modern conservation era. In many ways, the species represented a paradox. Although they were extremely valuable from the perspective of ivory trade and carving, military operations, labour in timber hauling, and cultural operations, elephants were a crop-raiding species whose numbers had to be controlled for the safety of forest fringe dwellers and their crops.[22]

While the government benefitted greatly in the form of revenues and elephant capture, sportsmen were eager to hunt elephants, and especially those engaged in extensive crop-raiding or killing people. In almost all operations involving elephants, be it capture, ivory extraction or timber hauling, local 'hillmen' communities were involved,[23] and the process entailed mutual benefits to the colonisers and the local communities. Most of these communities had long-term experience that allowed them to coexist relatively peacefully with elephants. However, from the early 1900s to the period preceding Indian independence, various interconnected, extra-local influences that may be classified under the broad definition of capitalist expansion[24] – such as colonial conservation, hydroelectric power projects, timber extraction and plantation agriculture, which mostly served to benefit the state or the British and Indian elite – brought about large-scale changes to local land use and to population dynamics, resulting in revised profiles of human–wildlife engagement, especially that with elephants. In the early 1940s, when the region was gripped by the Travancore famine, the hill forests that were leased out to the poor and the landless for temporary cultivation became sites of protracted conflict.[25] As the settler communities that moved in were unfamiliar with elephant forests, their agricultural practices, such as rice cultivation in the hills, attracted daily elephant raids. While some communities eventually reached a rather uneasy equilibrium, there are currently areas in the *ghats* where conflict with elephants continues.[26]

In the wider context of elephants in the subcontinent, a number of colonial elephant preservation laws from the late 1800 onwards were instrumental in increased state control over local responses and priorities (though crop-raiding elephants and rogues could still be killed under special circumstances).[27] Subsequently, overarching legal instruments such as the WLPA (which superseded previous legislation) resulted in cementing pockets of unresolved conflict that can be classified as zones of anomaly (e.g. parts of the Western Ghats, Bengal and Assam). The recategorisation of the Asian elephant from its original placement in Schedule II of the WLPA (which implied categorisation as 'Special Game') to Schedule I in 1977 was eventually carried out on account of its status as an endangered species as identified by CITES.[28] Reducing poaching and illegal trade in ivory and other elephant parts has been one of the key goals. In India, Project Elephant, a centrally sponsored scheme launched in 1992, has been responsible for formally implementing the Monitoring Illegal Killing of Elephants (MIKE) CITES programme. Akin to Project Tiger, Project Elephant is a large, nationwide effort to promote the conservation of free-ranging elephant populations and their habitats, especially their movement corridors.

In recent years there have been calls in scholarship from the developed world to promote elephants' companionship, conviviality and sentience.[29] While such sentiments are no doubt present in local human–elephant relationships (in fact India has a history of strong, organic coexistence frameworks),[30] it is important to note that several hundred people are killed each year by elephants. Neumann's caution against moral extensionism or the attribution of anthropomorphic equivalence to animals outside traditional frameworks is important, as this often results in violence against humans typically from marginalised communities.[31] In Kerala, a recent example of this played out after the tragic, yet accidental death of a pregnant elephant after it consumed fruit containing a concealed pig bomb. The intended target of the explosive-laden fruit was wild boar, which routinely visit smallholder farms along the forest fringe. Fuelled by comments on social media from sportspersons and celebrities, the international media and animal rights groups, public outrage against the smallholder farmers included calls for the murder of the offenders by stuffing explosives into their mouths, mowing them down with automatic weapons or drowning them using heavy chains tied to their legs. There were also attempts to communalise the issue by misreporting that the incident had occurred in an electoral constituency occupied by a minority community, as well as by demonising Kerala's ruling Left Democratic Front (LDF) alliance.

Crocodile Tears: Modern Third Rail Controversies in the Andaman Islands

Crocodiles too have their defenders. Owing to recent geopolitical and developmental priorities,[32] the Andaman and Nicobar Islands are believed to be on the cusp of large-scale environmental transformation. The plan to 'develop' and transform these remote islands into the next Singapore or Hong Kong has come under increased criticism from various quarters. At ground level, this brings one of the last remaining Indian habitats of the saltwater crocodile (*Crocodylus porosus*) into direct contact with tourism, the primary sector that is targeted for development in this region. In a bid to minimise undesirable interactions with visitors to the islands, the central government's focus on tourism is also purported to be accompanied by a plan for the removal of at least some crocodiles from tourism zones. When viewed from the perspective of recent attacks on settlers who live and work in the islands (often in occupations or spaces that put them in direct everyday contact with crocodiles), this can be perceived as a preferential strategy for tourists. It needs to be pointed out that though local forest authorities occasionally remove dangerous individuals through relocation – which exacerbates the problem as crocodiles have strong homing instincts and often return to their territories, even several hundred kilometres away, causing further conflict along the way – the typical response to petitions by local communities has been to point out that the animal is listed on Schedule I of the WLPA (which confers maximum protection for the species) and hence cannot be removed.

Here too, government departments (including in some instances the local Forest Department) and conservation groups holding middle-of-the-road positions that promote context-specific strategies are frequently opposed by extreme preservationist animal welfare lobbies with little knowledge, experience or presence in the area. The latter include urban animal lovers as well as individuals in the wildlife conservation community and government who object to the removal of problem individuals or to population management strategies, and are typically wholly opposed to the use of wild species. Fatalities from crocodile attacks are often compared to the larger number of everyday accidents such as drowning in a bid to trivialise the former, and without understanding human psychology in the context of catastrophic attacks, or acknowledging the risks associated with living in crocodile habitats. The objections of these extreme preservationist groups are not just against any actual removal or utilisation of crocodiles; rather, removal or sustainable use are often treated as third rail topics whose discussion itself is denounced. Although the islands offer a unique opportunity to explore coexistence strategies including the use of crocodiles in ways that are beneficial for local communities, the tussle between protectionists and groups in favour of sustainable use continues despite support for the latter from Indian experts such as Romulus Whitaker and those of international groups such as the IUCN. Most of the decision-making is not inclusive, and local communities continue to be excluded from these spaces, resulting in situations in which all stakeholders are likely to lose.

While crocodiles were hunted as well as killed as vermin during the colonial era,[33] the most significant threats to their existence across their range (including India) came in the form of international demand for skins.[34] Owing to extensive exploitation by the crocodile skin industry during the 1900s, crocodilian populations in India declined drastically. By the 1970s, the saltwater crocodile, the crocodilian species preferred on account of its small scales and large surface area, was on the brink of extinction. The Bhitarkanika and the Sunderbans in mainland India, and the Andaman and Nicobar Islands, were the saltie's last bastions. In response to these developments, crocodile conservation efforts were launched in the early 1970s, including wide-ranging surveys, the establishment of protected areas, and captive breeding. Robert Bustard, an FAO consultant and expert on crocodiles and marine turtles, was invited to design a recovery programme for all three species of crocodilians in India. The UNDP/FAO Crocodile Breeding and Management Project (Crocodile Project) was launched in 1975 in the state of Orissa (now Odisha). Prime Minister Indira Gandhi took a special interest in these matters. A non-governmental organisation, the Madras Crocodile Bank, was established in 1974; its founders Zai and Romulus Whitaker conducted extensive surveys for crocodiles, including the first such surveys in the Andaman Islands.[35] It was evident that poaching and nest destruction had led to greatly depleted populations even in these far-flung islands.[36]

Over the past two decades, conservation efforts, waterlogging, increased availability of swampland caused by the Indian Ocean tsunami, and naval surveillance (probably deterring poachers from South East Asia) are believed to have been favourable for population recoveries. A number of other regional scenarios across the range of the saltwater crocodile have shown population bounce-backs and recolonisation, with ensuing conflict typically in a twenty- to thirty-year period following protection.[37] In the Andaman Islands, circumstantial evidence points to increased numbers of crocodiles in at least some of the islands. This is corroborated by occasional crocodile attacks on livestock and people. In places such as Little Andaman, the victims have been mostly rural settlers, whereas fatal attacks in parts of Great Andaman have been on settlers as well as tourists. Crocodile attacks in the Nicobar Islands are fewer, and this is believed to be the result of their significance in cultural frameworks as well as traditional systems that occasionally remove problem animals or hunt them for food (parallels can be found in other regions, such as those in studies by Simon Pooley and colleagues).[38] For the most part, however, crocodiles are left alone, and Nicobari communities take the necessary precautions to avoid them, resulting in a scenario that is closer to the pragmatic end of the coexistence spectrum.[39] Unlike the situation with settler communities in the Andaman Islands, regional environmental laws permit these activities by the Indigenous Nicobari communities, who are allowed to continue their subsistence use.

A Birds-Eye View of Conservation in India

Colonial engagements with large free-ranging wildlife cannot be seen as operating separately from either the emergent capitalist paradigm or local specificities of the period. On the one hand, imperial agendas involving big game hunting and the 'war against dangerous beasts' can both be viewed as overarching instruments of power and subjugation within the subcontinent.[40] Hunting, or *shikar* in India, was more than just a pastime; in the case of tigers, elephants and even pigs it was a multidimensional political enterprise that yielded several benefits for the colonial administration, including those brought by surveillance and signalling, and knowledge-gathering of people, terrain and activities. Exclusionary sport hunting was a key aspect of the princely class in colonial India, such as the Rajputs and other 'martial races', as well as of the Mughals before them. At the same time, though they were asymmetrically beneficial to the colonisers, archival records and investigations on the ground show that communities routinely resisted, collaborated in, benefitted from and occasionally enjoyed hunting interactions. In some forms of sport such as pig-sticking, subaltern groups such as local shikaris and hunting tribes, including those who were prejudicially classified as 'criminal tribes', interacted extensively with sportsmen in an inversion of commonly accepted class hierarchies, and often stood to gain from the exercise of hunting; agrarian communities too displayed varied responses to hunting.[41]

Similarly, vermin control (a practice with a long history in Europe) elicited diverse responses from different parties. To take a culturally multifaceted species such as the tiger as an example, some communities wholeheartedly embraced and benefitted from this enterprise, others showed different levels of reluctance, and some engaged in active resistance.[42] The tiger held relatively less positive cultural significance for the British in comparison to their general respect for the lion. Even so, some colonial officials such as the naturalist G. P. Sanderson were against its removal as they argued that tigers provided valuable services in controlling crop-raiding herbivores.[43] Despite being an apex predator like the tiger, saltwater crocodiles (as well as their freshwater counterpart, the mugger) received much less protection and were often eliminated or sold for their skins. When compared to the trajectories of other species addressed here, we find that elephant populations were preserved in large part due to the economic and cultural significance of the species, though this was by no means homogeneous across their range. As their utility declined, so did regional and local efforts to capture and deploy elephants. In many places, when elephant capture ceased, their numbers began to expand, as did crop depredations; in others, agriculture, urbanisation and development reduced their numbers and fragmented their habitats, ushering in a need for protection.

According to M. K. Ranjitshinh, the architect of the WLPA, the most important motivating factor for wildlife preservation in the first half of the century was almost invariably *shikar*.[44] Although species preservation for its own sake was not on the agenda, declining heads of game, smaller trophies and eventually local extinctions were instrumental in the development of initial rules and game management plans. For wildlife, favourable conservation outcomes included the creation of hunting reserves and numerous restrictions on their utilisation. For all three species – the elephant, the tiger and the crocodile – demands for protection were also building up in the international arena during the 1970s. For local communities, this meant exclusion from forests and restrictions on hunting. Indigenous hunting, despite its rich diversity of cultural and ritualistic elements, came to be regarded as a quotidian exercise of hunting for the pot and was criminalised as 'poaching'; this in turn spawned rule-breaking and everyday acts of resistance, which had significant impacts on the ground.

In these revised ethics of colonial species preservation and game management, persuasion by 'sportsmen naturalists'[45] and 'penitent butchers' 'smitten with remorse'[46] became an important deciding factor. As pointed out by Sivaramakrishnan,[47] a conservation coalition comprised of colonial British as well as Indian elites (former aristocrats, naturalists, forest and wildlife officials) 'presented the usual litany of complaints about the illiterate, poor, rapidly multiplying farmers and hunters who knew not what was best for their body or soul'.[48] In the years following Indian independence, the home-grown contingent of this group (Salim Ali, Kailash Sankhala, Billy Arjan Singh, Lavkumar Kachar and M. Krishnan are examples) was more in favour of exclusionary conservation via protected areas and draconian conservation measures to protect important species such as the tiger, and 'tacked willingly between Indira Gandhi's populist energy and her authoritarian streak'.[49] Sivaramakrishnan[50] credits this coalition with facilitating 'a more explicit equation between nature and nation' and inciting 'a burgeoning nature-loving metropolitan middle class who mediated their personal transition from colonial elite to national elite through altered perspectives on nature and its place in a national heritage'. He adds that '[b]y the mid-1980s, Indira Gandhi and her group of advisors on wildlife policy had fully articulated an ecological nationalism that had natured the nation in the dominant colours of charismatic megafauna'.[51] For instance, Mrs Gandhi's advisors included key government bureaucrats such as Karan Singh and M. K. Ranjitsinh who took an anti-imperialist stance when it came to conservation. In tiger conservation, they not only succeeded in preventing large American organisations such as the Smithsonian Institution from establishing operations and field bases in India (as they had in other countries), they also resisted the continued influence of prominent international NGOs (such as the IUCN and WWF, which had initially mooted the idea of Project Tiger and offered financial support) and their Indian partners. The need to develop home-grown scientific and management expertise within the Indian Forest Service (and excluding Indian researchers and conservationists) was emphasised at several levels, even though there was arguably existing knowledge and wherewithal provided by American institutions, which their NGO counterparts in India were hoping to secure. In parallel, there were debates about identifying animals to serve as symbols for an independent India, resulting in the tiger, which had a more or less countrywide distribution – unlike the lion, which had a much narrower distribution – being picked as the national animal of India.

The impetus for the Indian Wildlife (Protection) Act, 1972, drafted by M. K. Ranjitsinh, came from none other than Mrs Gandhi, who, on re-election in 1971, requested the states (forests being a state subject until this point) to pass a resolution empowering the Government of India to pass central legislation for the protection of wildlife. Ranjitsinh carried out a study of several Indian states, as well as of other countries, especially developing countries in Asia and Africa, and found some of these useful for the Indian context. According to him, Maharashtra state (whose legislation already contained BNHS input) and countries such as Kenya, Indonesia and New Zealand could provide useful pointers. Ranjitsinh's framing of this legislation included a shift from conventional Indian jurisprudence, which was inquisitorial (i.e. innocent until proven guilty) to a more accusatorial one (the burden of proof rested on the accused as far as wildlife crimes were concerned), as well as disregard for the right of self-defence if the person was already committing an illegality (e.g. entry in a protected area). By playing on the prime minister's emotions on the sensitive topic of Indian princes (who had also lost their rights to hunt despite constitutional rights), Ranjitsinh also ensured that the hunting rituals and traditional practices of any tribal communities who opposed this aspect of the new legislation were also banned from most areas. This sleight of hand is best expressed in the following excerpt from his own memoir:

> The scowl on her [Indira Gandhi's] face was getting deeper, but the argument, especially with regard to the sensitive subject of Indian princes with its recent history, was reaching its mark. 'There's even a traditional right of hunting rhino, removing its heart and offering it to one's ancestors in the "tarpan" ceremony', I continued. This was practiced by the Ranas of Nepal, not the Indian princes, but if one just cites a grisly 'tradition' of some princes without stating the country to which they belong, it is not a complete lie, is it? If all these traditional rights are to be allowed, what is the purpose of having this Act, I asked?[52]

The lasting legacy of these individuals and the governmental and non-governmental organisations they are associated with has been a focus on charismatic mega-vertebrates, and a hardline protectionist stance in conservation that finds further support among urban, upper- and middle-class groups for criminalising the actions of local resource-dependent communities, especially those that hunt or harvest wildlife or that live in the midst of wildlife. Without giving adequate consideration to root causes or unforeseen consequences, these perspectives consider people's activities and encroachment into wildlife spaces as the primary catalyst of conflict, and are in turn alert to opportunities to relocate people from wild spaces.[53] Recent legislation such as the Scheduled Tribes and Other Traditional Forest Dwellers (Recognition of Forest Rights) Act, 2006, which aims to redress historical injustices such as eviction from forest areas, is therefore strongly opposed by protectionist conservationists. In tiger conservation, as we saw in the case of the Yavatmal tigress, this legacy still survives within the tiger versus tribal debates, as well as in support for the 'cosmopolitan tiger',[54] which is far removed from the reality of everyday fear and trepidation for people living in its range.

In recent years, emerging global discourse within conservation biology, environmental anthropology and related fields has further muddied the already murky waters of global–local conservation interdependencies. The attention to nonhuman subjects or the so-called 'animal turn', though welcome in academic debate, has found its way into applied scenarios. For instance, Kopnina's allegation of an 'anthropocentric bias in anthropology'[55] and the call for radical (eco-centric) anthropological advocacy to counter violence against nonhumans[56] are strongly aligned with neoliberal and radical exclusionary conservation ideologies such as Half Earth/Nature Needs Half[57] and with compassionate conservation.[58] A renewed focus on the rights of individual animals as espoused in simplistic ideologies such as compassionate conservation is also particularly problematic,[59] not least because of its flawed, Orientalist understandings of Indian concepts such as 'ahimsa'. Much of this literature is also characterised by its articulation of universal moralistic injunctions to feel compassion and grief, which ignore ground realities, and are enormously detrimental from the perspective of social justice.[60] For species involved in conflict along the forest fringe, such as elephants, the more-than-human turn has engendered discourse (and potentially action) championing conviviality and companionship in shared spaces.[61] As pointed out by Jepson et al.,[62] this could amount to trivialising the often catastrophic violence that people living in proximity to wild species have to contend with.

The intersection of heady urban-derived animal activism and the spillover of large free-ranging vertebrates like tigers and elephants into human-dominated spaces has been an emerging yet problematic issue in India. Deep ecological incitement that ignores local realities, the multiplicity of lived relationships and perceived inequalities have reinforced marginalisation and violence. These ideologies rarely find support within Indian academia but are restricted to ideologues at the extremes of the preservationist camp and urban animal rights and liberation groups. Scientists and pragmatic conservationists, especially those with long-term credentials of research on carnivores, are of the opinion that to keep the peace, it is necessary to dispose of man-eaters and other dangerous individuals before they kill again.[63] Similar opinions on selective removal or lethal control have been voiced by researchers in the context of elephants, crocodiles and other species that cause harm to human life. Much of this comes from long-term experience with communities and the realisation that tolerance is beginning to fray, in large part because of community opposition to environmental protection policies, especially those of exclusionary and coercive conservation. There has been some support in India for recent paradigms, such as that of 'coexistence',[64] that recommend moving beyond technical fixes and call for the reframing of mitigation strategies that highlight the multifaceted nature of human–wildlife engagements (including positive interactions and cultural entanglements). However, it is critical that these remain at the pragmatic end of the spectrum, especially in the context of species that cause catastrophic conflict, and that they not be subsumed by the lurking protectionist voices within this group of conservationists. For the larger subset of wildlife that interacts with human communities in ambivalent or less problematic ways, an alignment with different forms of 'reconciliation ecology'[65] might be more meaningful, as it offers opportunities that focus on conservation strategies in human-dominated landscapes modified by agriculture, development or other drivers that result in habitat loss.

Rangarajan points out that overarching instruments of British colonialism relating to environmental outcomes in India include state forestry and reservations, the development of modern irrigation systems and widespread state-sponsored campaigns to eliminate vermin.[66] While these operated in an interconnected fashion within the broad bounds of the capitalist project that was emerging at the time, they were also characterised by distinct specificities that set the South Asian region apart from other colonial contexts elsewhere. Regional and local specificities in turn make the South Asian context a complex one, and these complexities and specificities continue into the new millennium. Therefore, from the vantage point of the South and in terms of local histories and 'perspectives from below', we are afforded a nuanced, heterogeneous view that precludes clear dichotomies and incorporates discontinuities. At the same time, these include widespread commonalities, especially resistance to conservation and complex negotiations at the local level that cut across commonly accepted chronological divides. The same applies to conservation scenarios immediately after independence, as well as recent developments such as revitalised protectionist strategies and animal rights movements inspired by Northern, developed world discourse. Nature governance in India thus remains a complex project prone to tussles between various stakeholders, clashing influences and conflicting priorities.

Meera Anna Oommen is a member of the Board of Directors of the Dakshin Foundation Bangalore and of the Madras Crocodile Bank. She works on issues in environmental history, ecology and conservation science. Her current research takes a multidisciplinary approach to understanding the dynamics of human–wildlife conflict, social relationships with animals and the history of hunting. Her main fields of investigation are mountains (the Trans-Himalayas, Western Himalayas and Western Ghats) and island areas (Andaman and Nicobar).

Notes

1. Mahesh Rangarajan, 'The Raj and the Natural World: The War Against "Dangerous Beasts" in Colonial India', *Studies in History* 14(2) (1998), 265–99; Kalyanakrishnan Sivaramakrishnan, 'Thin Nationalism: Nature and Public Intellectualism in India', *Contributions to Indian Sociology* 45(1) (2011), 85–111.
2. Meera Anna Oommen, 'Beasts in the Garden: Human-Wildlife Coexistence in India's Past and Present', *Frontiers in Conservation Science* 2 (2021), 703432.
3. Ibid.
4. While the colonial phase extends well beyond this milestone to the 1700s, I consider 1858 (i.e. the transfer of East India Company lands and function to direct governance of the British Crown) to be an appropriate chronological starting point to analyse conservation action and legislation for the species under consideration. It also needs to be clarified that 'India' at this juncture included lands that were directly administered by the Crown, as well as princely states ruled by indigenous rulers under British paramountcy.
5. India became a free nation in 1947; a unified India that incorporated the former princely states became a reality only in 1956.
6. The analogy of flagships vs gunboats or battleships has been used in the following works: Kartik Shanker and Roshni Kutty, 'Sailing the Flagship Fantastic: Different Approaches to Sea Turtle Conservation in India', *Marine Anthropological Studies* 3(2) and 4(1) (2005), 213–40; Leo R. Douglas and Diogo Verissimo, 'Flagships or Battleships: Deconstructing the Relationship between Social Conflict and Conservation Flagship Species', *Environment and Society: Advances in Research* 4(1) (2013), 98–116.
7. I have relied on a number of media reports, public debates and personal communications with individuals involved in the incident to piece together the story of the Yavatmal tigress. A large number of newspaper reports in English as well as vernacular languages are available (e.g. Vivek Deshpande, 'After 53-Day Hunt, T1 shot, Decision Soon on Her Cubs', *Indian Express*, 4 November 2018, https://indianexpress.com/article/india/after-two-month-long-search-operation-maneating-tigress-t1-killed-in-yavatmal-forest-5432307/; Jaideep Hardikar, 'Chronicling the Killing of T1, the Tigress', *The Wire*, 6 November 2018, https://thewire.in/environment/avni-tigress-t1-killing), as are heated debates and interviews in the media (NDTV, 'We the People: Was Tigress Avni's Shooting Justified?', YouTube, 11 November 2018, https://www.youtube.com/watch?v=DJKTGXtkORQ&t=166s; Zee 24 Taas, 'Yavtamal Shooter Nawab Shafat Ali Khan Exclusive Interview on How to Hunting of Female Tiger', YouTube, 16 September, 2018, https://www.youtube.com/watch?v=5ZeePyb__eA, all retrieved 6 August 2024).
8. Several perfumes have been known to attract and induce cheek rubbing behaviour by jaguars, cheetahs, etc.
9. This terminology is opposed by animal activists and wildlife enthusiasts. As per its revised Standard Operating Procedures, the National Tiger Conservation Authority (NTCA) has replaced the term 'man-eater' with 'dangerous to human life'; National Tiger Conservation Authority (NTCA), 'Standard Operating Procedures and Tiger Safari Guidelines, November 2019', Ministry of Environment, Forest & Climate Change (New Delhi, Government of India, 2019), 54.

10. Rangarajan, 'The Raj and the Natural World'.
11. Ibid.; John MacKenzie, *The Empire of Nature: Hunting, Conservation and British Imperialism* (Manchester: Manchester University Press, 1988), 350; Oommen, 'Beasts in the Garden'.
12. Rangarajan, 'The Raj and the Natural World'; Ezra Rashkow, 'Resistance to Hunting in Pre-Independence India: Religious Environmentalism, Ecological Nationalism or Cultural Conservation?', *Modern Asian Studies* 49(2) (2014), 270–301; Ezra Rashkow, 'Making Subaltern Shikaris: Histories of the Hunted in Colonial Central India', *South Asian History and Culture* 5 (2014), 292–313. Peter Boomgaard's work from South East Asia also points to parallel relationships; Peter Boomgaard, *Frontiers of Fear: Tigers and People in the Malay World, 1600–1950* (New Haven, CT: Yale University Press, 2008).
13. The BNHS was constituted in the 1880s by English hunters and naturalists. Even in the initial decades after independence, this organisation's membership and activities continued to retain its elite, colonial flavour.
14. Thapar points out that Nehru's policies were themselves paradoxical as his tenure coincided with the initiation of some of the largest development projects in India: Valmik Thapar, *Battling for Survival: India's Wildlife over Two Centuries* (Delhi: Oxford University Press, 2003), 470.
15. Schaller finally succeeded in studying tigers in Kanha in Central India. In 1967 he published *The Deer and the Tiger*, considered the first work on tigers that went beyond traditional natural history to arrive at a 'scientific' account of the species and its prey in India.
16. The animosity of the Indian government to US interests and scientists was in part a response to the pro-Pakistan 'tilt' adopted by the United States in the India–Pakistan war of 1971: Michael Lewis, 'Indian Science for Indian Tigers? Conservation Biology and the Question of Cultural Values', *Journal of the History of Biology* 38(2) (2005), 185–207.
17. Ibid.; Jairam Ramesh, *Indira Gandhi: A Life in Nature* (New Delhi: Simon & Schuster India, 2017), 448.
18. Ranjan Chakrabarti, 'Local People and the Global Tiger: An Environmental History of the Sundarbans', *Global Environment* 3 (2009), 72–95.
19. In a close parallel with tiger conservation, elephant conservation has likewise been a traditionally male domain: Kartik Shanker, 'From Marine Turtles to Musth Bulls: Reviewing the Role of Conservation Icons in India' (forthcoming).
20. Federico De Romanis, 'Playing Sudoku on the Verso of the Muziris Papyrus', *Journal of Ancient History* 27 (2012), 75–101.
21. V. J. Varghese, 'Land, Labour and Migration: Understanding Kerala's Economic Modernity', Working Paper 420 (Thiruvananthapuram: Centre for Development Studies, 2009), 8. After 1858, most of these lands and relationships were transferred to the British Crown: M. Amruth, 'Changing Regimes of Forest Management: Institutional Change and Modes of Participation in the Western Ghats of Kerala' (Ph.D. diss., Mahatma Gandhi University, 2009).
22. Meera Anna Oommen, 'The Elephant in the Room: Histories of Place, Memory and Conflict with Wildlife along a Southern Indian Forest Fringe', *Environment and History* 25(2) (2019), 269–300.

23. Tribal communities living in or around forests such as the Urali, Kadar and Mala Arayar in Travancore and the Jenu Kuruba in Malabar.
24. Varghese, 'Land, Labour and Migration'.
25. Especially lands allotted for famine relief under the Grow More Food campaigns.
26. Oommen, 'The Elephant in the Room'.
27. The Madras Elephant Preservation Act of 1873, and its countrywide successor, the Elephant Preservation Act of 1879.
28. The Asian elephant was placed in Appendix 1 of CITES on 1 July 1975.
29. Jamie Lorimer, 'Elephants as Companion Species: The Lively Biogeographies of Asian Elephant Conservation in Sri Lanka', *Transactions of the Institute of British Geographers* 35(4) (2010), 491–506.
30. Tarsh Thekaekara and Thomas F. Thornton, 'Ethnic Diversity and Human–Elephant Conflict in the Nilgiris, South India', in Piers Locke and Jane Buckingham (eds), *Conflict, Negotiation, and Coexistence: Rethinking Human Elephant Relations in South Asia* (Oxford: Oxford University Press, 2016), 300–29.
31. Roderick P. Neumann, 'Political Ecology of Wildlife Conservation in the Mt. Meru Area of Northeast Tanzania', *Land Degradation and Rehabilitation* 3(2) (1992), 85–98.
32. In recent years, geopolitical shifts, especially those involving concerns about China's ambitions in the Indian Ocean region (the String of Pearls and the Malacca Dilemma), have prompted the Indian government to view the Andaman Islands as a strategically significant location for safeguarding India's interests in the Indian Ocean. Economic development, including opening the island to mainstream developmental activities such as tourism and an expansion in large-scale fisheries, are part of the new strategic vision for these islands: P. K. Roy and Aspi Cawasji, *Strategic Vision – 2030: Security and Development of Andaman and Nicobar Islands* (New Delhi: Vij Books India, 2017), 194.
33. Both the *mugger* (marsh crocodile, *Crocodylus palustris*) and the saltwater/estuarine crocodile (*C. porosus*) were hunted during the colonial period as vermin.
34. Rangarajan, 'The Raj and the Natural World'; Grahame Webb, Charlie Manolis and Matthew L. Brien, 'Saltwater Crocodile *Crocodylus porosus*', in Charlie Manolis and Colin Stevenson (eds), *Crocodiles: Status Survey and Conservation Action Plan* (Darwin: Crocodile Specialist Group, 2010), 99–113.
35. The contributions by the Whitaker family to the conservation of crocodiles and snakes in India are significant. Romulus Earl Whitaker is a naturalised Indian citizen credited with the establishment of several institutions including the Madras Snake Park and the Madras Crocodile Bank Trust (MCBT), and was also instrumental in setting up the Irula Snake Catchers' Industrial Cooperative Society and the Irula Tribal Women's' Welfare Society. Zai Whitaker is the daughter of Zafar Futehally, the pioneering Indian naturalist and conservationist of the BNHS. Zai and Romulus Whitaker conducted pioneering surveys of a number of endangered herpetofauna including crocodiles and snakes across the length and breadth of India. The MCBT continues to be one of modern Chennai's landmarks and a centre for education and research for India's little-known reptiles.
36. Romulus Whitaker and Zai Whitaker, 'A Preliminary Survey of Saltwater Crocodiles (*Crocodylus porosus*) in the Andaman Islands', *Journal of the Bombay Natural History Society* 76 (1978), 311–23; Binod C. Choudhury and Robert H. Bustard, 'Predation on

Natural Nests of the Saltwater Crocodile (*Crocodylus porosus* Scheider) on North Andaman Island with Notes on the Crocodile Population', *Journal of the Bombay Natural History Society* 76 (1980), 311–23.

37. Jan van der Ploeg, Francis Ratu, Judah Viravira, Matthew Brien, Christina Wood, Melvin Zama, Chelcia Gomese and Josef Hurutarau, 'Human-Crocodile Conflict in Solomon Islands. Penang, Malaysia: WorldFish. Program Report' (2019), 44.
38. Simon Pooley et al., 'Human–Crocodilian Interactions in Latin America and the Caribbean Region', *Conservation Science and Practice* 3(5) (2021), 351; Simon Pooley, 'The Entangled Relations of Humans and Nile Crocodiles in Africa, c. 1840–1992', *Environment and History* 22(3) (2016), 421–54.
39. Manish Chandi, personal communication.
40. MacKenzie, *The Empire of Nature*; Mahesh Rangarajan, 'Environmental Histories of South Asia: A Review Essay', *Environment and History* 2(2) (1996), 129–43.
41. Meera Anna Oommen, 'Colonial Pig-Sticking, Imperial Agendas and Natural History in the Indian Subcontinent', *Historical Journal* 64(3) (2021), 626–49.
42. Rangarajan, 'The Raj and the Natural World'.
43. George P. Sanderson, *Thirteen Years among the Wild Beasts of India* (London: W. H. Allen, 1879), 430. Ideas within the colonial machinery itself were heterogeneous. For instance, Dietrich Brandis, the first Inspector General of Forests, was a strong supporter of community forestry: Ramachandra Guha, 'The Prehistory of Community Forestry in India', *Environmental History* 6(2) (2001), 212–38.
44. M. K. Ranjitsinh, *A Life with Wildlife: From Princely India to the Present* (New Delhi: Harper Collins, 2017), 26.
45. Edward Percy Stebbing, *The Diary of a Sportsman Naturalist in India* (London: John Lane, 1920), 356.
46. 'Penitent butchers' was a term popularised by Fitter and Scott and refers to a category of late colonial-era sportsmen-turned-preservationists who played an increasingly significant role in deciding the future of conservation in the empire: Richard Fitter and Peter Scott, *The Penitent Butchers: The Fauna Preservation Society* (London: HarperCollins, 1978), 48.
47. Sivaramakrishnan, 'Thin Nationalism'.
48. Ibid., 93.
49. Ibid., 97.
50. Ibid., 95–96.
51. Ibid., 98.
52. Ranjitsinh, *A Life with Wildlife*, 124.
53. Ullas Karanth and Mysore Doraiswamy Madhusudan, 'Mitigating Human-Wildlife Conflict in Southern Asia', in John Terborgh et al. (eds), *Making Parks Work: Strategies for Preserving Tropical Nature* (Washington, DC: Island Press, 2002), 250–64; Krithi Karanth, Sahila Kudalkar and Shivangi Jain, 'Re-Building Communities: Voluntary Resettlement from Protected Areas in India', *Frontiers in Ecology and the Environment* 6 (2018).
54. Annu Jalais, 'Unmasking the Cosmopolitan Tiger', *Nature and Culture* 3(1) (2008), 25–40.
55. Helen Kopnina, 'Beyond Multispecies Ethnography: Engaging with Violence and Animal Rights in Anthropology', *Critical Anthropology* 37(3) (2017), 333–57.

56. Helen Kopnina et al., 'The "Future of Conservation" Debate: Defending Ecocentrism and the Nature Needs Half Movement', *Biological Conservation* 217 (2018), 140–48.
57. Edward O. Wilson, *Half-Earth: Our Planet's Fight for Life* (London: Liveright, 2016), 272.
58. Arian Wallach et al., 'Summoning Compassion to Address the Challenges of Conservation', *Conservation Biology* 32(6) (2018), 1255–65; Arian Wallach et al., 'Recognising Animal Personhood in Compassionate Conservation', *Conservation Biology* 34(5) (2020), 1097–106.
59. Wallach et al., 'Recognising Animal Personhood'.
60. Bram Buscher et al., 'Half-Earth or Whole Earth? Radical Ideas for Conservation, and Their Implications', *Oryx* 51(3) (2016), 407–13; Meera Anna Oommen et al., 'The Fatal Flaws of Compassionate Conservation', *Conservation Biology* 33(4) (2019), 784–87.
61. Lorimer, 'Elephants as Companion Species'.
62. Paul Jepson et al., 'Towards an Intra-Disciplinary Biogeography: A Response to Lorimer's "Lively Biogeographies" of Asian Elephant Conservation', *Transactions of the Institute of British Geographers* 36(1) (2011), 170–74.
63. Ullas Karanth and Rajesh Gopal, 'An Ecology-Based Policy Framework for Human–Tiger Coexistence in India', in Rosie S. Woodroffe, Simon Thirgood and Alan Rabinowitz (eds), *People and Wildlife (Conflict or Coexistence?)* (Cambridge: Cambridge University Press, 2009), 373–87.
64. See for instance Simon Pooley et al., 'An Interdisciplinary Review of Current and Future Approaches to Improving Human–Predator Relations', *Conservation Biology* 31(3) (2017), 513–23.
65. Michael Rosenzweig, 'Reconciliation Ecology and the Future of Species Diversity', *Oryx* 37(2) (2003), 194–206.
66. Rangarajan, 'The Raj and the Natural World'; Rangarajan, 'Environmental Histories of South Asia'.

Bibliography

Amruth, M. 'Changing Regimes of Forest Management: Institutional Change and Modes of Participation in the Western Ghats of Kerala'. Ph.D. diss., Mahatma Gandhi University, 2009.

Boomgaard, Peter. *Frontiers of Fear: Tigers and People in the Malay World, 1600–1950*. New Haven, CT: Yale University Press, 2008.

Buscher, Bram, Robert Fletcher, Dan Brockington, et al. 'Half-Earth or Whole Earth? Radical Ideas for Conservation, and Their Implications'. *Oryx* 51(3) (2016), 407–13.

Chakrabarti, Ranjan. 'Local People and the Global Tiger: An Environmental History of the Sundarbans'. *Global Environment* 3 (2009), 72–95.

Choudhury, Binod C., and Robert H. Bustard. 'Predation on Natural Nests of the Saltwater Crocodile (*Crocodylus porosus* Scheider) on North Andaman Island with Notes on the Crocodile Population'. *Journal of the Bombay Natural History Society* 76 (1980), 311–23.

Douglas, Leo R., and Diogo Verissimo. 'Flagships or Battleships: Deconstructing the Relationship between Social Conflict and Conservation Flagship Species'. *Environment and Society: Advances in Research* 4(1) (2013), 98–116.

Guha, Ramachandra. 'The Prehistory of Community Forestry in India'. *Environmental History* 6(2) (2001), 212–38.

Jalais, Annu. 'Unmasking the Cosmopolitan Tiger'. *Nature and Culture* 3(1) (2008), 25–40.

Jepson, Paul, Maan Barua, Richard J. Ladle and Kathleen Buckingham. 'Towards an Intra-Disciplinary Biogeography: A Response to Lorimer's "Lively Biogeographies" of Asian Elephant Conservation'. *Transactions of the Institute of British Geographers* 36(1) (2011), 170–74.

Karanth, Krithi K., Sahila Kudalkar, and Shivangi Jain. 'Re-Building Communities: Voluntary Resettlement From Protected Areas in India.' *Frontiers in Ecology and Evolution* 6 (November 23, 2018): 1–11.

Karanth, Ullas, and Mysore Doraiswamy Madhusudan. 'Mitigating Human-Wildlife Conflict in Southern Asia', in John Terborgh, Carel van Schalk, Lisa Davenport and Madhu Rao (eds), *Making Parks Work: Strategies for Preserving Tropical Nature* (Washington, DC: Island Press, 2002), 250–64.

Karanth, Ullas, and Rajesh Gopal. 'An Ecology-Based Policy Framework for Human–Tiger Coexistence in India', in Rosie S. Woodroffe, Simon Thirgood and Alan Rabinowitz (eds), *People and Wildlife (Conflict or Coexistence?)* (Cambridge: Cambridge University Press, 2009), 373–87.

Kopnina, Helen. 'Beyond Multispecies Ethnography: Engaging with Violence and Animal Rights in Anthropology'. *Critical Anthropology* 37(3) (2017), 333–57.

Kopnina, Helen, Haydn Washington, Joe Gray and Bron Taylor, 'The "Future of Conservation" Debate: Defending Ecocentrism and the Nature Needs Half Movement'. *Biological Conservation* 217 (2018), 140–48.

Lewis, Michael. 'Indian Science for Indian Tigers? Conservation Biology and the Question of Cultural Values'. *Journal of the History of Biology* 38(2) (2005), 185–207.

Lorimer, Jamie. 'Elephants as Companion Species: The Lively Biogeographies of Asian Elephant Conservation in Sri Lanka'. *Transactions of the Institute of British Geographers* 35(4) (2010), 491–506.

MacKenzie, John. *The Empire of Nature: Hunting, Conservation and British Imperialism*. Manchester: Manchester University Press, 1988.

Neumann, Roderick P. 'Political Ecology of Wildlife Conservation in the Mt. Meru Area of Northeast Tanzania'. *Land Degradation and Rehabilitation* 3(2) (1992), 85–98.

Oommen, Meera Anna. 'The Elephant in the Room: Histories of Place, Memory and Conflict with Wildlife along a Southern Indian Forest Fringe'. *Environment and History* 25(2) (2019), 269–300.

———. 'Beasts in the Garden: Human-Wildlife Coexistence in India's Past and Present'. *Frontiers in Conservation Science* 2 (2021), 703432.

———. 'Colonial Pig-Sticking, Imperial Agendas and Natural History in the Indian Subcontinent'. *Historical Journal* 64(3) (2021), 626–49.

Oommen, Meera Anna, Rosie Cooney, Madhuri Ramesh, Mike Archer, Dan Brockington, Bram Buscher, Robert Fletcher, Daniel J. D. Natusch, Abi T. Vanak, Grahame Webb and Kartik Shanker. 'The Fatal Flaws of Compassionate Conservation'. *Conservation Biology* 33(4) (2019), 784–87.

Pooley, Simon. 'The Entangled Relations of Humans and Nile Crocodiles in Africa, c. 1840–1992'. *Environment and History* 22(3) (2016), 421–54.

Pooley, Simon, et al. 'An Interdisciplinary Review of Current and Future Approaches to Improving Human–Predator Relations'. *Conservation Biology* 31(3) (2017), 513–23.

Pooley, Simon, et al. 'Human–Crocodilian Interactions in Latin America and the Caribbean Region'. *Conservation Science and Practice* 3(5) (2021), 351.

Ramesh, Jairam. *Indira Gandhi: A Life in Nature*. New Delhi: Simon & Schuster India, 2017.

Rangarajan, Mahesh. 'Environmental Histories of South Asia: A Review Essay'. *Environment and History* 2(2) (1996), 129–43.

———. 'The Raj and the Natural World: The War Against "Dangerous Beasts" in Colonial India'. *Studies in History* 14(2) (1998), 265–99.

Ranjitsinh, K. *A Life with Wildlife: From Princely India to the Present*. New Delhi: HarperCollins, 2017.

Rashkow, Ezra. 'Resistance to Hunting in Pre-Independence India: Religious Environmentalism, Ecological Nationalism or Cultural Conservation?' *Modern Asian Studies* 49(2) (2014), 270–301.

———. 'Making Subaltern Shikaris: Histories of the Hunted in Colonial Central India'. *South Asian History and Culture* 5 (2014): 292–313.

Romanis, Federico De. 'Playing Sudoku on the Verso of the Muziris Papyrus'. *Journal of Ancient History* 27 (2012), 75–101.

Rosenzweig, Michael. 'Reconciliation Ecology and the Future of Species Diversity'. *Oryx* 37(2) (2003), 194–206.

Shanker, Kartik, and Roshni Kutty. 'Sailing the Flagship Fantastic: Different Approaches to Sea Turtle Conservation in India'. *Marine Anthropological Studies* 3(2) and 4(1) (2005), 213–40.

Sivaramakrishnan, Kalyanakrishnan. 'Thin Nationalism: Nature and Public Intellectualism in India'. *Contributions to Indian Sociology* 45(1) (2011), 85–111.

Thapar, Valmik. *Battling for Survival: India's Wildlife over Two Centuries*. Delhi: Oxford University Press, 2003.

Thekaekara, Tarsh, and Thomas F. Thornton. 'Ethnic Diversity and Human–Elephant Conflict in the Nilgiris, South India', in Piers Locke and Jane Buckingham (eds), *Conflict, Negotiation, and Coexistence: Rethinking Human–Elephant Relations in South Asia* (Oxford: Oxford University Press, 2016), 300–29.

Varghese, V. J. 'Land, Labour and Migration: Understanding Kerala's Economic Modernity', Working Paper 420. Thiruvananthapuram: Centre for Development Studies, 2009.

Wallach, Arian, Marc Bekoff, Chelsea Batavia, Michael P. Nelson and Daniel Ramp. 'Summoning Compassion to Address the Challenges of Conservation'. *Conservation Biology* 32(6) (2018), 1255–65.

Wallach, Arian, Chelsea Batavia, Marc Bekoff, et al. 'Recognising Animal Personhood in Compassionate Conservation'. *Conservation Biology* 34(5) (2020), 1097–1106.

Webb, Grahame, Charlie Manolis and Matthew L. Brien. 'Saltwater Crocodile *Crocodylus porosus*', in Charlie Manolis and Colin Stevenson (eds), *Crocodiles: Status Survey and Conservation Action Plan* (Darwin: Crocodile Specialist Group, 2010), 99–113.

Whitaker, Romulus, and Zai Whitaker, 'A Preliminary Survey of Saltwater Crocodiles (*Crocodylus porosus*) in the Andaman Islands'. *Journal of the Bombay Natural History Society* 76 (1978), 311–23.

Wilson, Edward O. *Half-Earth: Our Planet's Fight for Life*. London: Liveright, 2016.

CHAPTER 12

Representing Space to Structure Time

Tropical Deforestation Fronts in the Light of Human–Territory Relations

Johan Oszwald

The landscapes along deforestation fronts – by which I mean wooded areas being conquered by mankind – all look very similar: a reddish track snaking through vast green forests. But the factors explaining the endogenous and exogenous mechanisms causing them are, for their part, highly various. In analysing these complex landscapes, sometimes on the edges of the ecumene, at the interface with other – often protected – areas, it is essential to adopt a multidisciplinary approach and envisage the place as an interconnected system under pressure from multiple protagonists – to be examined on several spatial and temporal scales. Geography is the natural locus for linking up various disciplines examining each environment as a socio-ecosystem to be apprehended in both its spatial and its temporal dimension.

Over the past fifty years, many deforestation fronts have advanced across tropical regions, especially in the three largest tracts of tropical forest in the world: the Asian tract (including islands and the continent), the African tract (across West and Central Africa) and the Amazonian tract. Although deforestation pioneer fronts are advancing in these vast spaces, mainly due to tropical forests being converted into farmland, experts at the FAO (the United Nations Food and Agriculture Organization) reckon that the rate of deforestation worldwide has dropped over the past ten years – while continuing apace in many countries.

It is estimated that 16 million hectares of forest were lost between 1990 and 2000, as against 13 million between 2000 and 2010, and 11 million between 2010 and 2020. Factoring in reforestation, the FAO calculates that the net loss of wooded areas stood at 5.2 million hectares per year between 2000 and 2010. These annual losses are associated with global changes in climate, including longer periods of drought, heightening the impact of seasonal fires, be they of natural or human origin. The consequences spill over onto the dynamics of these tropical spaces. With current climate change projections predicting a global rise in temperature, extreme events are expected to become more frequent in various parts of the wet tropical zone and the tropical belt (the zone lying between the two tropics). According to predictions for the twenty-first century by the Intergovernmental Panel on Climate Change (IPCC) and the Walker Institute, global warming will be more intense in Africa and Amazonia. Nevertheless, whatever the trajectory of climate change, including if rainfall levels remain unaltered, it will not prevent the dynamics of deforestation and agricultural development being played out on newly claimed lands.

This context of uncertain climate change is compounded by the recent and equally powerful and rapid transformations in the incorporation of tropical forest resources into the capitalist system by various components. The dynamics transforming rural societies in tropical areas are driven by strong climate constraints in a context of pronounced demographic growth, by modifications in rural–urban trade and by changes in economic and environmental policies around much of the world. These dynamics transforming rural societies generate new contexts, which are particularly complex to analyse given that their territorial scale is so variable and so dynamic. Furthermore, international institutions have appropriated the concept of sustainable development, ecosystem services or so-called 'alternative' development practices (such as agri-ecological modes of production and extractive reserves). This raises questions about their place in the political, social and territorial construction of these landscapes, which remain socio-ecosystems.

The concept of 'landscape' – at the overlap between the social sciences and the life sciences, or, more simply, an interface between nature and societies – leads us to apprehend a complex system with social and ecological components interacting and changing across time and space. In this respect, for geography, studying these landscapes involves a transdisciplinary approach drawing on history, anthropology and ecology to build a socio-ecosystemic approach to landscapes.

Writing in 1952, Pierre Monbeig, a geographer specialising in Latin America, pointed out that politics plays a primordial role in driving the institutional appropriation of land to reduce agrarian tension and 'avoid the need for agrarian reform while integrating entire swathes of the territory'.[1] In this context, the deforestation frontier may be described as a process functioning in several stages: land appropriation by clearing activities, initially the extraction of valuable wood, followed by farming; land speculation and legislation for appropriated lands; the spatialisation of land appropriation with the development of roads, and humanisation with the emergence of permanent settlements. The time between initial clearance and the emergence of settlements has dropped over recent decades.

This progressive spatial appropriation gives rise to the territorial reconfiguration of the frontier region. The conversion of land tends to be followed by various phases of development, in the form of sawmills, abattoirs or agricultural markets: it is a matter of humanising natural landscapes and hence integrating them into the world system. The frontier develops due to the interaction between public policy and private projects in a given territory, thus adding a spatial dynamic to a system involving various protagonists.

This frontier may take the form of collective social and territorial development, either organised by an agency implementing policy (as in Côte d'Ivoire), or as a result of actions arising from convergent individual interests (as in Gabon). Public appropriation projects often seek to limit agrarian tension by providing land to many farmers. These projects are thus characterised by a succession of small farms in a territory with many agricultural colonies. Conversely, private colonisation spaces are characterised by numerous vast agricultural holdings linked to transnational agro-industrial companies and integrated into international rationales pertaining to the globalisation of agriculture. These private appropriation systems, promoted by certain states, seek to generate a source of labour and deliver accelerated territorial development, as is often the case in Indonesia, Brazil and Malaysia, for instance.

There are many similarities between these two forms of colonisation, found irrespective of variations in territorial mode. In this chapter I examine two case studies, in Côte d'Ivoire and Gabon, to illustrate the territorial functioning, in the former, of the dynamics by which land is consumed, and, in the latter, of conservationist processes. These are analysed in the light of recent historical dynamics affecting the territories. To this end, I draw on cartographic approaches – based both on specific data, for example data gathered in situ for the climate part of my study, and on satellite data, especially Landsat satellite data covering the period since the 1980s – to map the landscape dynamic over the past fifty years.[2] This method throws light on the complex spatial-temporal functioning of these territories, stemming both from human–environment relations in spaces subject to differing legislation and forms of protection, and from the interplay between protagonists, which is often linked to global conservation issues.

In Côte d'Ivoire: Agricultural Frontiers Driven by a Globalised Economy

There is an extensive body of literature on the issue of deforestation and its (environmental) consequences in West Africa. Côte d'Ivoire is emblematic of human–environment relations in a context of rapid land consumption. It is one of the most deforested countries in the world, with several precisely documented frontiers following different trajectories. Analysing the development of deforestation fronts in Côte d'Ivoire in the second half of the twentieth century and early years of the twenty-first century involves examining at least three points: the issue of socio-environmental dynamics on a national scale (in a context of policies to promote agricultural frontiers); the place of protected territories within these dynamics (do they effectively protect land or rather act as reserves of land to be converted?); and the interplay of scales, protagonists and negative feedback loops (in a system operating on several spatiotemporal scales).

In international arenas, mainly Western experts and scientists long presented Côte d'Ivoire as a model of development after its independence in 1960. Nevertheless, this period was also characterised by growing instability, particularly after the mid-1980s, culminating in a major political crisis in the early 2000s.[3] To understand and make sense of this instability, we need to return to the origins of Côte d'Ivoire's development, and more specifically the spatial model stemming from French colonisation, a system largely built around exploiting and exporting the colony's agricultural and natural resources for the profit of the colonial metropoles.

Thus the former colony's development model was based on exporting unprocessed primary goods, especially mining products, wood and agricultural commodities. The economic development promoted by Félix Houphouët-Boigny, the country's first elected president, had three main objectives: perpetuating the model of specialised agricultural exports initiated in the 1920s; maintaining alliances with firms present on the colonial territory that were already structuring the economy; and welcoming the outside world, that is, foreign capital along with additional labour mainly from West Africa.

This triptych of proactive economic development was based around the wood extraction industry and certain agricultural perennials such as coffee, cocoa and palm oil. For this economic model to function, dense forests are of strategic importance for they are the privileged place for the plantations on which this agriculture-driven development model is based. And the areas of dense forestry lay in the southern half of Côte d'Ivoire. So this was where agricultural development policies were most pronounced, especially via the promotion of the slash-and-burn cycle.

Indeed, forest areas in southern Côte d'Ivoire experienced significant agricultural development in the 1950s, consuming vast quantities of land that had hitherto been little humanised. This dynamic stemmed from a proactive policy based on three pillars: a flexible and attractive investment code to attract foreign investors to Côte d'Ivoire; a guaranteed minimum price for raw agricultural commodities, negotiated with Côte d'Ivoire's trading partners, especially France; and a policy to attract migrants from neighbouring countries under pledges made by Félix Houphouët-Boigny in 1963 that 'the land belongs to those who farm it'. The state thus guaranteed access to land, public services and the right to vote for migrants coming to settle and work in Côte d'Ivoire.

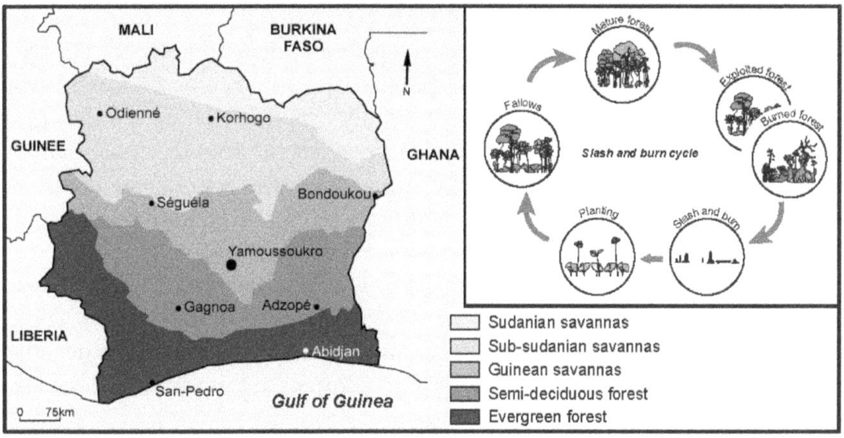

Map 12.1. Biomes in Côte d'Ivoire and the slash-and-burn cycle in tropical humid rainforests. © Johan Oszwald.

Implementing these proactive policies was made possible thanks to the favourable economic environment, with the regime protecting agricultural exports being maintained, in addition to aid from the European community. Indeed, the coffee and cocoa markets were regulated by international agreements between states, generating high levels of financial stability. Furthermore, the geopolitical context of the Cold War enabled Félix Houphouët-Boigny to negotiate privileged political treatment, especially by acting as a relay for France's Africa policy to other African partners.[4]

This economic environment secured by political negotiation bolstered the stability of Côte d'Ivoire's internal economic, political and social compromises, promoted in the media via highly personalised forms of rule. Nevertheless, this system started to falter in the 1980s and 1990s due to disruption caused by global geopolitics and economics, triggered by the emergence of a new worldwide paradigm instituted by the Washington consensus (a tacit agreement in which financial assistance to developing countries was conditional on their adopting good management practices mainly defined by the International Monetary Fund (IMF) and the World Bank).

This had a lasting impact on Côte d'Ivoire's economic equilibrium: the liberalisation of trade quickened, while the state withdrew from economic negotiations to the benefit of transnational firms. The markets responded to these upheavals with a steady drop in the price of coffee, cocoa and other agricultural commodities. And with an economy based primarily on extraction, Côte d'Ivoire was heavily dependent on international markets. This led to spiralling debts with the IMF and World Bank, causing an abrupt deterioration in Côte d'Ivoire's financial and economic situation.[5]

Faced with these changes, Félix Houphouët-Boigny tried to rescue the country's development model by taking a radical unilateral decision: between July 1987 and October 1989, he halted cocoa exports to bolster his negotiating position. But the international market rapidly adapted, turning to other producers such as Brazil or Colombia. This heralded the rout of Côte d'Ivoire's economy: the price guaranteed to farm managers was halved, the country's financial dependence on international donors accelerated and then the public banking sector was dismantled along with certain public services.[6]

Farmers responded by putting greater pressure on the land. To compensate for the financial losses caused by the drop in the guaranteed price of cocoa and coffee, they increased the surface farmed: this was the first of the negative feedbacks for Côte d'Ivoire's plantations. In addition to this increase in surface

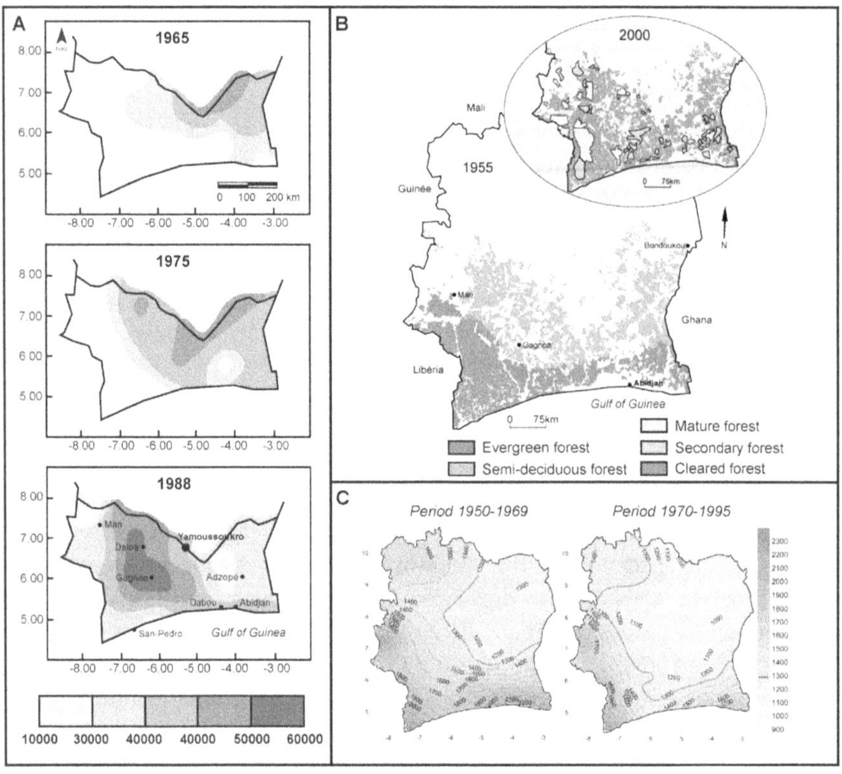

Map 12.2. Cocoa and forestry spaces in Côte d'Ivoire and the 1,300 mm isohyet. © Johan Oszwald.

A. Cocoa production (in tonnes) in Côte d'Ivoire.
B. State of forest areas in Côte d'Ivoire in 1955 and 2000.
C. Movement of the 1300 mm isohyet (cocoa development limit) in Côte d'Ivoire.

farmed to make up for financial losses, certain public services were dismantled, triggering a rare example of town-to-countryside migration, starting in the late 1980s.[7] This situation added to societal pressure on environments, with agricultural frontiers gobbling up land as they advanced across the south of the country.[8]

The cycle then continued. In the 1980s, rapidly advancing plantation systems – adding to pressure on forests due to the increase in farmed land – led to expanding cocoa supply and an increase in Côte d'Ivoire's stocks, causing prices to drop further.[9] This phenomenon then caused a new negative feedback, by accelerating the consumption of land by agriculture.

Data collection and mapping shows that forest areas have shrunk considerably since 1986, with consequences for regional rainfall variability.[10] In Map 12.1 above, the 1,300 mm rainfall isohyet corresponds to the optimal annual rainfall threshold for cocoa plantations or tropical forest. Beneath this, cocoa plantations suffer from water stress, impacting on the crop's profitability and weakening its resistance to predators and disease. Côte d'Ivoire's deforestation throughout the twentieth century probably had major repercussions on the water cycle. The amount of land where forest was felled brought about alterations in the region's climate conditions, making a drier climate more likely. This was because the expansion in plantation areas modified interactions between the soil and the atmosphere, disrupting the evaporation and evapotranspiration cycles. This rapid phenomenon triggered longer periods of drought, which also became more intense as compounded by the soil exhaustion caused by intensive monoculture.

Thus all the meteorological stations in Côte d'Ivoire tend to show a marked statistical shift in rainfall data in the early 1970s.[11] The 1,300 mm isohyet moved south-west, making cocoa cultivation less profitable in lands beneath this threshold, affecting the eastern and central regions in southern Côte d'Ivoire. This trend generated another negative feedback: to counter the dropping profitability of their plantations, planters tended once again to increase the area they farmed, as they had during previous economic crises. A vicious cycle of anthropo-environmental relations seemed to be at work in Côte d'Ivoire as of the late 1980s, threatening the country's entire forest areas.

In eastern Côte d'Ivoire in the region of Abengourou, one of the earliest farmed zones in the country, forest areas had already been converted into farmland in the 1980s. The only areas still available were zones protected by the state for conservation purposes or with a special protection status, such as the classified forests of Béki and Bossématié, forest reserves in the state's custody. Analysis of satellite images shows how heavily these zones were impacted, particularly between the late 1980s and early 1990s, for they were the last reserves of available forest land.

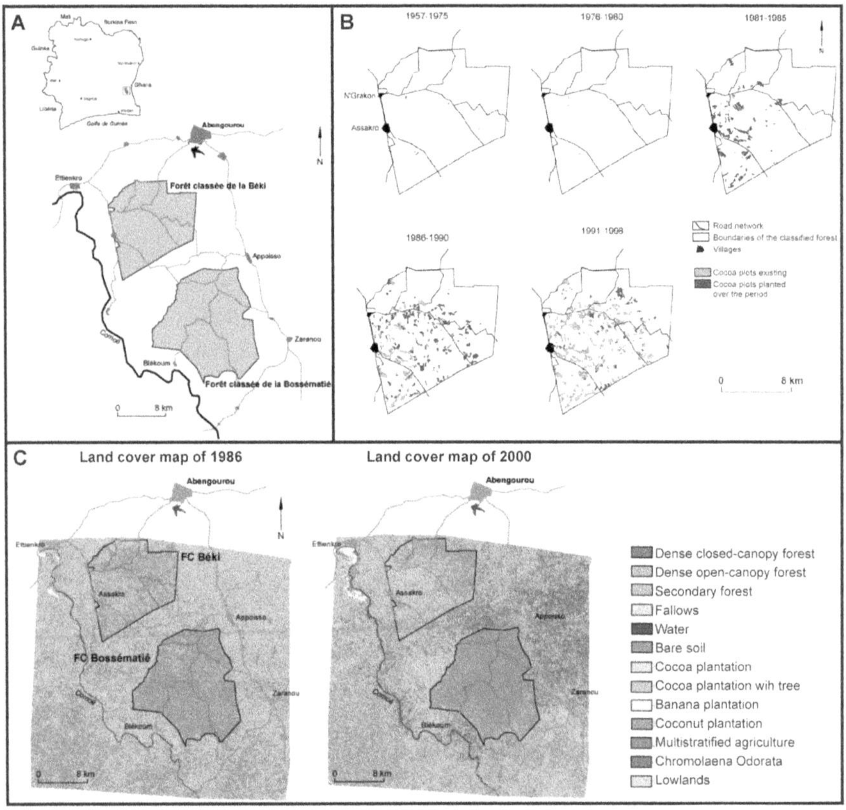

Map 12.3. New anthropo-environmental relations in Côte d'Ivoire. © Johan Oszwald.

A. The Abengourou area and the classified forests of Béki and Bossématié.
B. Illicit agricultural incursions in the classified forest of Béki.
C. Maps of the state of surfaces in the Abengourou zone in 1986 and in 2000 (SPOT HRV images).

As shown in the previous figure, all the negative feedbacks initially caused the farmed area to steadily increase, before farms then started to encroach on protected zones – the final reserves of farmable land. This phenomenon began in the 1970s, before accelerating as of 1984 when agricultural colonies were established in the classified forest of Béki, at an average rate of a hundred or so per year between 1984 and 1990: from 1977 to 1984, 145 farm managers settled in the classified forest, cultivating 1,564 hectares of land; from 1985 to 1990, 708 farm managers started operating in the classified forest, with an additional 5,867 hectares now being farmed; from 1990 to 1995, the number dropped, with 211 managers moving in to farm 1,395 additional hectares. The cultivated area covered a total of 8,800 out of the 16,100 hectares of the classified forest of Béki as a whole, amounting to nearly 55 per cent of the classified land.

In 2004, 87 per cent of the effective total of farm managers in Béki came from outside the region. 'Natives' were in a minority, and tended to cultivate land in the vicinity of the village to which they belonged. The more heterogeneous and dispersed plots mainly belonged to certain 'outsiders', especially Malians, Burkinabés and people from Côte d'Ivoire's Baoulé and Sénoufo ethnicities. This mix of planters is a clear illustration of the success on the ground of Félix Houphouët-Boigny's development policy. However, with a development model focusing on cash crops and with access to farmable land drying up as the 1980s crisis progressed, encroachment on forests only intensified, even on areas with special legal status, that is, covered by a form of protection.

The case of Côte d'Ivoire shows how complex human–environment relations are in spaces subject to differing legislation and forms of protection. Many causes explain the transformation of these forest tracts into areas given over to cash crop farming. The consequences of this phenomenon are equally varied, impacting on societal and environmental dynamics. It thus transpires that the interplay of protagonists is often linked to global economic and conservation issues, with a great diversity in situations at the local scale, all too often reduced to encroachment phenomena despite in fact stemming from eminently complex interactions.

In Lopé Park (Gabon): Conservation Policies and Landscape Dynamics

Like Côte d'Ivoire, certain West African countries such as Cameroon, Togo and Benin have chosen to develop by converting their forest areas. But others have adopted a diametrically opposed model, leading us to analyse conservation issues. In this respect, Gabon is a highly original case since it is one of the least populated countries on the African continent, with 1.53 million inhabitants, amounting to 5.7 pop/km^2 (against an average of 43 pop/km^2 in Africa in 2019).[12] This small population is concentrated in urban areas, meaning that many forest areas have been conserved, to the extent that dense rainforests still cover nearly 200,000 km^2 – 85 per cent of the territory – making it the second largest forest territory in Africa after the Democratic Republic of Congo.[13]

There has been commercial forestry in Gabon for over a hundred years, and the current government sees it as key to the country's economic development. The Gabonese government has thus divided the national territory into two zones. The first covers nearly one-third of the current forest area and lies to the west of the country, around Libreville and Port-Gentil. Here, the state encourages economic and trade activities, and is developing a relatively dense communications network to boost them.[14] This area is also home to 50 per cent of the country's population, and is subject to intense anthropic pressure, be it from agriculture, hunting or timber extraction. The second zone comprises 70 per cent of the remaining forest cover. Logging is less developed here, the communications network is less dense and of poorer quality and anthropic activities, which tend to be long-standing, are especially marginal as these areas have been heavily impacted by the rural exodus in Gabon over the past hundred years or more.[15]

The Gabonese government has been striving to adapt its forest policy over recent years to safeguard this economic, cultural and environmental asset for years to come. Thus since December 2001, new legislation has been in force based on three strategic development paths: the sustainable development of forest resources, the industrialisation of the logging industry and protecting ecosystems. To further this third strand, on 30 August 2002, in the wake of the Johannesburg Earth Summit, the government created thirteen national parks to be managed by the Agence nationale des parcs nationaux (ANPN), answering directly to the president. Thus 11 per cent of the territory – representative of Gabon's biological diversity – is now fully protected, with any extraction forbidden.

Among the country's various 'biodiversity sanctuaries', Lopé National Park (LNP) was the first protected area to be created, back in 1946. The 4,970 km^2 park lies in the centre of the country, in the northern part of the Chaillu Mountains. Before becoming an integral reserve, listed as World Heritage by UNESCO in 2007 due to its exceptional biological and ecological resources, the LNP underwent various changes in legal status.[16] This protected area had been a game reserve since 1946, before becoming Lopé-Okanda National Park in the 1960s. This new status did not prevent small-scale logging in the form of concessions during the 1980s.

This is no longer the case, the last logging companies having left the national park in the early 2000s. Thanks to conservation and economic tourism policies, the savannah areas in the LNP are maintained regularly in the dry season. Controlled burning is part of the local population's way of life and farming. Clearing growth makes it possible to spot game, prevents incidents with megafauna, keeps reptiles away, allows termite mounds to be harvested for fishing, ensures that edible mushrooms regrow and so on. Throughout history, the ecosystems in this area have been 'maintained' by local populations, and today it is the park managers who use controlled burning of the savannah. Hitherto these impacts have been little assessed, particularly in terms of landscape dynamics.

The ANPN uses Lopé Park to spearhead its media campaigns. It seeks to introduce wildlife and plant protection programmes, and its managers advocate

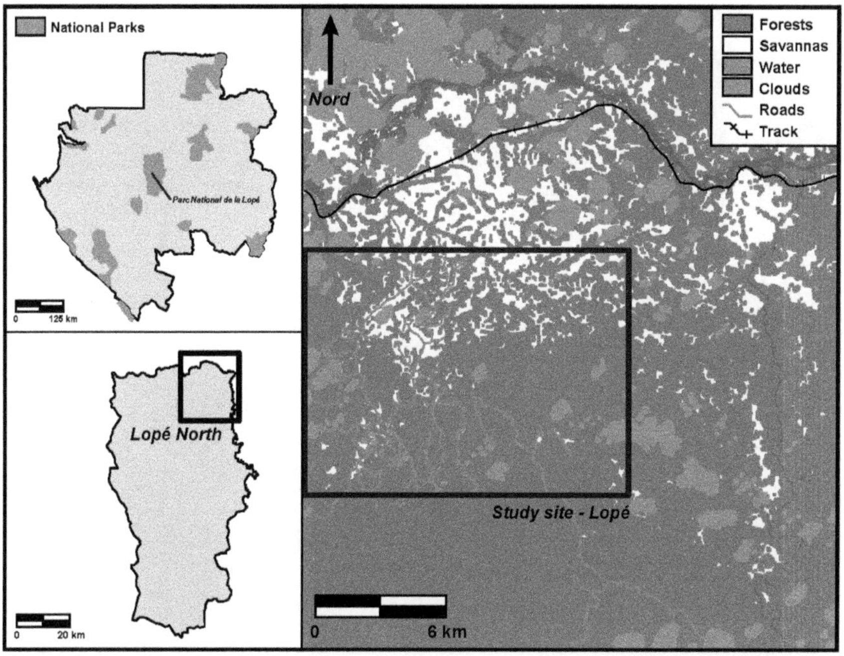

Map 12.4. Lopé National Park and the study zone. © Johan Oszwald.

developing a proactive policy to expand ecotourism in Gabon. Controlling anthropic pressure in the vicinity of the National Parks is currently the main objective of the Gabonese administration, and especially of the park custodians. Even though Gabon is underpopulated, illegal incursions to poach small game are a frequent occurrence. Based on maps obtained by photogrammetric and remote sensing analysis in 1982 and 2008, we may assess the impact of protection and conservation policies on environmental dynamics in Lopé National Park.

First, thematic land usage maps were drawn up for the forest–savannah transition zone for the years 1982 and 2006, based on remote sensing analysis of landscape elements. These maps show the significant diachronic vegetation dynamic. Indeed, they show a clear drop in savannah areas, down 24.5 per cent.

These land use maps were then compared to produce a diachronic forest versus non-forest overview. The resultant map gives a clearer picture of the scale of forest colonisation dynamics, enabling us to identify several processes: the expansion of forest areas by spatial contagion, especially around clearings; the expansion of forest areas along the river network (gallery forests tend to sprout up along temporary watercourses in the savannah, a phenomenon that tends to fragment the savannah near waterways, subsequently

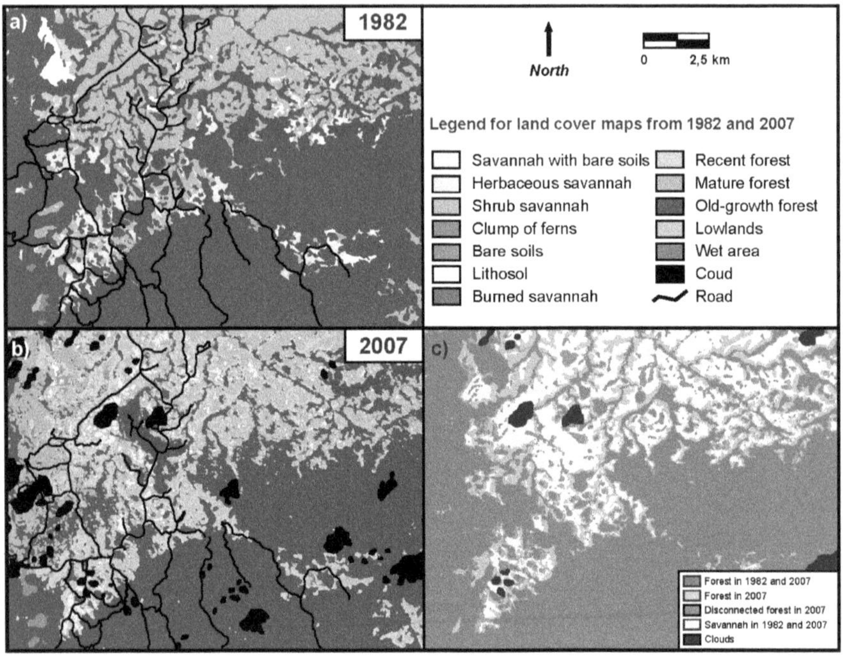

Map 12.5. The diachronic vegetation dynamic in Lopé Park. © Johan Oszwald.

A and B. Land use.
C. Forest–savannah dynamics (SPOT image, spatial resolution of 4 to 20 metres).

favouring the dilation of forest areas); and the appearance of many thickets in areas that are mainly savannah when rock formations or valley bottoms have been abandoned, leading to reforesting (a process that accelerates the connectivity of various areas, particularly isolated forest units in savannah areas in 1982).

To summarise all these observations, I created a model of the landscape dynamics, after surveying species in the field with the assistance of Lopé scientific staff.

The model of landscape dynamics in Lopé starts with the land being predominantly savannah. The grassy savannahs are composed predominantly of *Cyperaceae* and *Poaceae*. These savannahs often have bare areas of varying extent, forming mosaics. But as soon as the density of bare earth exceeds 40 per cent of the savannah, this is classified as a savannah–bare earth mosaic. The shift from grassy savannah to savannah with bare earth is generally caused by two factors. First, the intensification of annual slash-and-burn, intended to maintain the savannah, fragments the grass coverage engendered by age-old human occupation of the sites. Repetitive seasonal fires reduce the sub-

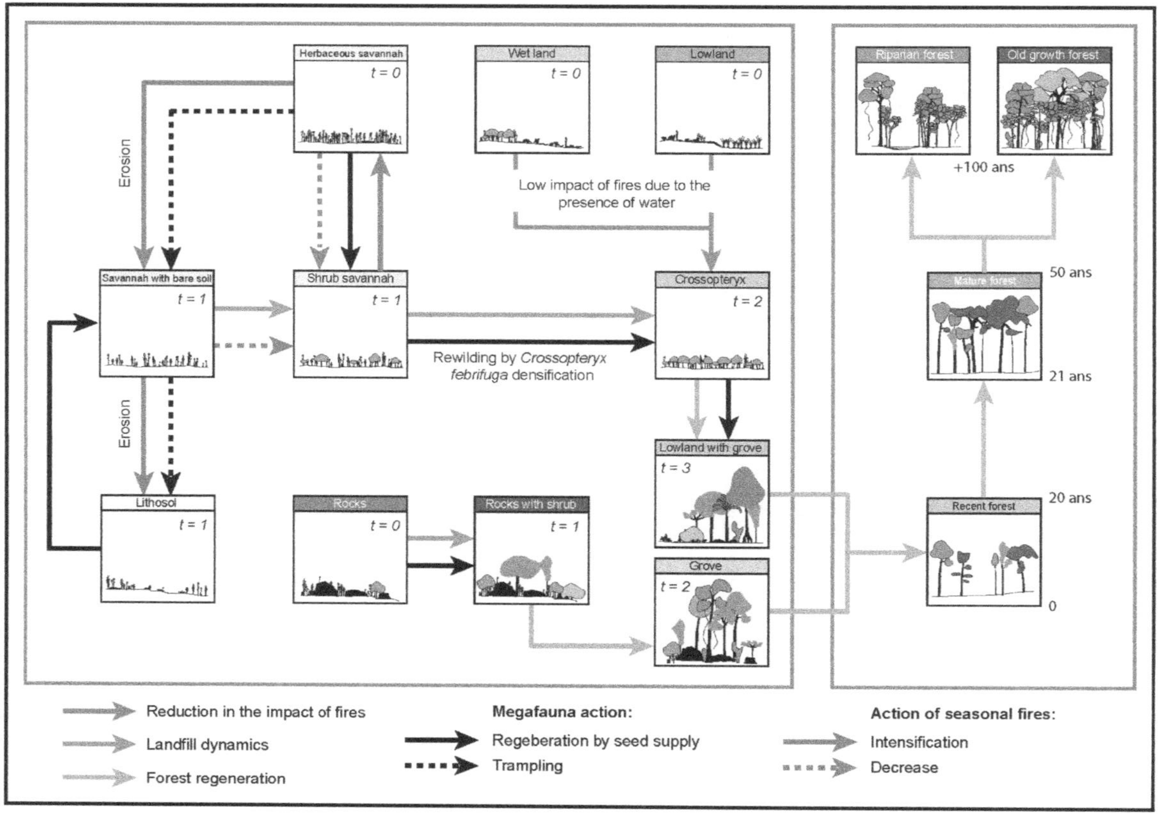

Figure 12.1. Forest dynamics in the forest–savannah mosaic in Lopé. © Johan Oszwald.

strate and plants, especially trunks and roots. Plants therefore grow back more slowly after a seasonal fire, leading to erosion during rainy seasons (either concentrated or sheet erosion). This was a significant factor before the zone was classified as a national park, when neighbouring populations used the site. Second, megafauna's repeated trampling of transit areas in the savannah leads to the appearance of expanses of bare earth. In the final stage, these processes trigger the appearance of lithosols, soils with sheet structures that develop along the upper parts of slopes.

Conversely, if burning occurs less frequently, grassy savannahs tend to evolve towards shrublands. Several decades ago, hundreds of zones were burned twice per year by villagers (during the short dry season from December to January, then during the long dry season from June to September). Over recent years since the zone was classified as national park, burning has only been carried out during the long dry season due to lack of staff, meaning that burning campaigns are less effective and certain areas are abandoned. Shrubland has the same species of grasses as grassy savannah, but certain shrubs take hold as there are fewer bushfires. These formations vary in density depending on seasonal fires. As shrubland lies close to humid zones, clumps of *Crossopteryx febrifuga* may also develop. These environments are severely disrupted by herds of buffalo passing back and forth, creating systems of grassy mounds surrounded by water pockets maintained by the repeated trampling of the ground.

Other elements in the landscape are also characterised by their progressively being abandoned, especially the rock piles. Rock formations accumulate sediments transported by erosive rains, leading to the formation of soils containing many nutrients. Additionally, these formations attract herds, especially buffaloes, who shelter there. Megafauna defecate in these areas, transporting many seeds. The sediments associated with the organic matter act as a seedbed for vegetation.

Landscape elements that have been left repeatedly abandoned give rise to forest landscape dynamics spreading over about 150 years: young forest transitions (0 to 20 years), with formations dominated by shrubland and a few isolated trees; mature forest transitions (21 to 50 years), with formations dominated by a few tree species within a mosaic of savannah (*Lophira alata*, *Uapaca guineensis*, *Pauridiantha efferata* and *Aucoumea klaineana*); and old forest transitions or dense rainforest (51 to 150 years), with formations dominated by very specific dense tree coverage (*Aucoumea klaineana*, *Lophira alata* and *Uapaca guineensis*).

From a strictly ecological point of view, Lopé National Park today is thus an example of a successful conservation policy, especially in terms of savannah areas being recolonised by forest.[17] This policy has led to a growing number of physical processes favourable to forest regeneration, and indeed, forest areas expanded by 24.5 per cent between 1982 and 2007 in the northern zone of Lopé, especially along permanent and temporary watercourses.

Most scientific studies tend to show that forest areas are continually shrinking in the tropics, despite being under state protection.[18] Indeed, the total surface area of forest concessions has increased markedly in Gabon over recent decades, despite the conservationist policies implemented in certain sectors. Nevertheless, in Lopé National Park there has been a notable increase in forest species, despite the presence of forest concessions within its perimeter. This increase is due to the ANPN's proactive policy to conserve and manage biodiversity and environments (especially via burning). To this end, it was not just logging companies that were banned from the park in the 2000s, but also populations living in or around it. A strict policy to preserve natural environments has thus been implemented, often to the detriment of local populations. Nevertheless, the same populations have given rise to these complex landscapes, characterised by highly diverse habitats. Such a situation raises questions about the concept of protection, and especially about the integration of local populations into protection actions, or in simpler terms, about the integration of socio-ecosystems into conservation policies.

These spatial analyses enable us to better apprehend anthropic pressure and conservation policies, and thus bring out the complexity of socio-environmental interactions. These society–environment interactions have been viewed here through multiscale approaches, drawing particularly on GIS and remote sensing, making it easier to understand social-environmental processes, connected either to market rationales and the development of public policy or to the implementation of policies to protect ecosystems.

As they are based on voluminous satellite images covering the entire planet, the mapping and quantification of changing modifications in forest cover or forest–farmland mosaics are reproducible. These approaches facilitate the monitoring of territories for which we often have little spatial information.

The transformation in forest cover occurring in most tropical regions has profound repercussions on environments and on how societies relate to them, especially when there is a shortage of land or when the last available spaces have a particular status. It thus seems essential to understand the dynamics at work, to characterise them and to determine the most influential factors for tropical forest areas as a whole, given that they are all subject not only to market rationales but also to the lengthy and equally changing history of nature conservation policies in the tropics.

These geographical approaches, which essentially aim to understand, through a systemic approach, the relationships between actors and policies at several spatial scales and the anthropic or natural territory, are an essential complement to the other approaches discussed in the previous chapters.

Johan Oszwald is Professor of Geography at the University of Rennes 2, where he is head of the Social Sciences Faculty. His research activities are carried out at the CNRS 6552 LETG (Littoral, Environnement, Télédétection, Géomatique). He has been working for more than twenty years on tropical deforestation fronts in Asia, Africa and America. He is a specialist in spatiotemporal geostatistics, remote sensing and mapping socio-ecosystems.

Notes

This chapter contains maps that are of a reduced resolution and that lack greyscale differentiation. Higher-resolution versions were unavailable for this publication. Due to the complex patterning of the satellite images, greyscale is unable to reproduce the differentiation conveyed in the colour versions.

1. Pierre Monbeig, *Pionniers et planteurs de São Paulo* (Paris: Cahiers de la FNSP, 1952), 28.
2. On these methods, see especially: Johan Oszwald et al., 'Analyse des directions de changement des états de surface végétaux pour renseigner la dynamique du front pionnier de Maçaranduba (Brésil) entre 1997 et 2006', *Télédétection* 9(2) (2010), 97–111.
3. Richard Banégas and Bruno Losch, 'La Côte d'Ivoire au bord de l'implosion', *Politique africaine* 87(3) (2002), 139–61.
4. Bruno Losch, 'Coup de cacao en Côte d'Ivoire: Économie politique d'une crise structurelle', *Critique Internationale* 9 (2000), 6–14.
5. Banégas and Losch, 'La Côte d'Ivoire'.
6. Losch, 'Coup de cacao'.
7. Jean-Louis Chaléard, 'Temps des villes, temps des vivres: L'essor du vivrier marchand en Côte d'Ivoire' (PhD diss., University of Nanterre, 1994).
8. Johan Oszwald, 'Dynamique des formations agroforestières en Côte d'Ivoire (depuis les années 1980 aux années 2000): Suivi par télédétection et développement d'une approche cartographique' (PhD diss., University of Science and Technology of Lille, 2005).
9. Éric Léonard and Marc Oswald, 'Une agriculture forestière sans forêt: Changements agro-écologiques et innovations paysannes en Côte d'Ivoire', *Nature Sciences Sociétés* 4(3) (1996), 202–16; Éric Léonard, *Différenciation et reproduction des exploitations agricoles dans le Sud-Ouest ivoirien: Une typologie des systèmes de production* (Abidjan: ORSTOM, 1993).
10. Sylvain Bigot, Johan Oszwald and Télesphore Brou Yao, 'Variations climatiques et changements d'occupation des sols en zones de forêt primaire: Le cas de la forêt classée du Haut-Sassandra (Côte d'Ivoire)', *Publications de l'Association Internationale de Climatologie* 15 (2003).
11. Télesphore Brou Yao et al., 'Récession pluviométrique et mutations socio-économiques en Côte d'Ivoire', *Publications de l'Association Internationale de Climatologie* 16 (2004).
12. *World Development Indicators*, annual report (Washington, DC: World Bank, 2002).
13. Benoît Mertens and Jan Makak, *Interactive Forest Atlas for Gabon* (Washington, DC: World Resources Institute and Ministère de l'économie forestière, des eaux, de la pêche et de l'aquaculture, 2009).

14. Ibid.
15. Roland Pourtier, *Le Gabon: Espace, histoire, société*, vol. 1 (Paris: L'Harmattan, 1990).
16. Norman Myers et al., 'Biodiversity Hotspots for Conservation Priorities', *Nature* 403 (2000), 853–58; Richard Oslisly, *Archéologie dans le parc national de la Lopé: Site mixte nature culture du Patrimoine mondial* (Paris: IRD and ANPN, 2010).
17. Jean Maley, 'Histoire récente de la forêt dense humide africaine et essai sur le dynamisme de quelques formations forestières', in Raymond Lanfranchi and Dominique Schwartz (eds), *Paysages quaternaires de l'Afrique Centrale Atlantique* (Paris: ORSTOM, 1990), 367–82; Miguel Leal et al., *The Biodiversity of Bai Djobo* (Gabon: Missouri Botanical Garden report, 2007).
18. FAO, *The State of the World's Forests (SOFO)* (Rome, 2012).

Bibliography

Banégas, Richard, and Bruno Losch. 'La Côte d'Ivoire au bord de l'implosion'. *Politique africaine* 87(3) (2002), 139–61.

Bigot, Sylvain, Johan Oszwald and Télesphore Brou Yao. 'Variations climatiques et changements d'occupation des sols en zones de forêt primaire: Le cas de la forêt classée du Haut-Sassandra (Côte d'Ivoire)'. *Publications de l'Association Internationale de Climatologie* 15 (2003).

Brou, Yao Télesphore, Johan Oszwald, Sylvain Bigot and Éric Servat. 'Récession pluviométrique et mutations socio-économiques en Côte d'Ivoire'. *Publications de l'Association Internationale de Climatologie* 16 (2004).

Chaléard, Jean-Louis. 'Temps des villes, temps des vivres: L'essor du vivrier marchand en Côte d'Ivoire'. PhD diss., University of Nanterre, 1994.

FAO. *The State of the World's Forests (SOFO)*. Rome, 2012.

Leal, Miguel, Daniel Nguema, Éric Mounoumoulossi and Patrick Bibbiemou. *The Biodiversity of Bai Djobo*. Gabon: Missouri Botanical Garden report, 2007.

Léonard, Éric. *Différenciation et reproduction des exploitations agricoles dans le Sud-Ouest ivoirien: Une typologie des systèmes de production*. Abidjan: ORSTOM, 1993.

Léonard, Éric, and Marc Oswald. 'Une agriculture forestière sans forêt: Changements agro-écologiques et innovations paysannes en Côte d'Ivoire'. *Nature Sciences Sociétés* 4(3) (1996), 202–16.

Losch, Bruno. 'Coup de cacao en Côte d'Ivoire: Économie politique d'une crise structurelle'. *Critique Internationale* 9 (2000), 6–14.

Maley, Jean. 'Histoire récente de la forêt dense humide africaine et essai sur le dynamisme de quelques formations forestières', in Raymond Lanfranchi and Dominique Schwartz (eds), *Paysages quaternaires de l'Afrique Centrale Atlantique* (Paris: ORSTOM, 1990), 367–82.

Mertens, Benoît, and Jan Makak. *Interactive Forest Atlas for Gabon*. Washington, DC: World Resources Institute and Ministère de l'économie forestière, des eaux, de la pêche et de l'aquaculture, 2009.

Monbeig, Pierre. *Pionniers et planteurs de São Paulo*. Paris: Cahiers de la FNSP, 1952.

Myers, Norman, et al. 'Biodiversity Hotspots for Conservation Priorities'. *Nature* 403 (2000), 853–58.

Oslisly, Richard. *Archéologie dans le parc national de la Lopé: Site mixte nature culture du Patrimoine mondial.* Paris: IRD and ANPN, 2010.
Oszwald, Johan. 'Dynamique des formations agroforestières en Côte d'Ivoire (depuis les années 1980 aux années 2000): Suivi par télédétection et développement d'une approche cartographique'. PhD diss., University of Science and Technology of Lille, 2005.
Oszwald, Johan, et al. 'Analyse des directions de changement des états de surface végétaux pour renseigner la dynamique du front pionnier de Maçaranduba (Brésil) entre 1997 et 2006'. *Télédétection* 9(2) (2010), 97–111.
Pourtier, Roland. *Le Gabon*: *Espace, histoire, société*, vol. 1. Paris: L'Harmattan, 1990.

CONCLUSION

Studying Nature, Networks and Power

What Next?

Guillaume Blanc

The imperial exploitation of forests and creation of forestry reserves in British Singapore; the conservation of elephants, elitist hunting and local usages of fauna in Kenya between the world wars; the myth of desertification and the expropriation of resources in 1940s Lebanon – colonialism brought capitalism and ecologism together. Over the course of this encounter, protecting nature was primarily a matter of protecting it from its inhabitants, in both Africa and Asia. This coercive and socially exclusionary dimension to conservation is anchored in practices that have continued since independences. The declinist discourse and criminalisation of the peasantry in 1970s socialist Tanzania; the alliance between international experts and national leaders to the detriment of certain populations in the Seychelles; the destruction of forests and creation of reserves in Côte d'Ivoire and Gabon since the 1990s; clashes between city-dwelling ecologists and rural anti-conservationists in India's protected areas in the twenty-first century – though conservationist doxa arose from the colonial phenomenon, it remained omnipresent in postcolonial days, in both Africa and Asia.

This occurred to such an extent that it has influenced nature policies even in countries that were never colonised. We may observe this in Ethiopia where, since 1961, a prevalent idea has it that the country's forest cover collapsed from 40 per cent in 1900 to 3 per cent 'today'. These figures stem from estimates made during the 1940s and 1950s by a Canadian and a German forester based on their own visual observations. But in 1961 an expert from the Food and Agricultural Organization (FAO) turned the 'fallacy of ubiquitous Malthusian narratives' into scientific fact; ever since, these figures have determined Ethiopian conservationist policy in general, and how nature has been turned into parks in particular.[1] For example, as stated in the introduction to this book, since the 1960s the population in Simien Park has been criminalised on the grounds that they are a threat to the local flora and fauna; in 2016 they were expelled by agents working for the Ethiopian national administration – who were thereby following UNESCO and IUCN recommendations to the letter. The neo-Malthusian nature of these practices and this knowledge does nothing to diminish the acute problems caused by human alterations to the country's environment. But it does reveal that nature management there is still based on the idea that an African national park should be devoid of inhabitants.

This has not prevented Ethiopia's leaders from using this ideal to serve specifically national objectives. In building a centralised nation state at a brisk pace since the onset of the postcolonial era in the 1960s, they have sought to win recognition for their nation from the outside world, the more effectively to impose it at home – in the Simien scrubland, for example, which has regularly opposed their power. That is why Ethiopia has appropriated international conservation norms: though stemming from colonial days, they bring the funds and legitimacy it lacked to plant the national flag in territories beyond its control.[2]

In Asia, Thailand presents a similar entanglement of nature and power, of colonial and postcolonial times. Like Ethiopia, the country has never been colonised. Yet as of the early twentieth century, the way nature was governed resembled the approach used by the British and French to complete their conquest of the region (in India, Burma and Indochina). As early as the 1900s, but especially as of the 1930s, the Siam monarchy took inspiration from the conservationist knowledge and norms being imposed elsewhere by Europeans, notably by calling on the British to train Thai civil servants working for the Royal Forestry Department. The monarchy used coercive state management of forests to claim and assert its territorial sovereignty across the entire kingdom – against the expansionist British and French empires, as well as against many local leaders who rejected its authority.[3]

Then in the late 1950s, as decolonisation in the region was coming to an end, with the beginning of the Cold War and a military junta now in power, the Thai state implemented an environmental policy that complied with 'modern' conservation standards. But this policy almost exclusively targeted the mountain lands of the *chao khao* populations, highlanders present throughout the country, especially along the border with Burma. This is where the authorities created 'national parks', 'protected forest zones' and 'nature reserves', in which populations were criminalised and controlled, with limits on available farmland, arrests for 'illegal encroachment' on the forest, and the forced displacement of villagers and villages. Throughout the 1960s and 1970s this undertaking was presented as scientific and moral. These 'peripheral ethnic groups' using agricultural practices deemed to be destructive (especially swidden cultivation) were accused of causing the disappearance of the kingdom's forests and natural heritage. This supposed damage to ecosystems by mountain populations thus stemmed from preconceived environmental notions dating from the colonial period and subsequently conveyed by international conservation institutions.[4] But the narrative of degradation enabled the national authorities to deny land use rights to 'mountain peoples' who rejected the modern Thai nation, or were at least hindering its construction. Once again – like in former colonies in Asia and Africa, as well as in Ethiopia, which had always been independent – the conservationist doxa was ultimately grounded in the colonial period, along with the model of the nation state originating from that period.[5]

Since the 1990s this has generated unprecedented forms of resistance. Many mountain peoples, especially the Karen, adopting the norms of the new 'community' conservation, have put forward the alternative model of nature management advocated by the *kananurak pamai* (forest conservation) movement. Faced with a central state wielding international conservation institutions' representations that are systematically declinist, the populations in turn wield the international community's representations of them as 'good savages' living in 'harmony' with nature.[6] Still, this has not prevented experts from international institutions recommending restrictions on local usage of protected areas, or the national authorities from implementing these recommendations to boost their control over their citizens.

Across both Africa and Asia, conservation is nowadays both neoliberal and communitarian. Yet it remains socially exclusionary, politically coercive and based on declinist discourse stemming from colonial beliefs and practices. That is why it generates perpetual conflict. This observation, which is central to many studies, runs throughout our book. It is no longer a matter of asking whether, in Africa, the Middle East and Asia, nature is or is not contested. The question is irrelevant for in fact 'nature' itself is fought over, with the (institutional) skirmish to seize territory, the (cultural) skirmish to impose a representation of the environment in the public sphere and the (material) skirmish to exploit a resource. So what we now need to do is focus on conservationist causes that produce nature. In other words, we need to simultaneously explore the causes that generate nature and the nature generated by these causes. To do this, four connected topics may be studied, together with the connections between them.

First, we need to continue studying international conservation institutions as 'calculation centres', or 'assemblage factors' as Bruno Latour puts it.[7] This has already been done by Simone Schleper for the IUCN for the period 1960–80.[8] Over these two key decades, it sent experts from its Swiss headquarters out into the field, pooled the data they collected, compiled it and then sent it back out as standardised norms. We now need to explore this history, starting from the earliest days of the IUCN with the founding of the International Office of Documentation and Correlation for the Protection of Nature in 1928, down to the present day. But we also need to produce other institutional studies by combing through the archives of UNESCO (in Paris), the WWF (in Gland),

Fauna & Flora International (in London), the Wildlife Conservation Society (in New York) and the zoological societies of Frankfurt, London and New York. This is no easy undertaking, for these institutions have not always catalogued their archives. Above all, researchers who criticise them are often denied access, as was the case for certain authors of this book at the IUCN's and the WWF's Swiss headquarters in Morges. Having said that, it is both possible and necessary to explore these archives. This would improve our knowledge of these institutions, enabling us to study their links and the debates within them. It would then be possible to retrace how, over the course of the long twentieth century, they have defined 'good' nature governance norms in general, and especially in Africa and Asia, two globalised continents where these institutions have had to accompany and adapt to the transition from colonial to postcolonial days.

Second, in the wake of Raf de Bont's history of 'nature's diplomats' from the 1920s to the 1960s,[9] we need to launch a series of biographical enquiries retracing the paths taken by international conservation experts over the course of the long twentieth century, shunning the superficial break between before and after independence. Mathieu Guérin has started this by retracing the imperial trajectory of Captain Ritchie, who worked in reserves in Kenya then Malaya. William Beinart, for his part, has looked at Joy Adamson's campaign for large East African wildlife in the years either side of Kenyan independence. And Joseph Hodge has traced the imperial and postcolonial career of Arthur Hugh Bunting, who was involved in Britain's policy to develop agriculture in East and West Africa as well as in South West Asia. This Afro-Asian trajectory seems to be the norm rather than the exception. David Anstey, John Blower, John Morton Boyd, Ian Grimwood, Peter Hay, Clive Nicol, John Stephenson, Patrick Stracey – one only has to compare Ethiopia's nature archives with those of Fauna & Flora International (previously the Society for the Preservation of the Fauna of the Empire) to see that from the 1950s to the early 1980s these experts worked in Uganda, Kenya and Ethiopia, but also in India, Burma and Thailand.[10] Taking care to 'treat actors seriously' and consider them as 'precious [yet] imperfect witnesses' of the singularity of a period,[11] it would be informative to produce numerous biographies and so conduct a genuine prosopography of conservation experts. This enquiry would enable us to see how 'good' nature governance norms were disseminated, applied and transformed, from the early days of the colonial period through to our present-day postcolonial era. Above all, such an embodied and situated enquiry would allow us to detect the reciprocal relationships and the tensions that developed between outside experts and the populations they met. We would thus be able to study in concrete terms the hybridity of 'global' nature that emerged in the contemporary period.

To do this we need to compare two types of archives: those held by the headquarters of United Nations agencies and colonial conservationist societies that have since become international NGOs, and those held in each colony and each postcolonial state where their experts worked. This would also enable us, third, to envisage conservationist conflicts from the point of view of 'other' official heritage-makers: the 'indigenous' then national rangers in protected areas working under the supervision of colonial wardens, then under decision-makers and international experts in postcolonial states. This is what Grégory Quenet and Rozenn Diallo have done in these pages, the former for the Seychelles, studied as it was moving from the colonial to the postcolonial period, and the latter for post-independence Mozambique. It is also what Pamela McElwee and Diana Davis have sought to do for the colonial period, the former in Indochina, the latter in Lebanon and Syria. They have nevertheless encountered two obstacles: the effacement of 'native' power players in colonial archives, and the dearth of written sources from the colonial period in the languages of the societies under study. But effacement and dearth do not equate to absence. Notification of fines issued in protected areas, accounts of scientific missions, media printed in the colonies, logbooks and diaries all provide material for extensive work in the archives, to be backed up by a great deal of long-term fieldwork to reveal the history of conservation as applied and experienced by colonisation's auxiliaries. And fieldwork on the ground is also key to uncovering the mechanisms via which postcolonial administrations played a part in global conservation. Several studies of this kind have already been conducted for South Africa,[12] Tanzania,[13] Kenya,[14] and Zambia, Uganda and Zimbabwe,[15] as well as for India,[16] Cambodia[17] and Malaysia.[18] For all these places, researchers have shown that although conservation rules were drawn up by the outside world, the less dominated of the colonised as well as most postcolonial administrations knew how to turn them to their advantage, or at least how to make do with them. Nevertheless, however innovative and valuable these studies are, most focus on one particular period (the early hunting reserves, the transition to independence or the recent financialisation of nature), leaving to one side the continuities and ruptures in contemporary conservation, from its wildlife bias in the late nineteenth century to its neoliberal contours in the twenty-first century. We now need to take this research a step further, and expand it to places whose conservation history is still poorly understood, such as Indonesia, Burma, Rwanda and the Central African Republic. This would enable us to explore colonial and national conservation histories, society by society, and observe how nature has been produced by negotiations and conflicts during which African and Asian societies, far from being *passive recipients*, also played a full part in *constructing* this globalised conserved nature.

This statement is true of all subalterns, or let us say of all anonymous people involved in the world of conservation. Admittedly, for those living in Africa and Asia, nature protection has generally meant injustice: they have been deprived of rights others enjoy (the right to use the land and to live on it), in the name of an ethic defined by others (according to which nature ought to be empty of or protected from occupants). But that does not mean that the inhabitants of this unjust nature play no part in its governance. They unquestionably do. The problem is to know how to find their trace in nature. To do so, all the archives already mentioned need to be studied: those of international institutions based in Europe and North America (development programmes, conference proceedings, official journals), those produced by the diaspora of conservation experts (mission reports, correspondence, proposed laws) and those produced by all power players, from imperial leaders to colonial park wardens, through to colonisation's auxiliaries and postcolonial leaders (laws, regulations, activity reports, wildlife censuses and notifications of fines). These sources generally mention 'natives' without naming them, or else essentialise them as 'ethnic groups', 'local populations' or 'local actors'. But by analysing this documentation 'against the grain we may detect traces of anonymous subalterns'.[19] To do this, we may define four categories of protagonists. First, those who resist, who may be sought at times of great revolts, such as the elephant hunters who took part in the 1905 Maji Maji rebellion in German East Africa,[20] or the expropriated peasants who joined the Mpondo insurgents in South Africa in 1960.[21] But we also need to identify those who resist through their everyday actions – who refuse to pay a fine, or illegally squat a plot of land, or encroach with their herd on a protected area, and so on. Behind such people stand those with power, and their history also needs to be sought in daily facts: the colonial ranger who seizes his neighbours' harvests, or, in postcolonial times, the park warden who turns a blind eye to small game hunting in exchange for recompense. Next come the workers who accept to be employed in a forest reserve to mark trees and count species, or who are employed to build tourist facilities, or else simply to accompany visitors to a protected area. Lastly are the wily, who flee on coming across a park warden, who accept to be photographed dancing naked in exchange for payment, who know how to approach scientists in the colony or international experts come to 'study' nature and in need of a villager to guide them for a day. All four figures may of course be deprived of their right to the land and to live in nature. But they nevertheless have their ways of partaking in its governance.[22] It is this history that we need to study today: the history of a nature produced through the increasing connections between calculation centres in the West, a diaspora of experts circulating between Africa and Asia, colonisation's auxiliaries then postcolonial leaders, and at the bottom of the conservation chain, the anonymous people who, while adopting, circumventing or contesting nature, bring it to life day after day.

Lastly, there is a fifth 'topic' to study, namely nature – the very stuff of conservation and of all conservation histories, but ultimately the hardest to historicise. How have Ethiopia's high plateaus, the Serengeti grasslands, Kenya's megafauna, Singapore's forests, Vietnam's mountains, Lebanon's deserts and the Seychelles' marine spaces evolved? The question needs to be asked, but answering it would require long-term studies covering vast territories and examining all the flora and fauna living there. Nature experts and managers have indeed conducted such surveys, whose results are held in archives: such-and-such a species was counted, such-and-such a mammal shot, in this reserve on that date. Yet there are not enough such surveys, and above all they are not exhaustive enough to allow us to draw any real conclusions, especially as they still focus too much on forests and megafauna, rather than on infinitely complex tropical ecosystems.[23] So we need to continue our research. First by continuing to study all the sources mentioned above, looking for the information they contain not only about wildlife numbers or the variety of flora, but also about ecosystems: watercourses, grasslands, plains and mountains. We then need to make far greater and far more extensive use of sources produced in the languages of the societies under study, looking for words used to designate the nonhuman world, that is, to name wild animal species, to classify plant species and to speak of destruction or preservation. Such lexicographical enquiry would allow us to apprehend the many African and Asian 'natures'. Furthermore, by studying how nature words change over time, we would doubtless better understand how nature has in fact evolved. Though achieving this would of course mean bringing together the life sciences and the social sciences.

This is what the author of this short conclusion has been trying to do since 2021, for example, with the 'Great African Rift' research unit, in which a hundred or so researchers are combining palaeoanthropology, geodynamics, history, palynology and botany in an attempt to pool their methods and so retrace this East African region's long-term environmental history. In a similar perspective, but applied to the Seychelles, Grégory Quenet launched the GENOHIST project in 2022, drawing on palaeogenomics: by sequencing ancient DNA and sedimentary DNA, the aim is to bring together historians and palaeogeneticists to compare environmental data with the archives and thereby think about how to write a non-anthropocentric history, while fully retaining all its comprehensive and social dimension. To build these histories of co-evolution, bringing together humans and nonhumans both in the words of the narrative and in the facts narrated, researchers need to relinquish their solitude and learn to truly work together. They also need to venture beyond their disciplinary silos and learn how to renew their questions and methods. But this, we wager, is a small price to pay for studying nature and humans as if they truly mattered equally.

Guillaume Blanc is Associate Professor of Contemporary History at the University of Rennes 2 and member of the Institut universitaire de France (2021–26), who works on the global governance of nature and people in Ethiopia and in contemporary East Africa. He directs the 'Environmental History' series for the Éditions de la Sorbonne and has recently published *La nature des hommes* (La Découverte, 2024) and *The Invention of Green Colonialism* (Polity, 2022).

Notes

1. James McCann, 'The Plow and the Forest: Narratives of Deforestation in Ethiopia, 1840–1992', *Environmental History* 2(2) (1997), 140–42.
2. Guillaume Blanc, *The Invention of Green Colonialism* (Cambridge: Polity, 2022).
3. Nancy Lee Peluso and Peter Vandergeest, 'Genealogies of the Political Forest and Customary Rights in Indonesia, Malaysia, and Thailand', *Journal of Asian Studies* 60(3) (2001), 761–812.
4. Katharine McKinnon, 'The Geopolitics of Shifting Cultivation in Thailand: A Brief History of the "Hill Tribe Problem"', in Malcolm Cairns (ed.), *Shifting Cultivation Policies: Balancing Environmental and Social Sustainability* (Wallingford: CABI, 2017), 151–66.
5. Pinkaew Laungaramsri, *Redefining Nature: Karen Ecological Knowledge and the Challenge to the Modern Conservation Paradigm* (Chennai: Earthworm, 2001).
6. Ibid. See also Pinkaew Laungaramsri, 'The Politic of Nature Conservation in Thailand', *Societies and Environments in Southeast Asia* (2017).
7. Bruno Latour, *La science en action: Introduction à la sociologie des sciences* (Paris: Gallimard, 1995), 515.
8. Simone Schleper, *Planning for the Planet: Environmental Expertise and the International Union for Conservation of Nature and Natural Resources, 1960–1980* (New York: Berghahn, 2019).
9. Raf de Bont, *Nature's Diplomats: Science, Internationalism, and Preservation, 1920–1960* (Pittsburgh, PA: University of Pittsburgh Press, 2021).
10. Guillaume Blanc, 'At the Roots of a Global Environmental History: An Ethiopian Loop in Nature's Archives', *Sources: Matériaux & terrains en études africaines* 4 (2022), 283–331.
11. Bernard Lepetit, 'L'histoire prend-t-elle les acteurs au sérieux?', *Espaces temps* 59–61 (1995), 112–22; Marc Bloch, *Apologie pour l'histoire ou métier d'historien* (Paris: Armand Colin, 1974), 138. I am very grateful to Wolf Feuerhahn for having shared these references and his thoughts about the role of 'actors' in history.
12. See for instance: Jacob Dlamini, *Safari Nation: A Social History of the Kruger National Park* (Athens: Ohio University Press, 2020).
13. See for instance: Roderick P. Neumann, *Imposing Wilderness: Struggles over Livelihood and Nature Preservation in Africa* (Berkeley: University of California Press, 1998).
14. See for instance: Reuben Matheka, 'The International Dimension of the Politics of Wildlife Conservation in Kenya, 1958–1968', *Journal of Eastern African Studies* 2(1) (2008), 112–33.

15. See for instance: Clark Gibson, *Politicians and Poachers: The Political Economy of Wildlife Policy in Africa* (Cambridge: Cambridge University Press, 1999).
16. See for instance: Ramachandra Guha and Madhav Gadgil, 'State Forestry and Social Conflict in British India', *Past & Present* 123 (1989), 141–77.
17. See for instance: Mathieu Guérin, 'France, the IUCN and Wildlife Conservation in Cambodia: From Colonial to Global Conservationism', *Journal of Southeast Asian Studies* 53(1–2) (2022), 1–26.
18. See for instance: Mathieu Guérin, 'Conserver la faune sauvage de la péninsule malaise, de la Malaya britannique à la Malaisie indépendante', *VertigO* 17(1) (2017).
19. Ann Laura Stoler, *Along the Archival Grain: Epistemic Anxieties and Colonial Common Sense* (Princeton, NJ: Princeton University Press, 2009).
20. Bernhard Gissibl, *The Nature of German Imperialism: Conservation and the Politics of Wildlife in Colonial East Africa* (New York: Berghahn, 2019), 109–40.
21. William Beinart, 'The Mpondo Revolt through the Eyes of Leonard Mdingi and Anderson Ganyile', in Thembela Kepe and Lungisile Ntsebeza (eds), *Rural Resistance in South Africa: The Mpondo Revolts after Fifty Years* (Leiden: Brill, 2011), 91–113.
22. Guillaume Blanc, *La nature des hommes: Une mission écologique pour 'sauver' l'Afrique* (Paris: La Découverte, 2024), 288–343.
23. Pierre-Michel Forget, Martine Hossaert-McKey and Odile Poncy (eds), *Écologie tropicale: De l'ombre à la lumière* (Paris: Cherche-Midi, 2014).

Bibliography

Beinart, William. 'The Mpondo Revolt through the Eyes of Leonard Mdingi and Anderson Ganyile', in Thembela Kepe and Lungisile Ntsebeza (eds), *Rural Resistance in South Africa: The Mpondo Revolts after Fifty Years* (Leiden: Brill, 2011), 91–113.

Blanc, Guillaume. 'At the Roots of a Global Environmental History: An Ethiopian Loop in Nature's Archives'. *Sources: Matériaux & terrains en études africaines* 4 (2022), 283–31.

———. *The Invention of Green Colonialism*. Cambridge: Polity, 2022.

———. *La nature des hommes: Une mission écologique pour 'sauver' l'Afrique*. Paris: La Découverte, 2024.

Bloch, Marc. *Apologie pour l'histoire ou métier d'historien*. Paris: Armand Colin, 1974.

De Bont, Raf. *Nature's Diplomats: Science, Internationalism, and Preservation, 1920–1960*. Pittsburgh, PA: University of Pittsburgh Press, 2021.

Dlamini, Jacob. *Safari Nation: A Social History of the Kruger National Park*. Athens: Ohio University Press, 2020.

Forget, Pierre-Michel, Martine Hossaert-McKey and Odile Poncy (eds). *Écologie tropicale: De l'ombre à la lumière*. Paris: Cherche-Midi, 2014.

Gibson, Clark. *Politicians and Poachers: The Political Economy of Wildlife Policy in Africa*. Cambridge: Cambridge University Press, 1999.

Gissibl, Bernhard. *The Nature of German Imperialism: Conservation and the Politics of Wildlife in Colonial East Africa*. New York: Berghahn, 2019.

Guérin, Mathieu. 'Conserver la faune sauvage de la péninsule malaise, de la Malaya britannique à la Malaisie indépendante'. *VertigO* 17(1) (2017). Retrieved 7 August 2024 from http://journals.openedition.org/vertigo/18503.

———. 'France, the IUCN and Wildlife Conservation in Cambodia: From Colonial to Global Conservationism'. *Journal of Southeast Asian Studies* 53(1–2) (2022), 1–26.

Guha, Ramachandra, and Madhav Gadgil. 'State Forestry and Social Conflict in British India'. *Past & Present* 123 (1989), 141–77.

Latour, Bruno. *La science en action: Introduction à la sociologie des sciences*. Paris: Gallimard, 1995.

Lepetit, Bernard. 'L'histoire prend-t-elle les acteurs au sérieux?' *Espaces temps* 59–61 (1995), 112–22.

Laungaramsri, Pinkaew. *Redefining Nature: Karen Ecological Knowledge and the Challenge to the Modern Conservation Paradigm*. Chennai: Earthworm, 2001.

———. 'The Politic of Nature Conservation in Thailand'. *Societies and Environments in Southeast Asia* (2017). Retrieved 7 August 2024 from https://nature.hypotheses.org/328.

Matheka, Reuben. 'The International Dimension of the Politics of Wildlife Conservation in Kenya, 1958–1968'. *Journal of Eastern African Studies* 2(1) (2008), 112–33.

McCann, James. 'The Plow and the Forest: Narratives of Deforestation in Ethiopia, 1840–1992'. *Environmental History* 2(2) (1997), 140–42.

McKinnon, Katharine. 'The Geopolitics of Shifting Cultivation in Thailand: A Brief History of the "Hill Tribe Problem"', in Malcolm Cairns (ed.), *Shifting Cultivation Policies: Balancing Environmental and Social Sustainability* (Wallingford: CABI, 2017), 151–66.

Neumann, Roderick P. *Imposing Wilderness: Struggles over Livelihood and Nature Preservation in Africa*. Berkeley: University of California Press, 1998.

Peluso, Nancy Lee, and Peter Vandergeest. 'Genealogies of the Political Forest and Customary Rights in Indonesia, Malaysia, and Thailand'. *Journal of Asian Studies* 60(3) (2001), 761–812.

Schleper, Simone. *Planning for the Planet: Environmental Expertise and the International Union for Conservation of Nature and Natural Resources, 1960–1980*. New York: Berghahn, 2019.

Stoler, Ann Laura. *Along the Archival Grain: Epistemic Anxieties and Colonial Common Sense*. Princeton, NJ: Princeton University Press, 2009.

Index

www.ingramcontent.com/pod-product-compliance
Ingram Content Group UK Ltd.
Pitfield, Milton Keynes, MK11 3LW, UK
UKHW021837150926
8656IPUK00004B/31

9 781805 398912